Optoelectronics for Technology and Engineering

Optoelectronics for Technicians and Engineering

ROBERT G. SEIPPEL

PRENTICE HALL CAREER & TECHNOLOGY, Englewood Cliffs, New Jersey 07632

LIBRARY OF CONGRESS
Library of Congress Cataloging-in-Publication Data

Seippel, Robert G.
Optoelectronics for technology and engineering / Robert G. Seippel.
p. cm.
Bibliography: p.
Includes index.
ISBN 0-13-638404-8
1. Optoelectronics. I. Title.
TA1750.S444 1989
621.38'0414—dc19 88-9923
CIP

Cover design: Diane Saxe
Cover photo courtesy: Coherent Inc., Palo Alto, CA, Everlase Laser Cutting Steel
Manufacturing buyer: Paula Benevento

Prentice-Hall, Inc.
A Paramount Communications Company
Englewood Cliffs, New Jersey 07632

Printed in the United States of America

10 9 8 7 6 5 4

ISBN 0-13-638404-8

Prentice-Hall International (UK) Limited, *London*
Prentice-Hall of Australia Pty. Limited, *Sydney*
Prentice-Hall Canada Inc., *Toronto*
Prentice-Hall Hispanoamericana, S.A., *Mexico*
Prentice-Hall of India Private Limited, *New Delhi*
Prentice-Hall of Japan, Inc., *Tokyo*
Simon & Schuster Asia Pte. Ltd., *Singapore*
Editora Prentice-Hall do Brasil, Ltda., *Rio de Janeiro*

Contents

Five Photodetection *95*

Six Laser Optics *135*

Nine
Laser *289*

Ten
Photometry/Radiometry *340*

Eleven
Safety *360*

Preface

Optoelectronics has suddenly become an area of heavy growth. From light emitting diodes and receiving devices through optical fibers to lasers, optoelectronics must now be dealt with as an emerging giant of the electronics and communications industry.

This growth has been spurred by many developments in materials and devices and each development has found a waiting market. Light sources, LEDs, and lasers are used for displays in numerous consumer products – VCRs, calculators, and compact disc players to name a few. Optical fibers carry heavy communication loads over long distances and past strong sources of electrical interference. Solar cells generate power both for satellites in space and for equipment in remote areas here on earth.

Solid state optical sensors are used in TV cameras and astronomical telescopes. Optical isolators remove delicate electronic circuitry from the dangers of high voltages.

Much of the credit for the phenomenal growth of fiber optics is due to the pioneers in the development of glass fibers, such as Standard Telecom Labs in England, Nippon Sheet Glass Company in Japan, and Bell Telephone Labs and Corning Glass Works in the United States. Since 1976, advancements in the fiber optics field have come so quickly that it is difficult to determine who developed what. Dozens of companies in the United States are creating new fibers and connectors, new light launching devices, and related electronics. A momentous new technological breakthrough seems highly unlikely, simply because industry is already there. The future promises improved manufacturing techniques that will tend to bring prices down to meet the needs of consumer economics.

While the world was thinking of the laser as a mysterious Buck Rogers type ray gun, Charles H. Townes of Columbia University was developing the first laser in 1954. Later in 1960, Theodore H. Maiman of Hughes Aircraft constructed the first ruby laser. In the closing months of 1960, A. Javin of Bell Labs created the first gas laser. Semiconductor lasers were devised simultaneously by General Electric, IBM, and Lincoln Laboratories in 1962.

These pioneers set the pace and companies such as Coherent, Molectron, and Spectra-Physics have followed by raising laser technology to star status in industry.

Today, lasers are no longer fictional aspirations. The broadest possible applications for lasers are now being realized. Some of these applications are in communications holography, medicine, direction finding, manufacturing, and light transmission.

Optoelectronics is old and new. It still has mystique but that mysteriousness is now mixed with reality. This book is dedicated to pioneers in the field and to the many who are dedicating their efforts toward its future.

* * *

A special thanks is given to the contributors of data and illustrations within this book. Those whose articles I have used as reference are listed in the bibliography. Illustrations have the standard courtesy lines. I sincerely hope that I have left no one out, for without the contributors' help this book would not be possible.

Robert G. Seippel, Ph.D.

Abbreviations Used in Optoelectronics

Abbreviation	*Identity*
AGC	Automatic Gain Control
AMI	Alternate Mark Inversion
APD	Avalanche Photodiode
AVC	Automatic Volume Control
BeO	Beryllium Oxide
BER	Bit Error Rate
CGS	Centimeter-Gram-Second System
cm	centimeter
CO_2	Carbon Dioxide
CVD	Chemical-Vapor Deposition
CW	Continuous Wave
E	Energy

Abbreviation	*Identity*
E-M	Electromagnetic
EMI	Electromagnetic Interference
EMP	Electromagnetic Pulse
FDM	Frequency Division Multiplexing
FO	Fiber Optics
HCN	Hydrogen Cyanide
He	Helium
HeCd	Helium Cadmium
HeNe	Helium Neon
Hg	Mercury
Hz	Hertz
IR	Infrared
Iv	Luminous Intensity
kW	kilowatts
LASER	Light Amplification by the Stimulated Emission of Radiation
LED	Light Emitting Diode
MKS	Meter-Kilogram-Second System
mm	Millimeters
Mv	Luminous Exitance
mW	Milliwatts
NA	Numerical Aperture
Nd	Neodyium
Ne	Neon
nm	Nanometers
NRZ	Nonreturn to Zero
PCM	Pulse Code Modulation
PCS	Plastic Clad Silica
PF	Packing Fraction
PIN	Positive Intrinsic Negative Layers
PPM	Pulse Position Modulation
PRF	Pulse Rate Frequency
PRM	Pulse Rate Modulation
PRR	Pulse Repetition Rate
PRT	Pulse Repetition Time
PWM	Pulse Width Modulation
Q	Quality of a Laser Beam
Qv	Luminous Energy
RFI	Radio Frequency Interference
RMS	Root Mean Square
RZ	Return to Zero

SNR	Signal-to-Noise Ratio
TDM	Time-Division Multiplexing
TEM	Transverse Electromagnetic Mode
UV	Ultraviolet
YAG	Yttrium Aluminum Garnet

Optoelectronics
for
Technology
and
Engineering

One Introduction to Optoelectronics

Optoelectronics is the marriage of several sciences—traditional optics with its lenses and prisms, and the new field of solid-state light sources and detectors and coherent lasers, thin optical fibers, and semiconductor electronics. New developments emerging from the laboratories would not by themselves create such a growth industry; applications had to be found. For optoelectronic devices this was easy. They offered solid-state reliability, improved efficiency, and low cost. The devices found applications in:

- Displays for calculators, home entertainment, and computer products
- Light beams from coherent sources used to measure distances
- Light beams in thin optical fibers carrying high volumes of telephone traffic with improved immunity to electrical interference
- Optical reading of compact-disc music
- Navigation with laser gyros
- Imaging with charge-coupled sensors for VCR cameras and deep-space astronomy
- Power generation with solar cells
- Electrical isolation with optical isolators
- Medical applications with lasers
- Credit-card security marking with holograms

OBJECTIVES

After studying this chapter and completing the self-check questions at the end of the chapter, the reader will be able to:

1. State several of the theories of light.
2. Explain several of the theories of electricity/electronics.
3. State the three basic bands of the frequency spectrum.
4. Calculate wavelength from several variable frequencies.
5. Define terms that relate to light rays.
   a. Propagation sources
   b. Action of light on material
   c. Photons
   d. Diffraction
6. Compare the two lenses, convex and concave.
7. Analyze the term *light separation* and its phenomena.
8. Expound on the term *fiber optics* to include its history and its applications.
9. Expound on the acronym *laser* to include its history and its applications.

PROGRESSION OF THE THEORIES OF LIGHT

The earliest theory of light was a theory of vision. It was thought that the eyes sent out invisible antennae and were thus able to feel or sense those things that were too distant to touch. This was called the "tactile" touching theory.

A second theory was the "emission" theory. This viewed all objects as emitting something which, when entering the eye, was sensed by some part of the eye, therefore the brain could see. Both the tactile theory and the emission theory were accepted by the Greek thinkers in 500 B.C. The tactile theory was dismissed, but the emission theory carried on and was used by Newton in developing his corpuscle theory.

Robert Hooke, a noted British scientist, published a book called *Micrographia* in 1665. In it he described light as small rapid vibrations and that variations in color are produced by these vibrations, thus originating the wave theory of light. The Royal Society of Great Britain honored Hooke by making him curator of experiments. Then in 1672, Sir Isaac Newton sent the society an account of his experiments in color separation with the prism. Shortly after the color theory, Newton developed his corpuscle theory and sent an account of it to the Royal Society. It was immediately turned down by the committee, led by Robert Hooke, because it conflicted with Hooke's wave theory.

As early as 1676, a Danish astronomer, Olaus Roemer, discovered that the speed of light was finite (measurable) and estimated its speed at 140,000 miles per second. It was not until 1926 at Mt. Wilson that Albert Michelson, an American, determined through more thorough experiments that the speed of light is 299,796 ± 4 kilometers per second. The speed has now been calculated by so many scientists that it would not be meaningful to present them all.

Along with the fact that light has velocity, experiments by Christian Huygens, a seventeenth-century Dutch scientist, proved that light also had polarity (or direction) depending on the plane. However, it was not until the nineteenth century that detailed experiments were made with light reflection and refraction. The two findings (polarity and finite speed of light), along with several wave theories, led further to the theory of transverse waves. It was thought that an elastic medium pervaded all space and changed only when matter was in its way.

In other words, light was a thing that moved at a finite speed, was polarized, and could be modified by refraction or reflection. In his famous principle in 1690, Christian Huygens stated that each point of a wavefront may be regarded as a source of new waves. This was very much like Maxwell's later electromagnetic theory of light. James Clerk Maxwell, a Scot, through experiments with propagation of electromagnetic waves in the nineteenth century, derived mathematical constants for electricity and magnetism. By these constants he was able to establish a frequency and wavelength electromagnetic spectrum. The Maxwell mathematical constants of the electromagnetic wave theory allowed Heinrich Hertz, a German physicist, to produce wavelengths by electrical means in 1887.

In 1900, Max Planck found that light energy is emitted by atoms in multiples of an energy unit. The unit was called a quantum and its magnitude depended on wavelength. Seven years later, in 1907, Einstein found concentrations of energy, which he called photons, when light released electrons from atoms.

Modern ideas of relativity, at least in the modern age are due to the work of Albert Einstein. The modern theory of gravitation is general relativity and was published by Einstein in 1915. General relativity pictures the world in four-dimensional space-time. Particles and light rays move along paths in four dimensional space in the straightest possible curves. The curves are dependent on the matter present in the space. All reference to absolute three-dimensional space are absent. General relativity is still largely theoretical. Special relativity asserts that all laws of physics are equally valid in all inertial frames of reference. In 1905, Einstein published this special theory of relativity. The importance of this publication is that the theory has some extremely striking practical implications. The first implication is that time and space separations are dependent on the selection of motion of the observer. The second prediction is that no body or physical effect travels faster than the speed of light. The third and probably the best known is the relation implication of mass m, energy E, and the speed of light c. The impact of the formula $E = mc^2$ is phenomenal. Technology has made great strides in areas such as atomic power and partical accelerators. Special relativity is by far the most practical and valid of all branches of physics.

Modern quantum mechanical light theory began around 1927 and incorporates the appropriate parts of the electromagnetic wave theory, the quantum theory, and the relativity theory. In Figure 1-1, the progression of major light theories is presented in chronological order. Minor theories have been omitted to preserve simplicity.

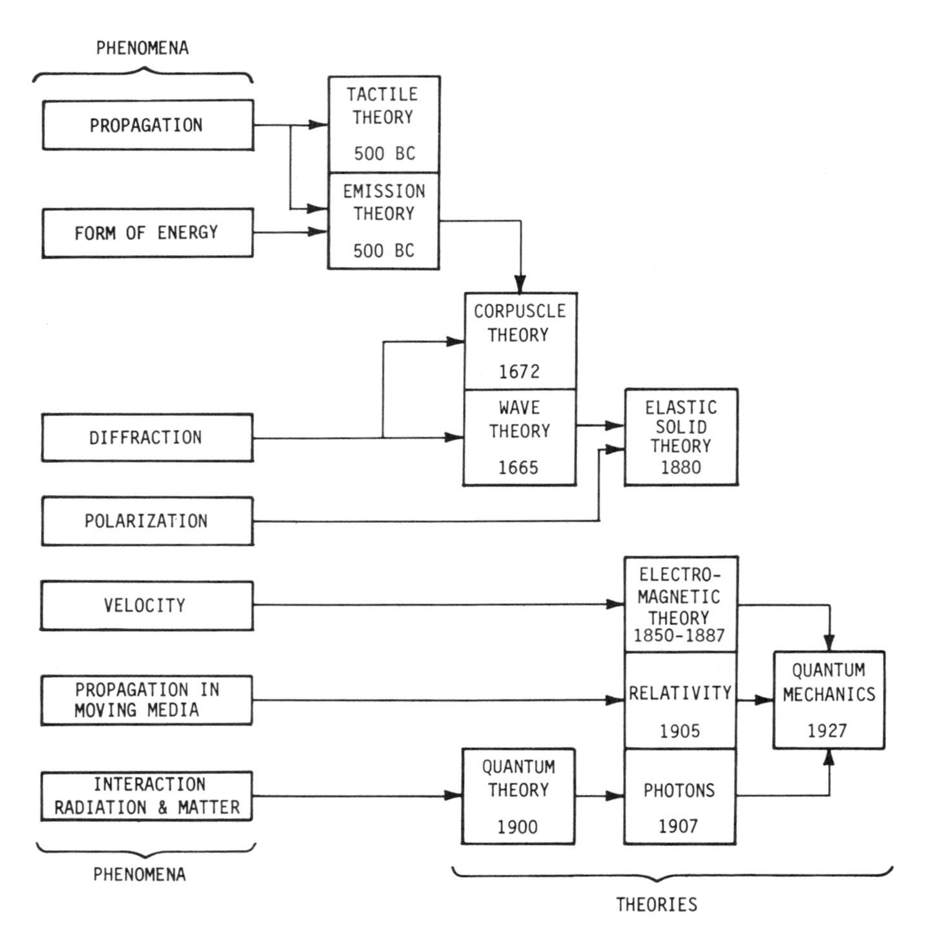

Figure 1-1 Progression of the theories of light (*Mike Anaya*).

PROGRESSION OF PHOTOELECTRICITY

The progression of photoelectricity necessarily follows the progression of light. The probable beginning of photoelectricity was with the discovery of the photovoltaic cell by Alexandre Becqueral in 1839. Becqueral found that a battery he was working with developed a better charge when exposed to light.

In 1888, Heinrich Hertz working on the invisible portion of the spectrum revealed the fact that certain metals produced electrical effects when exposed to ultraviolet light. The next steppingstone to photoelectricity was Willoughby Smith's discovery that the resistance of selenium changes when light is applied. Among other scientists experimenting with this phenomena in the late nineteenth century was R. E. Day.

Of extreme importance in photoelectricity history and indeed science in general was the discovery of radiation in 1896 by Antoine Henri Becqueral, son of Alexandre Becqueral the discoverer of the photovoltaic cell. The Becqueral family's work with light and electricty must be lauded as some of the greatest achievements of the time.

Just after the turn of the century, about 1905–1906, Max Planck developed his universal mathematical constant that light is emitted by atoms in multiples of a certain unit or quanta. The size of the unit being a quantum, which depends on the wavelength of the radiation. Einstein, in 1907, suggested that in order to give an adequate description of these localized pockets of energy developed from light exposure, that they had to be considered as particles separate from the atoms. Thus the concept of particles or photons of light energy was developed. In 1927, the photon theory was added to the quantum theory, the theory of relativity, and the electromagnetic theory, to establish the present-day light theory called quantum mechanics previously discussed. Figure 1–2 shows the development of the photoelectrical theory.

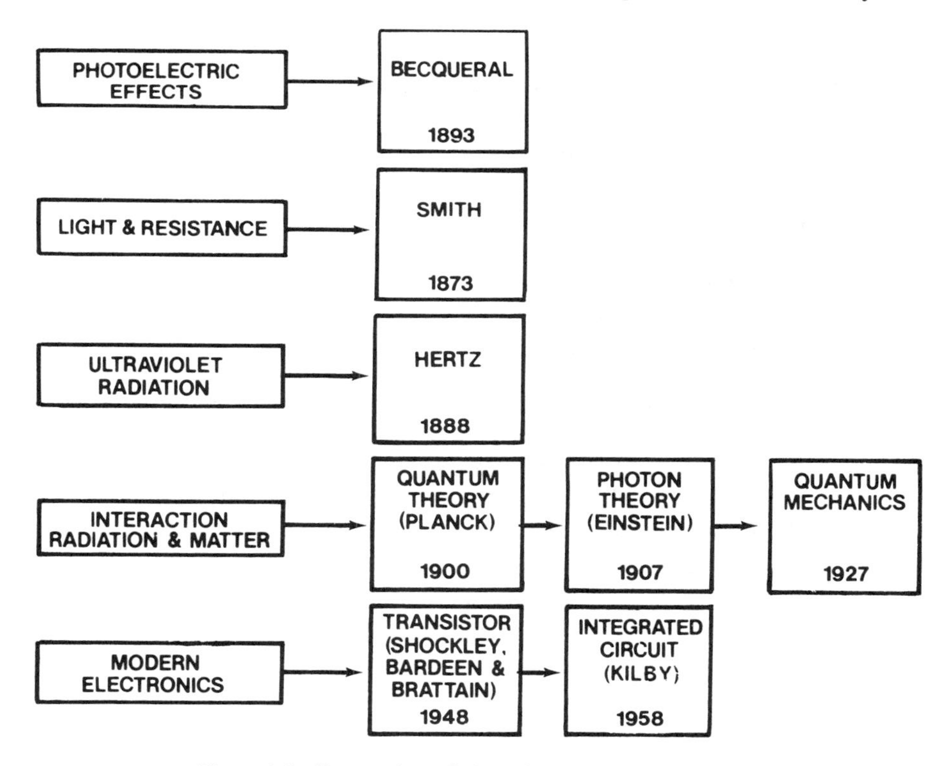

Figure 1–2 Progression of photoelectricity (*Mike Anaya*).

MODERN PROGRESS IN ELECTRONICS

Throughout the 1930s and 1940s many laboratories such as Bell Telephone and General Electric were developing photoelectric devices for commercial and military uses. The major breakthrough for the world of electronics was the invention of the transistor by Bardeen, Brattain, and Shockley of Bell Labs in 1948. This was the beginning of the age of electronics called *solid-state* or *semiconductor electronics.*

Since this discovery, electronics has been widely advanced by the design of integrated circuits. The principal designer was Kilby of Texas Instruments in 1958. Since integrated circuits are such a vast field in electronics today, we could not possibly give credit to everyone who is responsible for advancement in this area.

A stage in optoelectronics which seems appropriate to mention is the ultimate use of optoelectronics, the laser or "Light Amplification by Stimulated Emission of Radiation." In 1958, Charles H. Townes, an American, and Arthur L. Schawlow, a Canadian, wrote a paper in which they showed that it was possible to use stimulated emission for amplifiying light waves as well as microwaves (the maser). Two years later, in 1960, Theodore H. Maiman, a scientist working at Hughes Aircraft Company in California, built the first optical maser, otherwise known as the laser.

OPTOELECTRONIC OPERATIONAL SPECTRUM

Optoelectronics is the branch of electronics that deals with light. Electronic devices involved with light operate near the optical part of the frequency spectrum. The visible range is only a narrow portion of the optical spectrum. There are three basic bands of the frequency spectrum. These are:

1. *Infrared:* band of light wavelengths that is too long to be seen by the human eye
2. *Visible:* band of light wavelengths that the human eye responds to
3. *Ultraviolet:* band of light wavelengths that is too short for the human eye to see

Before we discuss each of the three basic bands of the optical spectrum, we shall explain several things that are involved in it.

A wavelength λ (the lowercase Greek letter lambda) is the amount of space occupied by the progression of one cycle of an electromagnetic wave. Calculation of the size of the wavelength is dependent on the frequency of the wave f and the velocity of light c.

$$\lambda = \frac{c \text{ (velocity of light)}}{f \text{ (frequency)}}$$

$$\lambda = \frac{c}{f}$$

The frequency of the wave f is the same frequency that originated the radiated wave. The velocity of the wave c is the velocity of light, which is approximately the same velocity for all electromagnetic waves, that is, 300,000,000 meters per second (186,000 miles per second). The velocity of light varies slightly in different materials.

An audio-frequency electromagnetic wave of 15,000 hertz (cycles per second) would have a wavelength of

$$\lambda = \frac{300{,}000{,}000}{15{,}000} = 20{,}000 \text{ meters}$$

A radio frequency of 300 megahertz would have a wavelength of

$$\lambda = \frac{300{,}000{,}000}{300{,}000{,}000} = 1 \text{ meter}$$

Higher frequencies such as 3000 megahertz would have a wavelength of 0.1 meter.

Even higher frequencies such as 300,000 megahertz would radiate a wavelength of 0.001 meter, which could also be written as 1000 microns. It may be convenient to discuss wavelengths in microns because of the size of the frequency involved.

In the optical spectrum another value, the angstrom (Å), is used to express wavelength.

$$1 \text{ angstrom (Å)} = 0.0001 \text{ micron } (\mu\text{m})$$

or

$$1 \text{ micron } (\mu\text{m}) = 10{,}000 \text{ angstroms (Å)}$$

In terms of scientific notation

$$1 \text{ angstrom (Å)} = 10^{-10} \text{ meter}$$

$$1 \text{ micron } (\mu\text{m}) = 10^{-6} \text{ meter}$$

The optical spectrum operates from wavelengths of 0.005 to 4000 microns (50 to 40,000,000 angstroms). In frequencies, these are extremely high values (6×10^{16} to 7.5×10^{10} hertz) (see Figure 1-3).

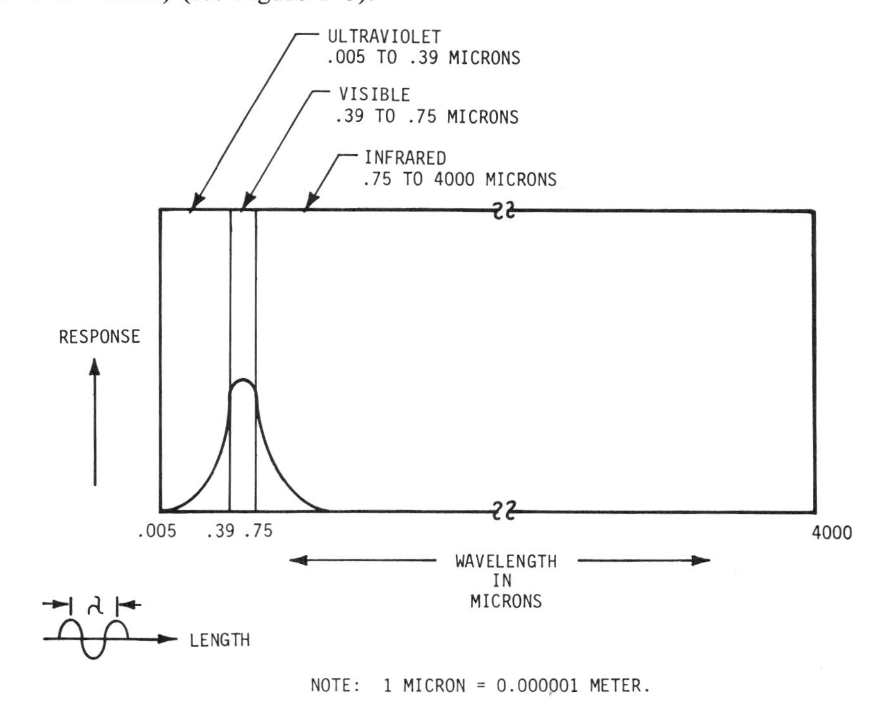

Figure 1-3 The optical spectrum, including the portion visible to the human eye (*Mike Anaya*).

Ultraviolet Wavelengths

Wavelengths for ultraviolet radiated frequencies range from 0.005 to 0.3900 micron (50 to 3900 angstroms). In frequencies, these are 6×10^{16} to 7.69×10^{14} hertz.

Visible Wavelengths

Visible radiated wavelengths range from 0.3900 to 0.7500 micron (3900 to 7500 angstroms). In frequencies, these are 7.69×10^{14} to 4×10^{14} hertz.

Infrared Wavelengths

Wavelengths for infrared radiated frequencies range from 0.7500 to 4000 micron (7500 to 40,000,000 angstroms). In frequencies, these are 4×10^{14} to 7.5×10^{10} hertz. Infrared frequencies are too low and ultraviolet frequencies are too high for human eye response.

Wavelengths are in inverse proportion to frequency; therefore, infrared wavelengths are too long for human-eye response and ultraviolet wavelengths are too short for human-eye response. Another point worth mentioning is that the wavelength range for infrared (0.75 to 4000 microns) is much greater than either visual or ultraviolet wavelength ranges.

Color

The human eye sees violet (approximately 0.43-micron wavelength) on one side of the color spectrum and red (approximately 0.68-micron wavelength) on the opposite side of the spectrum. In between these extremes, the eyes see blue, green, yellow, and orange. Two beams of light that have the same wavelength are seen as the same color. Two beams that are seen to have the same color usually have the same wavelength. However, a mixture of two colors such as green and red will match a beam of blue. Therefore, the wavelengths are not necessarily the colors they may seem to the eye. Matches of colors are only correct to the person who has extremely good vision.

TABLE 1-1 OPTICAL COLORS AND THEIR WAVELENGTHS

	Color	Wavelengths* (Microns)	Frequencies (Hertz)
	Ultraviolet	0.005–0.39	6×10^{16}–7.69×10^{14}
Visual response	Violet	0.40–0.45	7.5–6.6 $\times 10^{14}$
	Blue	0.45–0.50	6.6–6.0 $\times 10^{14}$
	Green	0.50–0.57	6.0–5.27 $\times 10^{14}$
	Yellow	0.57–0.59	5.27–5.01 $\times 10^{14}$
	Orange	0.59–0.61	5.01–4.92 $\times 10^{14}$
	Red	0.61–0.70	4.92–4.28 $\times 10^{14}$
	Infrared	0.70–4000	4.28×10^{14}–7.5×10^{10}

*Approximates only, with overlapping wavelengths and frequencies.

The graduation of colors and their wavelengths from one end of the color (visual) spectrum have overlapping wavelengths depending on the observer. Table 1–1 shows typical colors along with their estimated wavelengths and radiated frequencies.

LIGHT RAYS

Light travels in beams or rays. A *ray* is the path of the light. In the seventeenth century, propagation of light was simply explained and represented by rays. In Figure 1–4, light from a pinpoint-size hole is interrupted by an opaque (not pervious to light) disk in its path. The opaque disk prevents light from passing through. This causes a distinct, sharp-edged shadow on a backboard receptor or movie screen. The small hole shadow is called an *umbra.* This figure is an extremely good example to argue the facts that light travels from a source to a receptor on beams or rays and secondly, that the rays are straight lines. Rays interrupted by opaque objects cannot get to the receptor. Newton's first laws of motion stated that particles travel in a straight line unless moved by some force. Newton therefore determined that light particles were weightless and furthermore the particle's speeds were so fast that gravity had no effect on them. This is easy to accept, as illustrated in Figure 1–4.

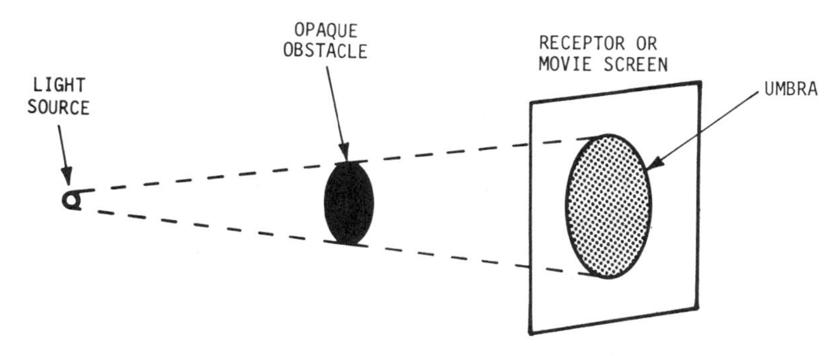

Figure 1–4 Formation of shadows in support of straight-line propagation of light (*Mike Anaya*).

Propagation Sources

Light rays or beams are propagated at various wavelengths. Some sources radiate a much narrower bandwidth than others (see Figure 1–5). Monochromatic light is radiation that has one wavelength (or an extremely narrow band of several wavelengths). Heterochromatic light is radiation that has several very distinct wavelengths. Panchromatic light is radiation that includes wavelengths in a great range of the spectrum.

Action of Light on Material

All material does not "see" light in the same manner. There are basically three different light-receptor materials. These are opaque, transparent, and translucent. Material that does not allow any light to travel through it is called *opaque.* If you

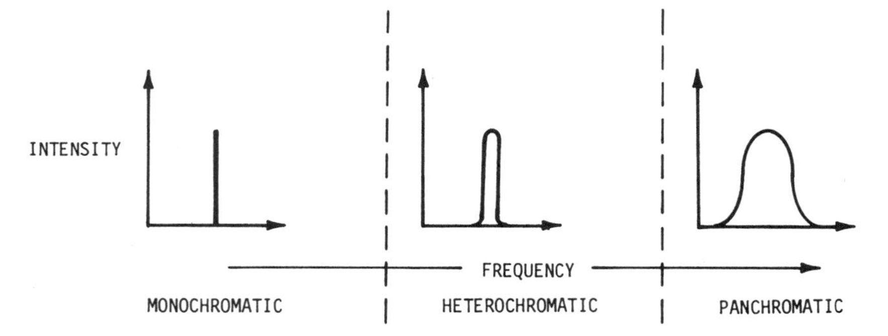

Figure 1–5 Radiation from sources of light rays (*Mile Anaya*).

can see partially through a substance but not clearly, the substance is called *translucent*. Finally, *transparent* material allows you to see through it clearly.

Photons

Response of the human eye to light is determined by the energy of the radiation. Radiant energy travels in energy groups known as *photons*. The frequency of the radiation determines the strength of the photons. The higher the frequency of the radiation, the higher the energy of the photons. As an example, wavelengths in the visible range of the spectrum near the ultraviolet range have short wavelengths, higher frequencies, and higher proton energy. Furthermore, wavelengths in the visible range of the spectrum near the infrared range have long wavelengths, lower frequencies, and lower photon energy. When an electron is freed by absorbing the energy of a photon, the result of the combination is called a *photoelectron*. Again the energy of the photoelectron is dependent on the frequency of the radiation. The energy of a photon is expressed in watts and can be calculated by the following equation:

$$W = hf$$

where W = energy in watts

h = Planck's constant (6.63×10^{-34} joule-second)

f = frequency of the photon wavelength in hertz

Diffraction

Up to this point we have assumed that light travels in straight lines. This is not altogether true. In experiments by Newton called *Newton's rings,* he discovered that lights does spread, although to a very small extent, from the edges of the light beam. The shadow of an opaque obstacle is not perfectly sharp, but some light does indeed penetrate into the shadow area. The fringes near the opaque shadow shown in Figure 1–4 appear sharp, but under magnification would show interference or what we now call *diffractions*. In conclusion, we might add that light does not always correspond to its visual appearance.

LENS SYSTEM

A lens is a piece of glass or other transparent device such as plastic, with two curved or one curved and one flat surface. The purpose of the lens is to bring together or spread apart light rays as they pass through it. There are two basic lens shapes, *convex* and *concave.* All lenses are made in the form of one of these shapes or both in combination. We shall illustrate the more simple forms and describe their operation. Lens discussion is purposely brief as the intent is to provide basic operation only. Lenses and their operation could easily fill a text of their own.

Convex Lenses

The convex lens is a converging lens. That is, it causes light to come together. In Figure 1-6A, a plano-convex lens accepts parallel light rays and converges them into a focal point. In Figure 1-6B, a plano-convex lens accepts light rays from a converged source then collimates the rays or makes them parallel. In Figure 1-6C and D, a double convex lens does essentially the same thing.

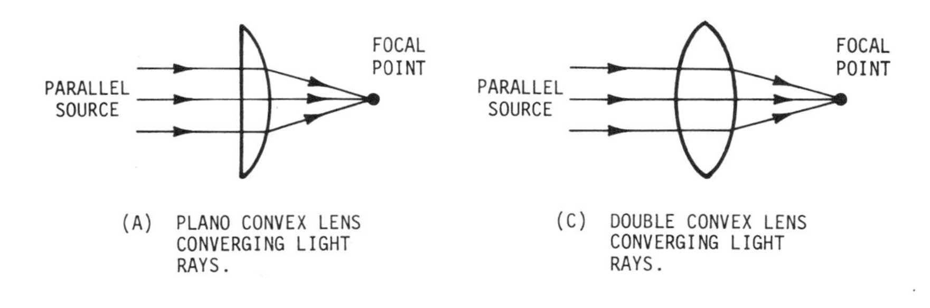

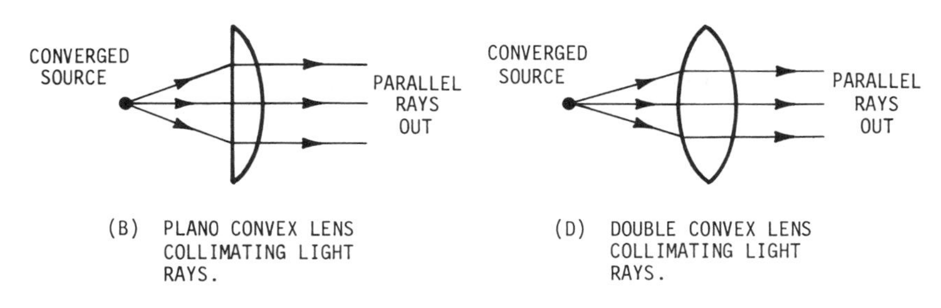

Figure 1-6 Convex lenses (*Mike Anaya*).

Concave Lenses

The concave lens is a diverging lens. That is, it causes light to spread apart. In Figure 1–7a, a plano-concave lens causes parallel light rays to diverge or spread out. In Figure 1–7b, a plano-concave lens causes rays that are spread out to become parallel. In Figure 1–7c and d, a double concave lens does essentially the same thing.

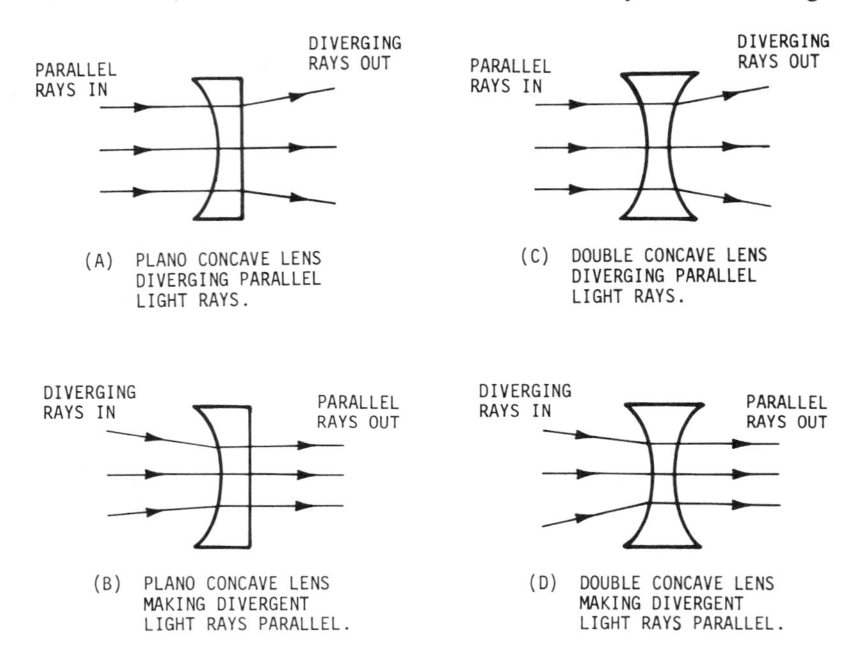

Figure 1–7 Concave lenses (*Mike Anaya*).

Lenses in Combination

Often it is necessary to utilize two lenses to provide a final result. In Figure 1–8, a concave lens is used to convert divergent light rays to parallel light rays. The parallel lines are then directed through a convex lens which converges them into a focal point.

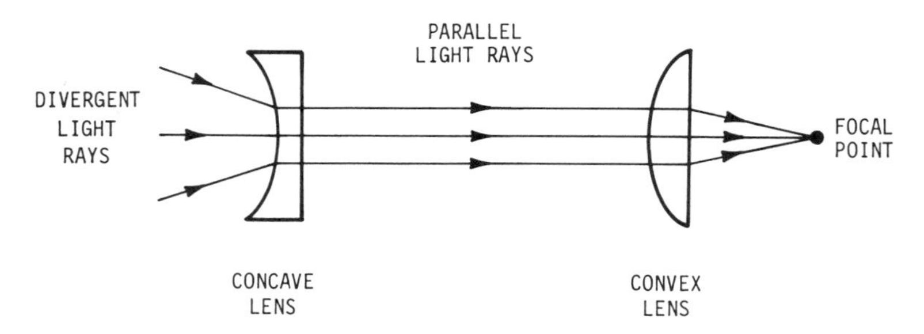

Figure 1–8 Lenses in combination (*Mike Anaya*).

LIGHT SEPARATION

In 1672, Isaac Newton developed his theory of color. He allowed a thin beam of white light to pass through a shutter. He directed this thin beam through a prism. The prism produced colors of the rainbow: red, orange, yellow, green, blue, indigo, and violet as the light passed through. From this he determined that white light is made up of the colors of the spectrum. He further decided that red was bent the least, orange a little more, and so on, to the violet, which was bent the most. Thus he concluded that the angle of refraction through a prism varies with the color.

In Figure 1–9a, white light passes through a shutter and a prism to separate the colors of the spectrum. Newton's second thought was to reconstruct the colors

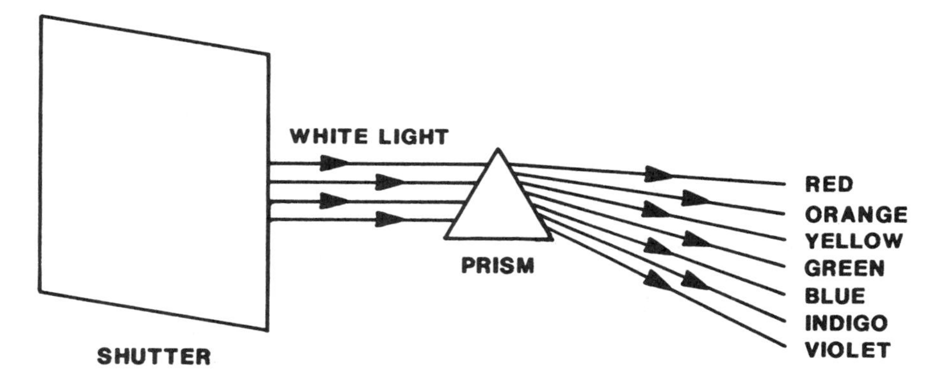

(A) WHITE LIGHT CONVERTS TO COLOR SPECTRUM.

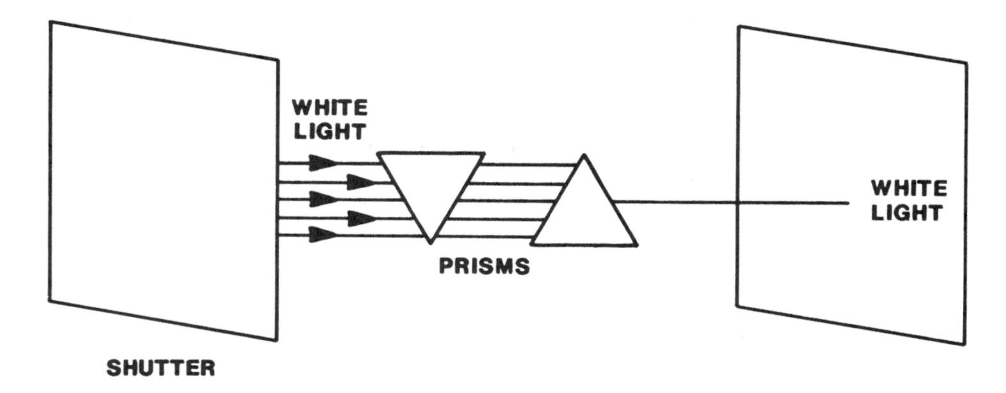

(B) TWO PRISMS CONVERT WHITE LIGHT TO COLOR TO WHITE LIGHT.

Figure 1–9 Light separation.

back to white light. He placed a second prism next to the first. The colors were changed back to white light as shown in Figure 1-9b. Simple lenses are similar to prisms and can focus different-color light at different points, a serious problem in color photography. The problem can be corrected with lens grouping and special coatings.

INTRODUCTION TO FIBER OPTIC ELECTRONICS

Fiber Optics Defined

Fiber optics is the science that deals with the transmission of light through extremely thin fibers of glass, plastic, or other transparent material. Optical fibers are dielectric waveguides for electromagnetic energy at optical wavelengths. The fibers provide a path for a single beam of light or in multiples, such as the transposition of a complete image. The fibers are provided as a single fiber or a cable bundle. They may be bent or curved (within limits) to meet the needs of special routing.

History of Fiber Optic Electronics

Probably the founder of fiber optic electronics was the British physicist John Tyndall. Tyndall found in 1870 that light will follow water into a container and flowing from the container. It was, however, not until after World War II and into the 1950s that major corporations began study in the area of light and fiber optics. In 1968, Standard Telecom Labs in England developed silica glass with an extremely low transmission loss. Since then the Corning Glass Works and Bell Telephone Labs in the United States and Nippon Sheet Glass Company in Japan have developed glass fibers with typical loss factors of 1 to 2 decibels per kilometer (dB/km) and even much lower (0.1 dB/km) under lab conditions. These low-loss capabilities lend themselves to the development of low-cost, high-efficiency fiber optic communication systems. After 1976, advancements have come so quickly that it is difficult to determine who developed what. Dozens of companies in the United States are creating new fibers and connectors, new light launching devices, and related electronics. A momentous new technological breakthrough seems highly unlikely, simple because industry is already there. The future looks to manufacturing techniques that will bring prices down to meet the needs of consumers.

Optical Fiber Frequency Spectrum

Most optical fibers are manufactured for use in the visible frequency spectrum. Frequency responses tend to fall off on the ultraviolet and the infrared ends of the spectrum. Modern fiber optics have been manufactured to operate in the 4000-angstrom-wavelength range on the ultraviolet range of spectrum. On the infrared end of the spectrum, wavelengths of 10 microns are now common.

Why Use Optical Fibers for Electronic Communication?

The advantages of using fiber optics in communication networks are almost as limitless as a designers ability to create. The optical transmission path provides electrical input and output isolation. To a designer this means total freedom from ground loops, along with lightning-safe installation. Bandwidths are not a problem because they are independent of the cable size. Extremely long-distance cables are available, along with an endless variety of bandwidth products. This means greater data rates at longer distances than wire or coaxial cabling.

One of the major problems encountered in communication electronics is EMI (electromagnetic induction). With fiber optics there is no EMI susceptibly, no induced noise, and no crosstalk. Finally, the fibers and cables are extremely lightweight and low in cost compared to wire. These reasons add up to lower cost in installation and less maintenance time.

Fiber Optics in Operation

In cooperative experiments with Bell Labs in 1973, Western Electric, in its Atlanta, Georgia facility designed and developed light-guide fibers for telecommunication cabling. The fibers were arrayed into ribbons and placed in experimental cables at Atlantic Bell Labs. These cables were shipped to Chicago for testing. In May 1978 AT&T reported a successful one-year trial of a full-service lightwave installation in Chicago. The Chicago test involved a 1.5-mile link under downtown streets between two Illinois Bell Telephone switching stations and between one of the stations and an office housing a number of customers.

In late 1979, AT&T set three more installations into operation. A Connecticut hookup used technology in a portion of telephone network between the telephone company central office and local terminals located near customer premises. At about the same time a Florida installation provided service to a power station. In Arizona, Mountain Bell supplied service to a government agency near a power station. These experimental stations provided proof that lightwave systems are immune to electrical interference.

Probably New York Telephone set up the most comprehensive and innovative telecommunication systems to date. Two major lightwave installations began operating in 1980. The temporary one in upstate New York was a 2 1/2-mile lightwave link that operated successfully at Lake Placid during the Winter Olympic Games. The second, a permanent installation, links telephone company offices in White Plains with those on East 38th Street in downtown Manhattan. These systems carry incredibly large volumes of information, such as voice, video, and computer data, and in substantially less space than standard cable and wire systems. Both projects involved Bell Telephone Laboratories, AT&T, Western Electric, and New York Telephone.

In October 1978, General Cable Corporation announced that it would furnish the first optical fiber cable for use with the railroad industry at Union Pacific's Denver, Colorado, terminal. The optical fiber cable was designed and manufactured by General Cable to link a remote CCTV video camera with terminal office equipment to record information, car numbers, and reporting marks printed on passing freight cars. This information is used to check the accuracy of computerized advance makeups of arriving and departing trains, thereby verifying the makeup of the trains as well as the location of all the cars on the trains. The television also provides a check of all cars received from and sent to other railroads in the area. By having accurate information available promptly, switching can be prearranged and speeded up, thereby improving service to shippers. The optical fiber cable is an aerial installation and was continuously monitored for a year in order to assess its performance under conditions of vibration, ice and wind loading, and significant temperature changes. The optical fibers were manufactured by Corning Glass Works, Corning, New York.

Not only must the transmission medium be able to generate reliability over a broad range of harsh environment with a minimum of maintenance but also it must be economically feasible and, ideally, be immune to electrical and electromagnetic interference as well as offer a broad bandwidth of transmission capacity available for other railroad services.

Another optical fiber cable was designed, manufactured, and installed by General Cable for the U.S. Air Force's Arnold Engineering Development Center in Tennessee. The cable connects rocket engine test sites with a central data processing facility for real-time data analysis at the free world's largest rocket test facility. This was the first optical fiber cable fully engineered with splices, terminals, and other components for a working high-speed data transmission link—up to 150 megabits per second (Mb/s).

For NASA, General Cable designed, manufactured, and installed at Cape Kennedy an optical fiber cable for transmission of high-speed digital data and video signals between locations at the center. They also designed, developed, and manufactured for General Telephone & Electronics Corporation, the world's first optical fiber cable to provide regular telephone service to the public. It was placed in operation in California. Also, a similar optical fiber cable, providing telephone service in Brussels, Belgium, was manufactured by General Cable.

Teleprompter Cable Television Incorporated hooked up an 800-foot fiber optic link from an antenna on a Manhattan roof to head-end equipment on the bottom floor of a 34-floor building. The U.S. Navy's Airborne Light Optical Fiber Technology (ALOFT) program successfully demonstrated the use of fiber optics in flight tests on board an A–7 test aircraft. Point-to-point optical links were used. Northrup Corporation and ITT Cannon Electric on a joint research program developed a fiber optic bus to interconnect avionics equipment on board an F–5E tactical fighter. The bus is used on the F–16 and F–18 aircraft.

For several years now a fiber optic telephone system has operated with little or no failure on the *USS Littlerock*. Sonar links are now installed on U.S. navy submarines. The U.S. Army is using optical fiber cables in ground tactical and strategic telecommunications (deployable). General Motors has developed a fiber optic harness system to transmit a control signal to dashboard instruments. Sperry Univac and Digital Equipment Corporation are using fiber optics to connect mainframes and peripheral equipment on their computers. As early as April 1976, Rediffusion Engineering Limited in England installed a color TV cable (fiber optic) over a 1.5-kilometer stretch in Hastings, Sussex. Significant advances have and are being made in fiber optic technology in all electronics fields.

New Innovations in Fiber Optics

Vehicles used to lift personnel into high-tension lines may use fiber optics to reduce the chance of power spikes and the risk of injuring personnel by moving them into high-tension lines. Information could be taken to miners by fiber optics and therefore reduce the chance of spark generation. An ocean diver's heartbeat, breath rate, and body temperature could be monitored all on one fiber. Life-support systems and patient monitoring systems can be developed with clear safety from external interference.

Basic Operation of Fiber Optic Systems

A typical fiber optic system is illustrated in Figure 1–10. The operation of this simple system is defined in simple terms here and will be described in detail in Chapter 8. A system consists of the following:

1. A transmitter accepts an electrical signal and converts it to a current to drive the light source.
2. The light source launches the optical signal into the fiber.
3. The optical fiber provides a path for the optical signal.
4. A light detector detects and converts the optical signal to an electrical signal.
5. A receiver produces low noise and large voltage gain from the power detector signal.
6. Various connectors and splices interface the system.

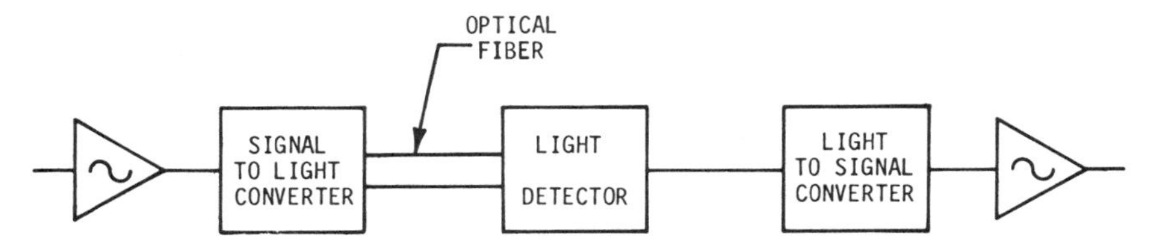

Figure 1–10 Typical fiber optic system (*Mike Anaya*).

INTRODUCTION TO LASER TECHNOLOGY

Laser Technology Defined

Laser technology is the science that deals with the concentration of light into small but powerful beams. The word *laser* is an acronym, that is, it is derived from the initials of "Light Amplification by Stimulated Emission of Radiation." The acronym was chosen when the technology of the maser was shifted from microwaves to lightwaves.

Maser is an acronym for "Microwave Amplification by Stimulated Emission of Radiation." The function of the maser is to concentrate atoms in an active medium such as a crystal or a gas at microwave frequencies, raise them to a higher level of energy by excitation, and then emit this radiated energy ito a narrow, intense beam. The function of the laser is the same, except that the operating frequency of the laser is in the optical spectrum, whereas the maser operates in microwave frequencies. The wavelength of microwaves range from 1 millimeter to around 30 centimeters (and probably greater). Optical wavelengths range from 0.005 to about 4000 microns. (A micron is equal to 10^{-6} meter.)

History of the Laser

Discussion of maser technology began in the early 1950s, but the first maser was not built until 1954. C. H. Townes, a professor at Columbia University, and students conceived and constructed the first maser. It was developed in an ammonia medium. Solid-state masers using paramagnetic ions embedded in crystal were developed late in 1957.

The maser researchers were involved in extending the use of masers into the optical wavelength. In 1958 a classic article was published on optical masers by C. H. Townes and A. L. Schawlow, a Canadian. The article triggered enthusiasm which led to the development of the laser. The first laser was created by Theodore H. Maiman, a scientist working for Hughes Aircraft Company in California. Maiman directed a beam of light into ruby crystals with a xenon flashlamp and measured the emitted radiation from the ruby. He discovered that an increase in input radiation beyond threshold caused emitted radiation to become extremely intense and highly directional. This was ironic, since maser scientists had rejected the ruby as a poor material for masers.

Uranium lasers were developed near the end of 1960 along with other rare-earth materials. Also near the end of 1960, A. Javin of Bell Laboratories created the helium neon laser. Semiconductor lasers (injection laser diodes) were simultaneously manufactured in 1962 by three American companies, General Electric, IBM, and Lincoln Laboratories. Therefore, it is difficult to give credit to one.

Laser Types

There are basically four laser types: gas lasers, liquid lasers, solid lasers, and semiconductor lasers. The gas laser uses gas as an active medium. A mixture of helium and neon is enclosed in a glass tube. An electrical current is discharged into the gas and a continuous flow of coherent (one frequency) waves is issued through the output coupler. The continuous-wave (CW) output is monochromatic (one color).

Liquid lasers use organic dyes enclosed in a glass tube for an active medium. A pump circulates the dye in the tube. A powerful pulse of light excites the organic dye. Solid lasers use a crystal such as the ruby for an active medium. The ruby crystal is solid and cylindrical. Each end is polished and parallel. A tungsten lamp tied to an alternating-current power supply excites the ruby (active medium). The output from the laser is continuous wave (CW) and coherent.

The semiconductor laser is made from semiconductor PN junctions. It is usually called an injection laser diode (ILD). The excitation mechanism (a direct-current power supply) controls the amount of current to the active medium (the ILD). The output light beam is easily modulated.

Laser Characteristics

All lasers have some common characteristics. These are an active material to convert energy into laser light, a pumping source to provide power or energy, optics to direct the beam repeatedly through the active material to become amplified, and optics to direct the beam into a narrow powerful cone of divergence. Other characteristics common to all lasers are a feedback mechanism to provide continuous operation and an output coupler to transmit power out of the laser. These characteristics are covered in detail in Chapter 9.

Laser Light Properties

Laser light has unique properties. The light is one color (monochromatic). Other properties are radiance, directionality, and coherence.

The radiance of the laser is extremely intense and is often compared with the intensity of the sun. The laser, when focused to a fine, hairlike beam, can concentrate all its power into the narrow beam; therefore, it has directional properties. If the beam were allowed to diverge or spread out, it would lose its power. Unlike normal beams of light, a coherent beam from a laser will travel long distances with very little divergence.

A laser beam is of one color (monochromatic) and one frequency, and the light beams are in phase with each other (coherent); that is, they are in time step.

Basic Laser

As an introduction to laser technology let us take a cursory look at a basic laser, the ruby laser (see Figure 1-11). As was previously discussed, the ruby laser was the first developed and was created by Theodore H. Maiman, a scientist at Hughes Aircraft Company. The ruby laser is constructed of an active medium, an excitation source, a feedback mechanism, and an output coupler. The active medium in the ruby laser is a tubular crystalline aluminum oxide structure with a small amount of

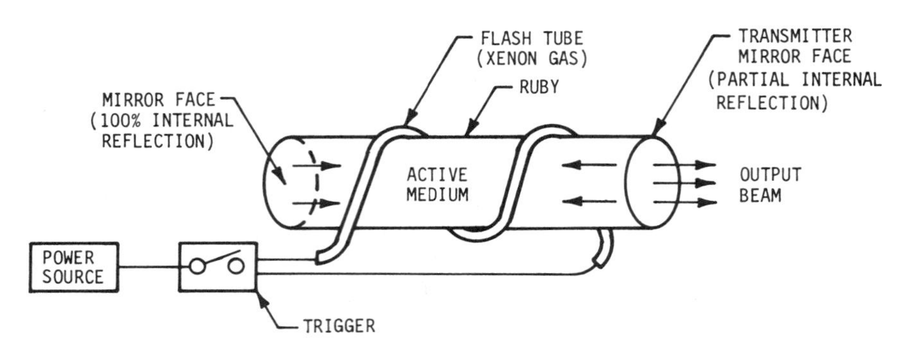

Figure 1–11 Basic laser (*Mike Anaya*).

chromium dissolved or doped into its lattice structure. The crystalline structure is grown in the same manner as other solid-state material, such as silicon. The ends of the crystal are highly polished and are cut parallel to each other. These ends are coated with a reflective coating. One end of the ruby reflects 100% of the light internally. The other end acts as a transmitter, allowing only a small portion of the light to pass outward while reflecting most of the light back into the ruby for stimulated emission. The excitation source is usually a flashtube. The tube is fitted with a trigger that emits a burst of light to start the process.

Let's now go through the operation and consider how it all comes together. The flashtube is energized by a trigger and produces a high-level burst of light similar to a camera electronic flash. The flash, being adjacent to the ruby cylinder, causes the chromium atoms within the crystalline structure to become excited. The process of pumping raises the chromium atoms from ground energy state to excited energy state. The ions then decay and fall to a lower or intermediate energy state. When the population of ions in the intermediate state is greater than the ground state, population inversion is present, that is, there are now more excited atoms than ground-state atoms.

When the population inversion is present, laser action (lasing) can take place. In time, the excited chromium atoms will fall to the ground energy state. When they do, photons are emitted. A photon is a pocket of light caused by radiant energy. The frequency of the energy determines the strength of the photons. Higher frequencies cause greater-strength photons.

Now let us consider the term *stimulated emission*. Chromium atoms are pumped to an excited energy state, then go through a spontaneous decay to an intermediate energy state where an accumulation of excited atoms creates a population inversion. As time progresses, the excited atoms fall from the intermediate energy state to the ground state. When they do, photons are emitted. As the photons are emitted they strike excited atoms and two other identical photons will emerge. These two photons in turn strike other excited atoms and stimulated emission occurs. The more often this happens the greater the intensity of the light beam.

When the beam of photons strikes either mirror ends of the ruby, the reflection of the light beam into the ruby causes an even greater beam intensity because of photon action. The beam is red in color because the laser is amplifying at a wavelength of 6943 angstroms (0.6943 micron). The white light of the flashtube produces many wavelengths; however, only one wavelength can excite the atoms from ground state to excited energy state. The amplified light is fed partially out of the laser by a transmitting mirror. The transmitting mirror reflects most of the light back into the ruby crystal to attain an ongoing process of stimulated emission and therefore amplification. If too much energy is coupled out of the transmitting mirror, the stimulated emission will decay and die. Therefore, the excitation mechanism must constantly pump energy into the laser.

This is just a cursory look at the operation of a laser. In Chapter 9 we provide a more in-depth look at laser types, characteristics, properties, and operation.

Lasers in Operation

The laser has become one of the most productive devices in present-day commerce and industry. It has been used in communications, holography, medicine, direction finding, manufacturing, and many other fields. It has always been a fictional aspiration to use a laserlike death ray or an antiballistic weapon that could destroy aircraft or missiles. The advances that are being made today allow the laser to operate with more power, greater distance, and greater control. These fictional thoughts are not as far from reality as they were a decade ago.

In the communications field, lasers are utilized in audio, radio, and television transmission. In television, the scene being shot is often illuminated by laser-light scanning. Laser beams have a narrow bandwidth, high directional capability, and low divergence. These properties are extremely important in communications using modulated light. Lasers are used in production of three-dimensional movies and television in a science called *holography*. Holography is also used in manufacturing processes, where it is used to detect vibrations and locate strains. Holography is as new as laser technology and its future just as promising.

In medicine there is no technology that has advanced to the extent of the laser. Ruby lasers for years have been used for eye surgery. Argon ion lasers are useful as scalpels. The body's skin, fat and muscle are transparent to selective laser wavelengths; therefore, cancer cells can be destroyed without harming the other parts of the body. Decayed areas of the teeth can be destroyed without harm to healthy teeth.

Distance measuring for surveying and for military fire control are other uses of the laser. Laser radars are now far past their infancy.

In manufacturing, the laser is used with holography to detect strains and to measure irregular objects. Powerful lasers are used to cut stacks of cloth, drill fine holes accurately, weld, and to align objects such as aircraft wings.

In light transmission, injection laser diodes are utilized to launch light into small-diameter optical fibers. Indeed, laser technology and the fiber optic industry are moving rapidly to the forefront in all industry and especially in conjunction with the electronics industry.

SELF-CHECK QUESTIONS

Self-check questions allow students to evaluate how well they have learned what they are studying.

1. Name several of the theories of light.
2. Who discovered the photovoltaic effect? In what year?
3. What part of the spectrum did Heinrich Hertz find to have special effects on metal?
4. What discovery did Willoughby Smith present to light science?
5. Who discovered radiation? In what year?
6. Who developed the quantum theory?
7. Who were the inventors of the transistor?
8. Integrated circuits were first designed by what company and designer?
9. Who was the first person to develop the concept of a laser?
10. Who built the first laser?
11. What are the three basic bands of frequency in the optical spectrum?
12. What is a wavelength?
13. What is the velocity of light?
14. What is the range of ultraviolet wavelengths?
15. What is the range of visible wavelengths?
16. What is the range of infrared wavelengths?
17. What is the visual color range of the human eye?
18. What is Newton's first law of motion in regard to light?
19. What are the three terms developed for the radiation of light of various bandwidths?
20. What are the three basic material types called light receptors?
21. What is a photon?
22. What is diffraction?
23. What are the two lens types?
24. How can you separate white light into multiple colors?
25. Define the term *fiber optics*.
26. Who were some of the developers of fiber optics?
27. In what wavelength range do fiber optics operate?
28. What are some of the advantages of using fiber optics?
29. Of what does a basic fiber optic system consist?
30. The word *laser* is an acronym. For what do the letters of the acronym stand?
31. What are the four basic laser types?
32. What are the laser's unique properties?
33. Of what does a basic laser consist?
34. Name some of the uses of the laser.

Two

Optoelectronics Terminology

Light is either leaving a source or arriving at a detector or sensor. At all other times it is in motion between the two. This chapter deals with optoelectronics terminology and the sourcing, transmitting, and receiving of light.

OBJECTIVES

After studying this chapter and completing the self-check questions at the end of the chapter, the reader will be able to:

1. State the definition of solar constant.
2. Explain the several energy states of the atom to include decay from one energy state to another.
3. Define the term *stimulated emission* and its relation to laser technology.
4. Compare the two light sources, point and extended.
5. Explain the two terms *point source intensity* and *area source intensity.*
6. Define the terms that are representative of light transmitting.
   a. Electromagnetic waves
   b. Polarization
   c. Coherence
   d. Beam divergence
7. Define the terms that are representative of light reception.
   a. Interference effects
   b. Incidence

c. Reflection
 (1) Diffuse
 (2) Fresnel
 (3) Surface
d. Scattering
e. Dispersion
f. Absorption
g. Photodetection
h. Electroluminescence

GENERAL

The sun and the stars are the natural sources of light. The aurora is another natural source of light. Reflective sources of light include planets and moons. The amount of light energy received from the sun (solar irradiance) outside the earth's atmosphere is called the *solar constant* and is equal to 140 milliwatts per square centimeter. The atmosphere absorbs much of this. If we consider a clear day with no fog, smog, or cloud cover, the amount of solar irradiance reaching the earth is somewhat less, say 100 milliwatts per square centimeter. The amount of light energy received from the sun is an inverse curve that is dependent primarily on the distance from the sun.

Light is the form of electromagnetic radiation that acts upon the retina of the eye and the optic nerve. This makes sight possible. The term *light* has been altered by technical advances to include electromagnetic wavelengths not visible to the human eye. The entire frequency spectrum that deals with wavelengths visible and adjacent wavelengths is called the *optical frequency spectrum.*

There are three basic levels in the optical frequency spectrum:

1. *Infrared:* band of light wavelengths that are too long for response by the human eye
2. *Visible:* band of light wavelengths to which the human eye responds
3. *Ultraviolet:* band of light wavelengths that are too short for response by the human eye

Optoelectronics deals with this entire spectrum. You may wish to review the section "Optoelectronic Operational Spectrum" in Chapter 1 for details of the spectrum. Wavelength ranges for these areas of the spectrum are 0.005 to 0.3900 micron for the ultraviolet, 0.3900 to 0.7500 micron for the visible, and 0.7500 to 4000 microns for the infrared. These values will be repeated at significant places throughout this book.

Since light is a measurable source, two scientific fields have been created to measure it. Photometry is involved with the visible electromagnetic wavelengths. Radiometry deals with those wavelengths that are too short or too long for the human eye to perceive. In Chapter 10 we explain these two scientific fields of measurement and describe techniques and applications for measurement.

The theory of light called the quantum theory defines the photon. The photon is an uncharged particle and has an energy level of its own which is dependent on its frequency or wavelength. The higher the frequency of the energy, the greater the strength of the photons.

Energy States

Each atom has several energy states (see Figure 2–1). The lowest energy state is the ground state. Any energy state above ground represents an excited state and can be called by names such as first or second energy states or intermediate energy states. If an atom is in one of its energy states (E_4) and decays to a lower energy state (E_2), the loss of energy (in electron volts) is emitted as a photon. The energy of the photon is equal to the difference between the energy of the two energy states. The process of decay from one energy state to another is called *spontaneous decay* or *spontaneous emission.*

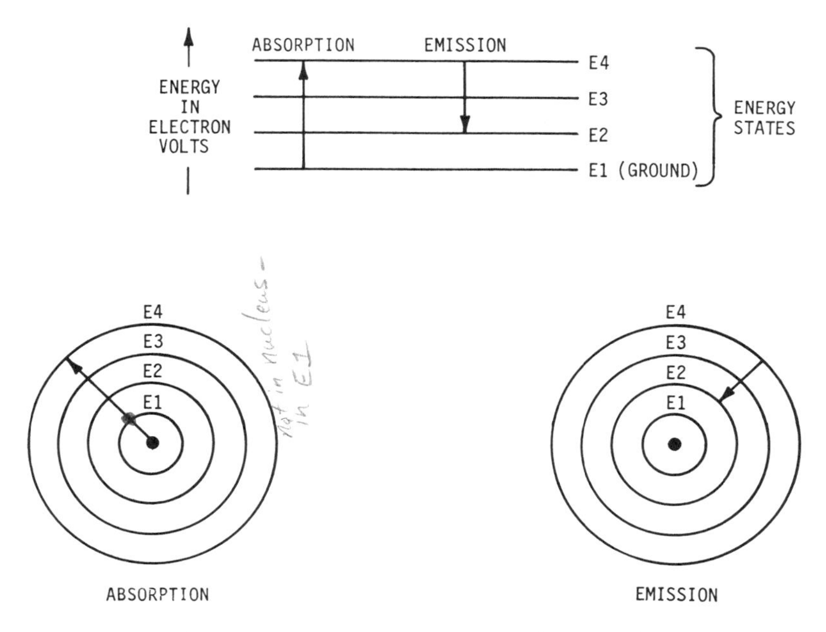

Figure 2–1 Energy states in atoms (*Mike Anaya*).

An atom may be irradiated by some light source whose energy is equal to the difference between the ground state and an energy state. This can cause an electron to change from one energy state to another by absorbing light energy. This transition from one energy level to another is called *absorption.* In making the transition the atom absorbs a packet of energy called the *photon.* This is similar to the process of emission.

The energy absorbed or emitted (photon) is equal to the difference between the energy states. For instance,

$$E_2 - E_1 = E_p \text{ (energy of photon)}$$

Also,

$$E_p = hf$$

where h = Planck's constant

f = frequency of the radiation

The value of h has a dimension of energy and time. The value is 6.625×10^{-34} joule-second. The value of f is in cycles per second (hertz). Photon energy is said to be directly proportional to frequency.

Photon energy may also be expressed in terms of wavelength. You may recall that wavelength is equal to the speed of light in meters per second divided by frequency,

$$\lambda = \frac{c}{f}$$

where λ = wavelength, in meters

c = speed of light, in meters per second

f = frequency of the electromagnetic wave, in hertz

The formula can also be stated in the following manner:

$$c = f\lambda \qquad \text{and} \qquad f = \frac{c}{\lambda}$$

If we substitute this value of f in the formula for the energy of a photon, we have a second formula that relates the energy of the photon to wavelength.

$$E_p = hf$$

$$E_p = h\left(\frac{c}{\lambda}\right)$$

$$E_p = \frac{hc}{\lambda}$$

Stimulated Emission

You may recall from the last discussion on energy levels (or states) that spontaneous emission occurs when an atom decays from one energy state to a lower energy state (see Figure 2–2). The power that originally caused the atom to move into a higher level of energy was an electric current or a powerful light. Another way for an atom to change energy levels is to absorb a photon. This can happen only when the energy difference between two levels of an atom, the level it is in and the level to which it is being excited, are equal to the energy of the photon.

Stimulated emission is a laser process. Let's consider an atom that is in, say,

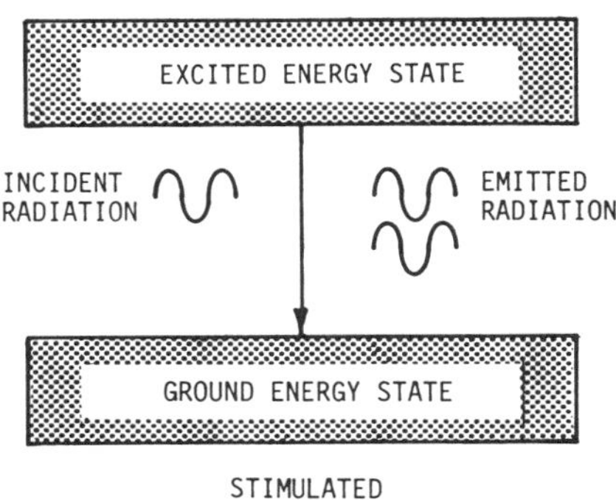

Figure 2–2 Stimulated emission (*Mike Anaya*).

a third-level energy state and will probably decay to a lower energy state. While the atom is still in the higher energy state an external source of radiation is applied that is extremely close to the frequency of the atom in the excited state. Then there is a good chance that the external source will stimulate the atom into moving from that excited state to the ground level. In the case of stimulated absorption the incident radiation moves the atom to a higher energy level. In the case of stimulated emission the atom is moved to a lower level and emits two wave packets (photons of precisely the same frequency).

The stimulated emission and the incident frequency are the same and move in the same direction and are in step (in phase) with each other. Therefore, they are coherent. The time that an atom remains in an excited state is extremely short, about 10^{-8} second. Stimulated emission must take place when the atoms are in this excited state. Each atom has several energy states and moves between these states. The longer the time between these energy states, the better the atoms are adapted for stimulated emission. Long-duration energy states are called *metastable states* and have energy state times of 10^{-6} to 10^{-3} second. When excitation has taken place to cause absorption, the population of the excited-state atoms exceeds the ground-state atoms; when this happens the process is called *population inversion.*

You may recall the Bohr theory of the atom from other basic atomic studies. The theory describes the atom as having a nucleus consisting of neutrons and protons. Electrons rotate around the nucleus in shells or rings. The outside ring holds valence electrons. These electrons are the farthest away from the nucleus and therefore are not held as tightly in place as the electrons closer to the nucleus. If there is enough radiant energy applied to the atom, the electrons can be removed from their orbits. The electrons can then escape from the force applied by the nucleus and become free carriers. The amount of energy required to remove the electron from the valence ring (excite the valence electron into an excited state) is called the *energy gap*. Listed in Table 2–1 are some of the solid-state materials used in optoelectronics along with their energy gaps.

TABLE 2-1 OPTOELECTRONIC MATERIALS AND THEIR GAPS

Atom	Energy Gap in (eV)
Cadmium sulfide (CdS)	2.4
Gallium phosphide (GaP)	2.2
Cadmium selenide (CdSe)	1.7
Gallium arsenide (GaAs)	1.4
Silicon (Si)	1.1
Germanium (Ge)	0.7
Lead sulfide (PbS)	0.37
Lead telluride (PbTe)	0.29
Lead selenide (PbSe)	0.26

LIGHT SOURCING

Basic Light Sources

A point source is theoretically a dimensionless (no length, width, or height) point in space from which light is emitted equally in all directions (see Figure 2-3a). In reality this cannot be, for any point or light source must have a finite size. Point sources can, however, appear to be a defined point such as a star that is light-years away and a laser beam. In the illustration, rays that leave point P arrive at point A in parallel lines of the direction P to A. Likewise, rays leave point P to points B and C in the same manner.

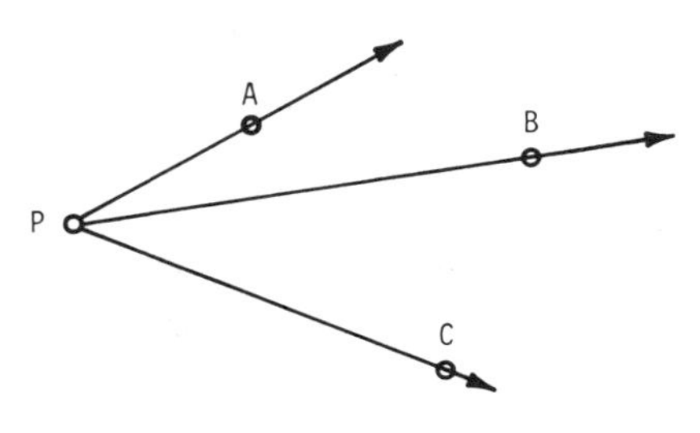

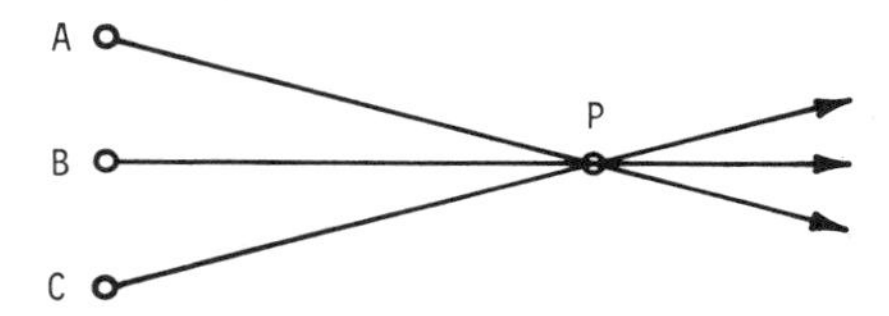

Figure 2-3 Basic light sources (*Mike Anaya*).

In Figure 2–3b, an extended source of light is shown. This figure depicts a point in space that is illuminated by light being radiated from several different directions. Points, *A, B,* and *C* (the extended source) are illuminating point *P* from several directions. The point *P* then is illuminated by each source point. The farther you move from point source (*A, B, C*), the light lines become more nearly parallel. From an extreme distance the extended source appears to be a point source. For example, a flashlight viewed from a distance of 5 feet would appear to be an extended source that radiated from many points on its face. That same flashlight when viewed from 500 feet would appear to be a point source.

Point Source Intensity

Intensity is the amount of radiant or luminous flux per unit solid angle that is diverging from the light source. Photons of light energy combine to create radiant power called *flux.* The steradian is a solid angle of a sphere equal to the area of the sphere divided by the radius squared (see Figure 2–4a).

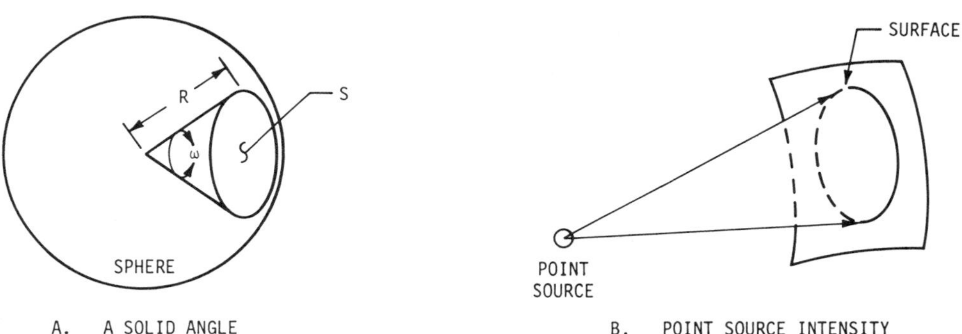

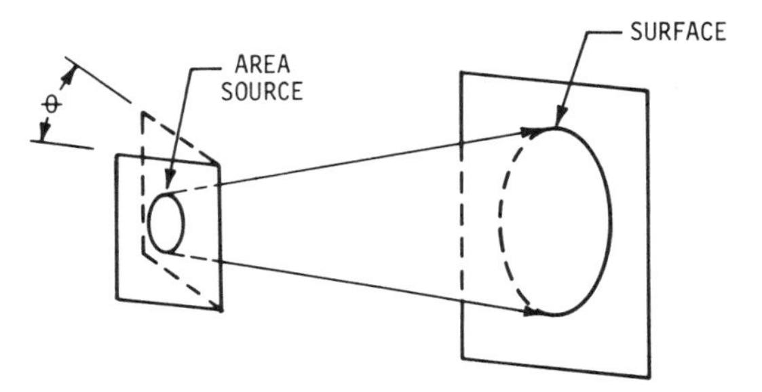

Figure 2–4 Source intensity (*Mike Anaya*).

$$\omega = \frac{A}{r^2}$$

where
ω = solid angle = 1 steradian
A = curved surface area of the sphere
r = radius of the sphere

There are 4π steradians within a sphere. That is, there are 4π solid angles projected from a point source at the center of the sphere.

Point source intensity in radiometric terms is measured in watts per steradian. Radiant intensity (the light power of the point source) is the amount of power from a point source generated in a unit solid angle (see Figure 2–4b).

$$I_R = \frac{P}{4\pi}$$

where
I_R = radiant intensity, in watts per steradian
P = radiant flux, in watts
$4\pi = 4 \times 3.1416$

Point source intensity in photometric terms is measured in candelas or lumens per steradian. Luminous intensity is the amount of power from a point source generated in a solid angle (see Figure 2–4b).

$$I_L = \frac{F}{4\pi}$$

where
I_L = luminous intensity, in candelas or lumens per steradian
F = luminous flux, in lumens
$4\pi = 4 \times 3.1416$

Note that the solid angle is applicable to both radiometric and photometric measurements. Measurement of light intensity along with other light measurement is covered in Chapter 10 (see Figure 2–4c).

Area Source Intensity

Intensity, once again, is the amount of radiant or luminous flux diverging from a projected source area or light source on a finite area. This flux per unit area is also called *exitance.* Area source intensity in radiometric terms is measured in watts per centimeter squared.

$$B_R = \text{W/cm}^2$$

where
B_R = radiance
W = watts
cm^2 = area of source

Area source intensity in photometric terms is measured in candelas or lumens per square centimeter.

$$B_L = L/cm^2$$

where
B_L = luminance
L = lumens
cm^2 = area of source

An area source whose angular distribution is a circle is called a *Lambertian source.* As you will see later on in the book, a flat LED (light-emitting diode) chip is an excellent approximation of the Lambertian area light source.

LIGHT TRANSMITTING

Electromagnetic Waves

Light energy travels in a definite energy pattern called an *electromagnetic wave*—the same as radio waves but at a much higher frequency. As the name suggests, there are two components to the wave: one electrical, the other magnetic. The two components pass energy back and forth between them, so that the total energy content is constant. For example, when the electric field is strongest, the magnetic field will have just reached zero.

If an observer is stationary in space and watches a wave pass by (they have a habit of passing very fast, i.e., at the speed of light), he or she will see the electric field "slowly" increasing in strength as the wave passes. After reaching a peak, the strength would die back to zero. At this instant the magnetic field will have reached its strongest point. The electric field will then build up with the opposite polarity and the magnetic field would shrink to zero. Each buildup and decay of energy follows the sine-wave pattern of Figure 2-5.

Any electromagnetic wave has four fundamental properties: its frequency, velocity, polarization, and strength (amplitude). From frequency and velocity, a fifth, its wavelength, can be calculated. The *period* of an electromagnetic wave can also be calculated. This is the amount of time a stationary observer measures for the wave to cycle through one complete set of changes.

$$T = \frac{1}{f}$$

where
T = period of one oscillation, in seconds
f = wave's frequency, in hertz

The *velocity* of light, although very high, does depend on the properties of the material it is passing through—its dielectric constant in particular. In empty space light travels at 300,000,000 meters per second. In any other material it travels slightly slower because the dielectric constant will be higher. In these materials the speed also depends on the wavelength, so that blue light travels slower than red. The speed changes are important because they explain how lenses, prisms, and optical fibers work.

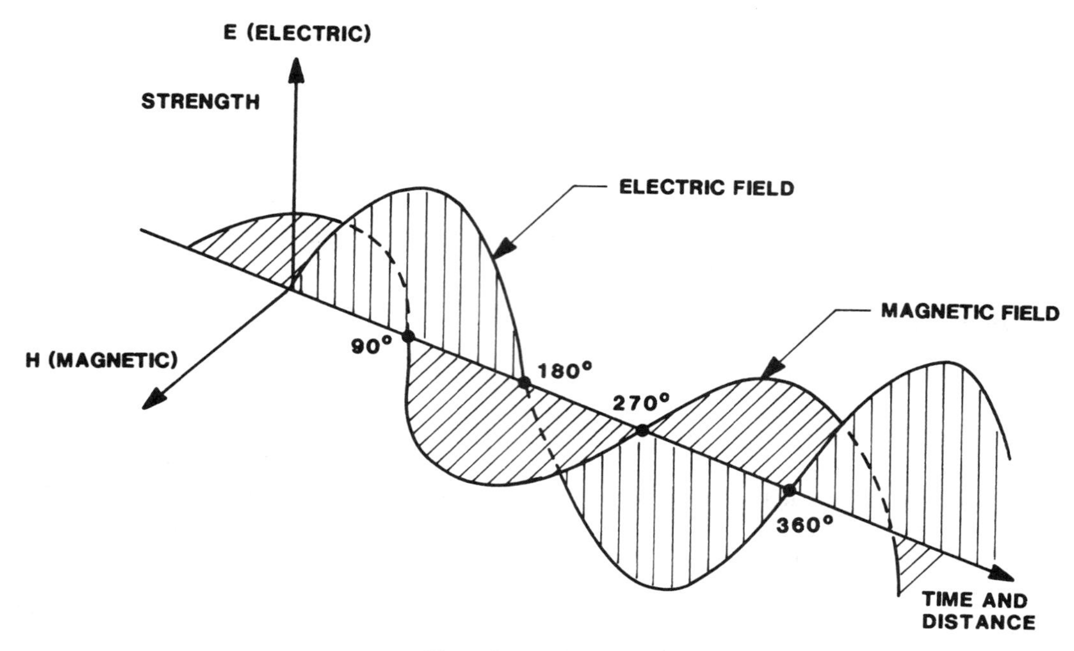

Figure 2–5 Electromagnetic waves.

The physical length of the wave between any two identical points is its *wavelength.* As one wave of light passes through different materials it will always have the same frequency of vibration, but because the velocity may change slightly, the wavelength will change in direct proportion.

$$\lambda = \frac{v}{f}$$

where λ = wavelength, in meters
v = velocity, in meters per second
f = frequency, in hertz

What humans call *color* is simply another expression for frequency. It is the sensation that our optical and nervous systems experience when light of specific wavelengths hits our eyes. Single or tightly grouped frequencies could be experienced as red, orange, yellow, green, blue, violet, or any shade in between. Energy at all frequencies will be viewed as white. As mentioned previously, the eye could experience the color blue either from a single-frequency "blue" light or from two frequencies, "red" and "green." Remember that this is a mixing property of the eye that other electronic detectors will not possess.

Polarization of Waves

All light energy has a polarization. This describes how the electric field of the E/M wave is oriented (i.e., vertical, horizontal, or some other angle). Some sources, an incandescent lamp for example, emit light with a random assortment of polarizations. This light is then said to be unpolarized even though each individual component does have some specific polarization. A special case occurs when two separate beams with the same strength and wavelength travel together. If one is polarized vertically and the other horizontally, they will appear to have a rotating electric field and the result is called *circular polarization* (see Figure 2–6).

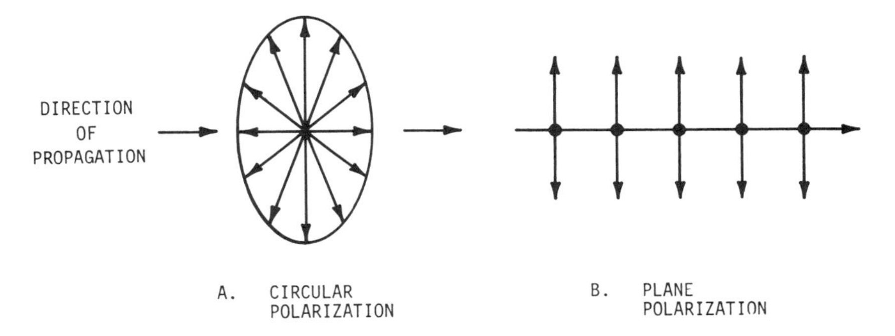

Figure 2–6 Polarization of waves (*Mike Anaya*).

Polarization is very important in many areas of optoelectronics. Within optoelectronic systems such as the laser, linear polarizers are used to separate out all rays that are not parallel to the propagation axis. The beam can be tested for polarization by rotating the polarizer by 90°. This should cancel the beam by orienting the polarizer so that its axis is perpendicular to the plane of the transmitted beam. In Chapter 6 we illustrate and describe the optic polarizer.

Coherence

Light sources such as fluorescent lamps do not radiate coherently. The light from these lamps is radiated in many frequencies and just as many phases. As a result the light beams do not have uniform direction and certainly are not polarized. This incoherency can be compared with the light of a laser beam, which is highly directional and polarized. Its wavefronts are defined and its frequency along with phase relationships are constant. In effect, the beam is monochromatic (one color) and coherent.

The coherency of the light waves is imperative to good laser operation. To have coherency the waves must have a consistent relationship between their wave troughs and wave crests. Wave troughs are points of minimum vibration and wave crests are points of maximum vibration. Coherency is divided into two major relationships,

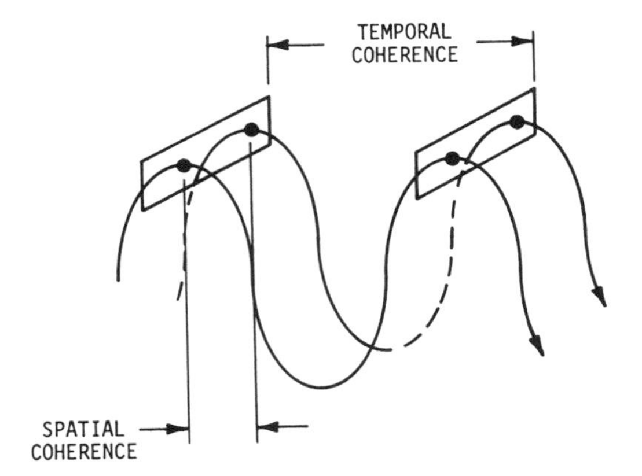

Figure 2–7 Coherence (*Mike Anaya*).

spatial and temporal (see Figure 2–7). To be spatially coherent, waves must have phase correlation across a wavefront at any point in time. The spatially coherent waves also maintain their shape in relation to time. To be temporally coherent, waves must be frequency constant so that phase correlation is in the direction of propagation. The phenomenon of coherence allows the laser beam to be focused to a minute unit area. A conventional source cannot be focused without loss of power. Furthermore, coherent light can be made to produce constructive interference of waves such as those used in holography.

Beam Divergence

All light beams tend to spread out as they travel great distances. The amount is directly proportional to the wavelength and inversely proportional to the initial diameter of the beam. Laser beams are no exception.

In Figure 2–8 a typical beam is illustrated. D_1 represents the diameter of the beam at its aperture (output from the laser). D_2 represents the diameter of the beam at the detector, receiver or somewhere in space. L is the length of the beam at that point. Angle θ is the angle of divergence or fanning out. The angle in the illustration

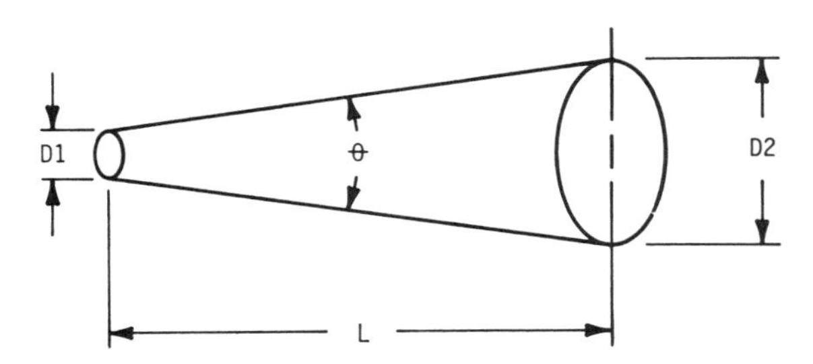

Figure 2–8 Beam divergence (*Mike Anaya*).

is extremely exaggerated for the divergence of a laser beam is a very narrow cone. The calculation for the diameter of beam divergence is as follows:

$$D_2 = L \angle\theta + D_1 \qquad \text{(for very small angles)}$$

where

D_2 = diameter of beam divergence
L = length of projected beam
D_1 = diameter of the projection at the aperture
$\angle\theta$ = angle of divergence, in radians

The angle θ must be converted to radian measure for these calculations. You may recall from your geometric studies that a radian is equal to 57.3°. Two pi (2π) radians are equal to 360° angular rotation about a point. Since the angle of divergence is quite small, the radian measure of that angle is usually in milliradians. One milliradian is one-thousandth of a radian or 0.0573°, which represents 3.44 minutes of arc.

Often, the beam divergence is specified in half-angle divergence. In this case the half-angle would be θ divided by 2, with minor considerations given to the aperture diameter. Typical laser beam divergence is 1 milliradian for gas lasers and 10 milliradians for dye lasers. Semiconductor lasers have large beam divergence angles such as 500 milliradians.

Since it is imperative that the beam be thin with low divergence angles, it behooves the designer to plan for focusing to preclude large divergence angles. Focusing is accomplished with lenses. The ability of a lens to focus is dependent on its focal length and the beam source divergence angle. There are limits to which the beams can be focused. These limits are approximately equal to the square of the beam wavelength. For large apertures, lenses may not be needed.

LIGHT RECEPTION

The reception of light is completely dependent on what use or application is intended. If the desire is to light a room, the light should reflect diffusely from most objects in the room with absorption dependent on the materials and colors in the room. If the light were to be used for fiber optic applications, fiber ends would reflect at zero percent if possible and the light refracted into the fiber would be maximum. The amount of absorption within the fiber would be held to a minimum. If the light were to be used for photodetection, the amount of absorption would be extremely high (100% if possible). The amount of reflection would be held to a minimum. Thus the application decides the use of materials that are low or high absorbers, reflect or do not reflect, and refract or do not refract.

Interference Effects

The study of interference and its peculiarities is an extremely broad field. The field does have applications in optoelectronics, primarily in laser technology and holography. Interference effects may be produced by coherent light. Broadly defined, interference is the interaction between two beams of light. The two beams should be coherent for the interaction cannot take place with incoherent light. The interaction produces interference fringes. The fringes are variations of resultant amplitudes of the interaction as the waves vary from point to point in a field of view. The variations are due to path differences where the waves reinforce each other at certain points and oppose each other at other points. These interference patterns allow the relative coherency of a light source. The measurement is made by analysis of the interference fringe pattern.

The first attempt to observe interference patterns was made by Francesco Grimaldi in the seventeenth century. He placed a monochromatic light from a single source in front of a screen with two slits in it. He did observe interference, but it was not clear. It was later shown by Thomas Young that the interference could only be seen clearly if the light source is small (see Figure 2–9). Young's experiment involved a small light source radiated through a slit in a screen which he called a *source slit*. A second screen with two slits, each equidistant from a horizontal line extending from the source slit, was placed between the source and the viewing screen. Some waves from the source are intercepted by slit 1. Slits 2 and 3 sample these waves at two points. If the phase relationship of the two wave fronts is fixed, the interference pattern will appear on the screen as bright and dark lines or fringes. The fringes may be analyzed thusly. If the path difference (D) between the two path lengths is equal to half a wavelength, the waves arrive at point P, 180° out of phase with each other. Similarly, if the path difference (D) between the two paths is equal to a full wavelength, the waves arrive at point P in phase.

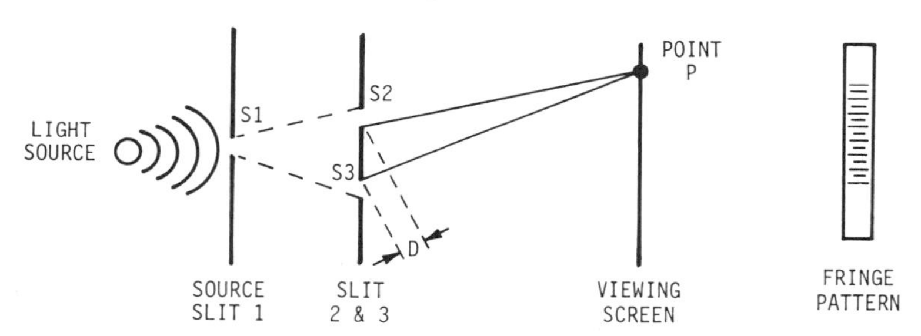

Figure 2–9 Interference effects (Young's experiment) (*Mike Anaya*).

If the interference patterns projected on the screen are dark fringes, the two waves are out of phase. Out-of-phase interference patterns are called *destructive interference*. In Figure 2–10a, interference patterns 180° out of phase are illustrated. The wave pattern is a standing wave such as is found in laser cavities. However, if the interference patterns projected on the screen are bright fringes, the two waves are in phase. In-phase interference patterns are called *constructive interference*. In Figure 2–10b, these in-phase waves are summed to a higher-amplitude resultant.

Incoherent light does not produce spatial or temporal coherence. The wavefront is not clearly defined and the phase relationships are constantly shifting; therefore, incoherent light sources cannot be used for the formation of interference fringes.

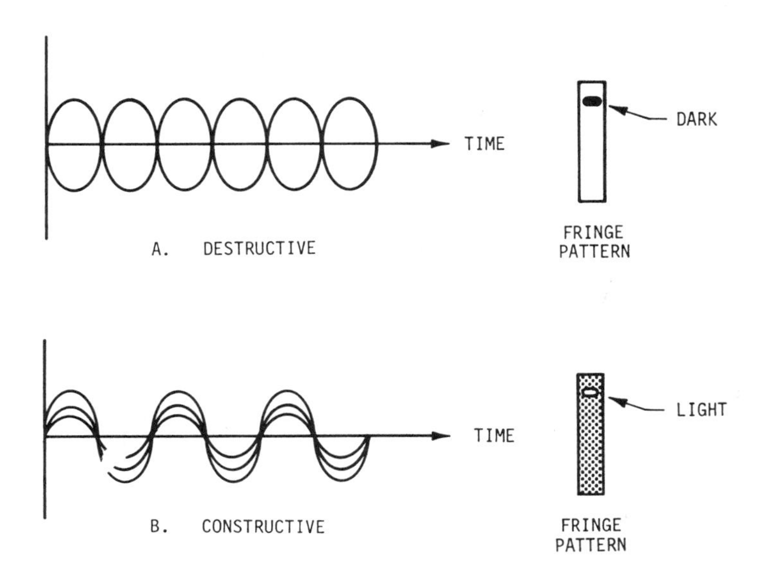

Figure 2–10 Interference patterns (*Mike Anaya*).

However, if the incoherent light source is projected through a narrow-band filter it may be utilized. The filter tends to isolate monochromatic waves of the light. That light may contain some degree of temporal coherence.

Incidence

Incidence is the amount of flux per unit area that is normal or perpendicular to a surface or detector. If the flux is not normal (not perpendicular), the normal component of the angular flux is the incidence. In radiometric terms, incidence is called *radiant incidence* or *irradiance.* Irradiance (E_e) is measured in watts per square meter ($E_e = W/m^2$). Incidence in photometric terms is called *luminous incidence* or illuminance (also called illumination). Illuminance is measured in lux (lx) or lumens per square meter (lm/m^2).

When light falls on a surface it is partly reflected from the surface and partly absorbed. It may also be converted by the surface into some other form of energy. In this case it becomes part of an active surface such as a photodiode or phototransistor. In any event, incidence is how much flux falls on the surface area. From a point source, the law of illumination of light falling on a given area is called the *inverse square law* (see Figure 2–11). You will note that 1 lumen of flux illuminates 1 square foot of area from 1 foot away at 1 footcandle (1 lumen/square foot). At 2 feet from the point source, an area of 4 square feet is illuminated to 1/4 footcandle. Finally, at 3 feet from the point source, an area of 9 square feet is illuminated to 1/9 footcandle. The distance away from the source is squared to provide the illuminated area. The illumination of that area is the inverse to the square of the distance.

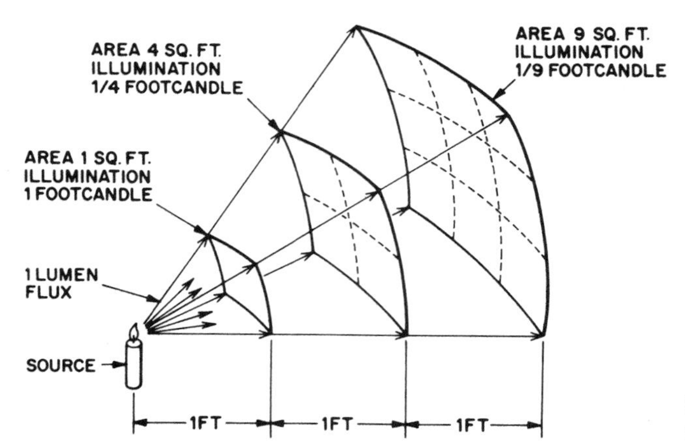

Figure 2–11 Point source illumination (the inverse square law).

Illumination from an area source must be looked at from a different perspective. The area source must be assumed to be an extended source. That is, each point or element on the surface contributes to the luminous intensity of the source. The luminous intensity in a given direction, divided by the projected area of the source is called the *luminance* of the source. In Figure 2–12, area source illumination is illustrated. An angle is made between a normal line (perpendicular to the surface) and the incident flux lines. The angle is used to calculate the effective area presented to incident flux. The effective area is equal to area times cosine $\angle \theta$. The illumination in lumens per square feet is equal to the area of the receiver times cosine $\angle \theta$ times the incident flux. Refer to Chapter 10 for more details on photometry/radiometry.

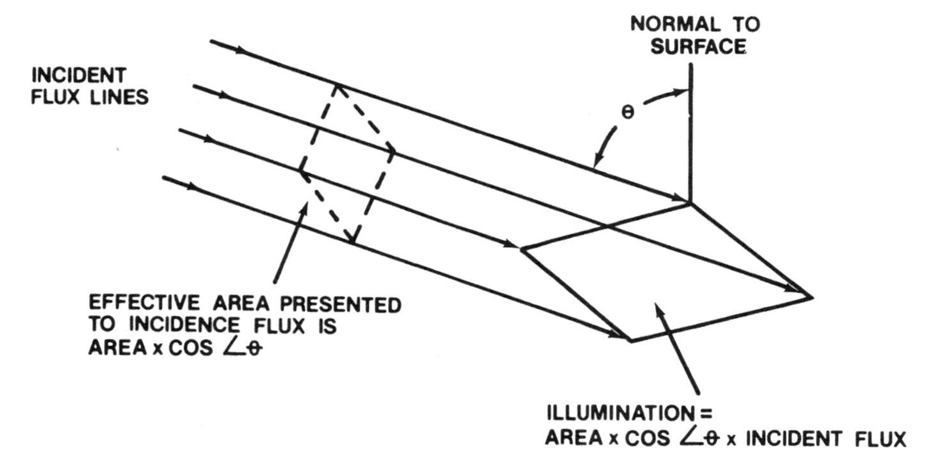

Figure 2–12 Area source illumination (the cosine law) (*Mike Anaya*).

Reflection

When light beams or rays strike a surface, they reflect from that surface either in a diffuse manner or a specular manner or in some cases both. The amount of reflection is largely due to the type of material on the surface of the reflector (see Figure 2–13).

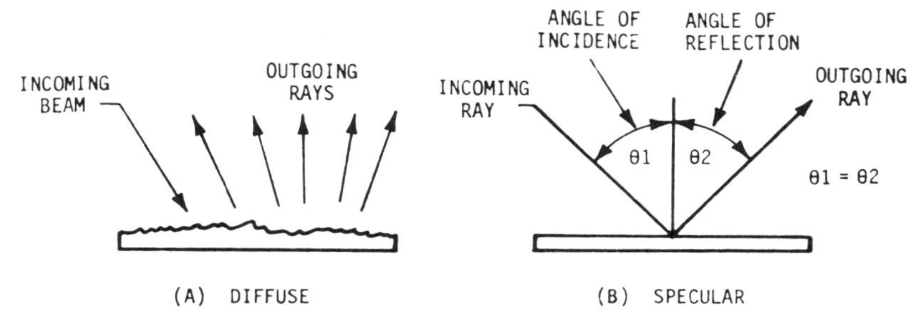

Figure 2–13 Reflection (*Mike Anaya*).

Diffuse reflection. Diffuse reflectors scatter the light outward in several beams depending, of course, on the surface material and the amount of light the material absorbs. Virtually all material absorbs some light. In Figure 2–13a, the principle of diffuse reflection is illustrated. A smooth surface such as a freshly painted house would diffuse light in rather equal and alike reflected rays. Another surface such as a rock-covered roof would cause light rays to be reflected in scattered, nondescript lines.

Surface or specular reflection. Whenever a beam of light from one medium, such as air, strikes a second mirror-like medium such as glass, part of the beam is reflected (see Figure 2–13b). The angle that the incident rays strike the second medium is called the *angle of incidence.* This angle is the angle made by incident beam and a line normal (perpendicular) to the boundary of the two media. Part of the incident beam is reflected at an angle called the *angle of reflection.* This angle is made by the reflected beam and a line normal (perpendicular) to the boundary of the two media:

$$\text{angle of incidence} = \text{angle of reflection}$$

If the angle of incidence varies, so does the angle of reflection by the same amount. The angle of incidence and the angle of reflection lie in the same plane.

Fresnel reflection. Fresnel reflections become Fresnel reflection losses when dealing with fiber optics. Fresnel losses are losses that are a result of the differences between refractive indexes of the glass fiber core and its cladding. Variations of Fresnel reflections are related to unpolarized light, light polarized perpendicular to the plane of incidence, and light polarized parallel to the plane of incidence.

Surface Refraction

Whenever a beam of light passes from one medium, such as air, to another medium, such as water, the beam separates at the intersection. Part of the beam is reflected back into the incident medium (air) and part is refracted into the second medium (water). The angle made by the incident beam and a line normal (perpendicular) to the intersection is called the *angle of incidence.* The angle made by the reflected beam and the line normal to the intersection is called the *angle of reflection.* The angle of incidence and the angle of reflection are equal. The angle made by the refracted beam and a line normal (perpendicular) to the intersection is called the *angle of refraction* (see Figure 2–14).

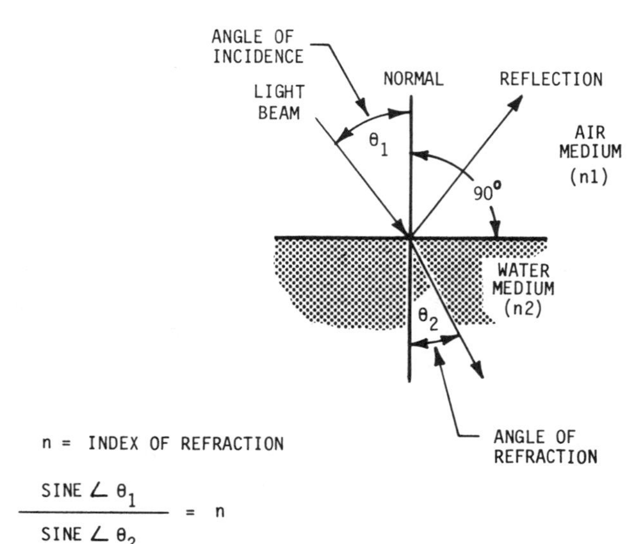

Figure 2–14 Surface refraction (*Mike Anaya*).

The ratio of the indexes of refraction of the two indexes of refraction is thus

$$n = \frac{n_2}{n_1}$$

or

$$n = \text{index of refraction}$$

The ratio of the sine of the angle of incidence to the sine of the angle of refraction is equal to the index of refraction.

$$n = \frac{\text{sine} \quad \angle \theta_1}{\text{sine} \quad \angle \theta_2}$$

Therefore,

$$\frac{n_2}{n_1} = \frac{\text{sine} \quad \angle \theta_1}{\text{sine} \quad \angle \theta_2}$$

or

$$n_1 \text{ sine } \angle \theta 1 = n_2 \text{ sine } \angle \theta_2$$

Furthermore, the ratio is a constant ratio. Whenever the angle of incidence changes, the angle of refraction changes to retain the ratio. The law is *Snell's law.* Snell was a Dutch astronomer and professor of mathematics at the University of Leyden in Holland. In fiber optic applications it is imperative that the light incident to the fiber end be refracted into the fiber core. The amount of light refracted is determined by the refractive indices of the air and the fiber core. Refer to Chapter 7 for fiber optic transmission angles.

Scattering

Scattering is differentiated from absorption in the following manner. With true absorption, the intensity of the beam is decreased in calculable terms as it penetrates the medium. Light energy absorbed in the material is converted to heat motion of molecules. Consider a long tunnel where you could only see light from one end. As you walked nearer that end the light gets brighter. The light is absorbed as it travels through the tunnel at the absorption rate of the air medium. If the tunnel were then filled with a light cloud of smoke, the smoke would scatter some of the light from the main beams; therefore, the intensity of the light from a fixed distance would decrease. You may observe scattering effects by watching dust particles as the sun shines in a window. Parts of the rays are scattered by the dust particles.

A scattering type known as *Rayleigh scattering* is caused by micro-irregularities in the medium. A wave passing through the medium strikes these micro-irregularities in the mainstream of the waves. The waves reflected from the microparticles are spherical and do not follow the main wave, but scatter. Therefore, the intensity of the beam is diminished. In fiber optic application the composition of the glass must be considered to ensure low Rayleigh scattering. Silica has low scattering losses.

Dispersion

Whenever a beam of light enters matter its velocity decreases within that medium. The subject of dispersion deals with the speed of light in a medium and the variation in conjunction with wavelength. The speed of light in a medium can be calculated by its refractive index. The refractive index (n) of any material is the ratio of the speed of light in a vacuum to its speed in the medium.

$$n = \frac{c}{v}$$

where n = refractive index of a medium
c = speed of light in a vacuum
v = speed of light in a medium

Then the speed of light (v) in the medium may be calculated by

$$v = \frac{c}{n}$$

The refractive index for air is 1.000, whereas optical glasses have values of 1.520 to 1.720 for refractive indexes. Any change in refractive index would represent a change in velocity.

The dispersion of each material is different. Curves are supplied by manufacturers that plot wavelength against refractive index. As the wavelength decreases the refractive index increases and the dispersion ($dn/d\lambda$) increases. The rate of increase becomes greater at shorter wavelengths. Dispersion is the spreading of light rays. In the propagation of light rays through a medium such as an optical fiber the problems of dispersion are multiplied. With fiber optics there are two types of dispersion, intermodal and intramodal. Intermodal dispersion is caused by the propagation of rays of the same wavelength along different paths through the fiber medium. This results in the wavelengths arriving at the opposite end of the fiber at different times. Intramodal dispersion is due to variations of the refractive index of the material that the fiber was made from. Dispersion is measured in nanoseconds per kilometer (ns/km) in fiber optic applications.

Absorption

Whenever a beam of light enters matter its intensity decreases as it travels further into the medium. There are two types of absorption, general and selective. General absorption is said to reduce all wavelengths of the light by the same amount. There are no substances that absorb all wavelengths equally. Some material such as lampblack absorbs near 100% of light wavelengths. With selective absorption the material selectively absorbs certain wavelengths and rejects others. Almost all colored things such as flowers and leaves owe their color to selective absorption. Light rays penetrate the surface of the substance. Selective wavelengths (therefore colors) are absorbed, whereas others are reflected or scattered and escape from the surface. These wavelengths appear in color to the human eye. In optoelectronics it is imperative that near 100% of incident beams be absorbed in devices such as solar cells and detectors. On the reverse side, zero percent absorption is important in the transmission of light waves in an optical fiber cable and lenses.

Electroluminescence

The most modern of the optoelectronic effects is *electroluminescence.* This effect involves the application of electrical energy to a semiconductor material, causing it to emit light (radiated) energy. Current is caused to flow through a load from a negative terminal through negatively charged metal. Electrons are released and are drawn through positively charged metal to the positive terminal. An energy gap is developed between the negatively charged metal and the positively charged metal at their junction. It is at this junction that electrons combine with positive charges to create photons. These photons cause the device to radiate light through a glass envelope similar to a small lightbulb.

Combinations of the two metals used provide different light wavelengths. Material such as gallium arsenide added to silicon provides one wavelength. Adding phosphorus to the combination provides another wavelength. Different materials will cause radiated light to appear to the human eye as different colors, such as green or red. Other materials will radiate invisible light in the ultraviolet or infrared ranges of the spectrum. This is an oversimplification of the process. Much more detail is provided in Chapter 4.

Photodetection

Photodetection includes three phenomena: photoemission, photoconduction, and photovoltaic action. Photoemission is defined as the emission of electrons by light energy. The effect of light on certain materials causes electrons to be displaced from the surface. The emitted (radiated) electrons are called *photoelectrons.* Light is applied to the cathode of a basic phototube. The light causes photoelectrons to be drawn through a load by the positive anode. Thus the light determines the current flow and the voltage drop across the load resistance. Current flow is also controlled by the potential applied to the anode.

Operation through photoconduction involves a change in resistance of the photosensitive material. A wafer of photoconductive material is placed underneath a glass window to protect it from exposure. The photoconductive material is tied to a load resistance and a power source. The clear glass window allows light radiation to strike the photoconductive material freeing valence electrons. The resistance of the photoconductive material decreases, causing current through the load to increase.

The photovoltaic cell uses dissimilar metals to generate an electromotive force in response to radiated light. A light-sensitive material is placed beneath a thin layer of transparent metal and next to a dissimilar metal. The light-sensitive material is exposed to radiation through the thin transparent metal that acts as a filter. When exposed, free electrons are removed from the light-sensitive material, causing electrons to flow to the dissimilar metal. This creates current flow and a difference of potential between the two terminals connected to the load. Photodetection is described in detail in Chapter 5.

SELF-CHECK QUESTIONS

Self-check questions allow students to evaluate how well they have learned what they are studying.

1. What is light?
2. What is the meaning of *optical frequency spectrum?*
3. Name the three energy states of an atom.
4. Define the term *spontaneous decay.*
5. Define the term *absorption.*
6. The formula for wavelength is $\lambda = c/f$; what do the terms of the equation represent?
7. Define the term *stimulated emission.*
8. Define the term *population inversion.*
9. Define the term *energy gap.*
10. What are the two types of light source geometry?
11. What is light source intensity?
12. Name the five basic quantities or measurements of an electromagnetic wave.
13. Rays of light have been grouped into numbers of wavelengths for technical terminology use. What are these groups?
14. Define the term *polarization.*
15. When are light beams said to be coherent?
16. What is beam divergence?
17. What is meant by the term *interference effects?*
18. Define the term *incidence.*
19. Name the three types of reflection.
20. Define the term *refraction.* What relation does incidence have with refraction in Snell's law?
21. Define the term *scattering.*
22. Define the term *dispersion.*
23. Define the term *absorption.*
24. Name the three types of photodetection.
25. Define the term *electroluminescence.*

Three

Semiconductor Physics

Most devices used in optoelectronics are semiconductors. It therefore seems necessary to provide the reader with a review of the basics of semiconductor physics. This chapter is a short discussion of these elements.

OBJECTIVES

After studying this chapter and completing the self-check questions at the end of the chapter, the reader will be able to:

1. Describe the atom, its makeup, atomic number, and weight.
2. Examine a two-dimensional model of the atom.
3. Explain the fundamental structure of the atom caused by doping.
4. Present the interface between a space relationship such as a valence model, and an energy model of an atom.
5. Discuss in general terms the quantum theory and the energy states of the photon.
6. Define *radiant energy.*
7. Compare the differences between intrinsic and extrinsic materials.
8. Analyze the energy-level schematic of extrinsic semiconductor donor material.
9. Analyze the energy-level schematic of extrinsic semiconductor acceptor material.
10. Compare the energy-level schematic of biased and unbiased PN junctions.
11. Explain the effects of sunlight on a PN junction.

SEMICONDUCTOR PHYSICS

The Atom

The atom is the smallest particle of a material that retains the identity of the material. For instance, a copper atom is the smallest part of a piece of copper wire. Most of you are probably familiar with the two-dimensional model of a typical atom. Let's review it once more to reestablish some pertinent terms. The atom consists of a nucleus with electrons in orbit much the same as the planets orbit the sun. Within the nucleus are the proton and the neutron. The number of protons equal the number of orbiting electrons. The number of neutrons are not necessarily equal to the number of protons. The atomic weight of the atom is equal to the number of protons plus the number of neutrons. The weight may or may not be twice the number of electrons in orbit. For instance the atom sodium has an atomic number (number of electrons) of 11 while its atomic weight is 22. In comparison, an atom of aluminum has 13 electrons while its atomic weight is 27. The electron is a very small thing. It is said to be 1/1840 of the mass of a proton or a neutron.

Valence Bond Model

A valence bond model (see Figure 3–1) has been established by research. No matter how many electrons are involved, the maximum number of electrons in any orbit follows the same pattern. For instance, the *K* shell has two electron spaces possible, but no more. The *L* shell has possible two electron positions in its first ring and six electron spaces in its second ring. The *M, N, O, P, Q* shells follow this same pattern. For explanation purposes let's consider the copper atom. The copper atom has 29 electrons, 29 protons and 35 neutrons. Its atomic weight is approximately 64 (see Figure 3–2). Its *K* shell has 2 electrons. Its *L* shell has 8 electrons. Its *M* shell has 18 electrons and its *N* shell has but one electron in its valence. This, of course, adds

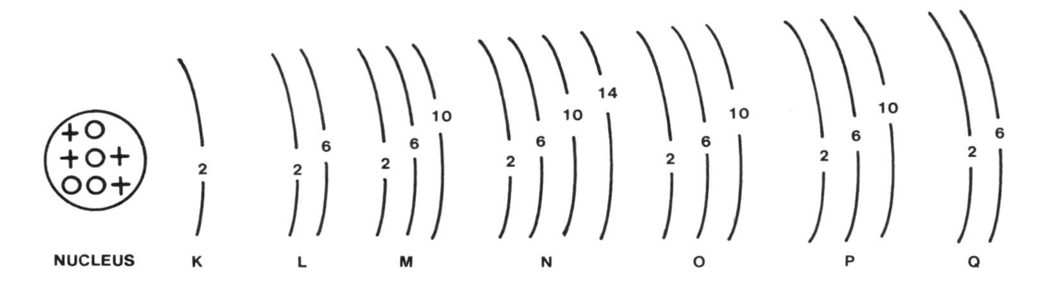

Figure 3–1 Valence bond two-dimensional model.

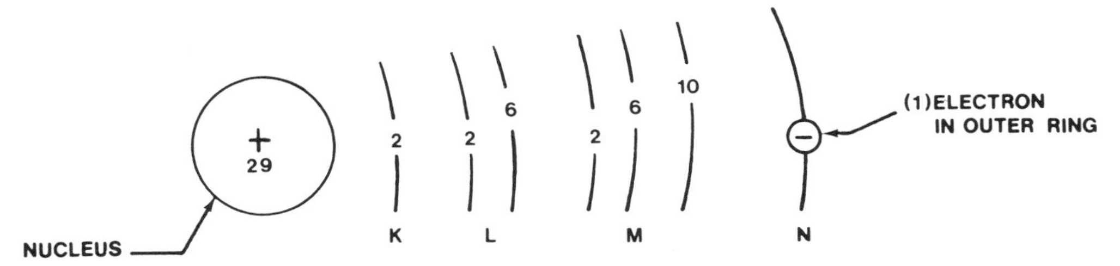

Figure 3–2 Valence bond two-dimensional model of copper atom.

to a total of 29. The outer ring is called a *valence*. In the two-dimensional illustration, it appears that electrons follow the same orbit. In actuality, the electrons all have their own orbit. Other good *conductors* in the table of elements (aluminum, silver, and gold) have only one electron in the valence. A good conductor has this structure. It is possible to move the electron from the valence, therefore producing electron flow (electricity). Some force is used to remove the electron, such as a magnetic field, friction, or heat. Indeed, in the case of photovoltaic material the electrons are removed from orbit by radiant light. They are called *free electrons*.

Just as some materials are good conductors, other materials are poor conductors. Poor conductors of electricity are called *insulators*. Insulators are usually compounds such as glass, mica, ceramics, and other substances. Since the atoms of these substances generally have a large number of electrons in the valence, the attraction of the electrons to the nucleus is strong.

Some special atoms are neither good conductors or good insulators. These atoms are called *semiconductors*. There are two semiconductor atoms in common use, silicon and germanium. The silicon atom has 14 electrons and the germanium atom has 32 electrons. Each of these atoms has 4 electrons in its outer shell, 2 in one ring and 2 in its valence ring. In Figure 3–3, a two-dimensional model of the silicon atom is illustrated. In Figure 3–3a, the atom of silicon is shown to have 14 electrons, 2 in its *K* shell, 8 in its *L* shell, and 2 apiece in the outer rings of the *M* shell. Semiconductors such as silicon or germanium have crystalline structures. Each silicon atom, as shown

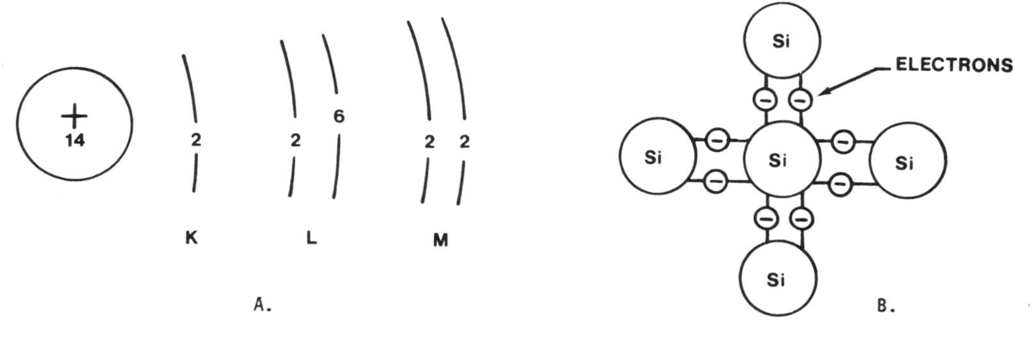

Figure 3–3 Two-dimensional model of silicon atom.

in Figure 3–3b, shares one of its electrons with each of its neighboring atoms. This is called *diamond sharing, covalent bonding,* or *valence bonding*. Pure silicon has this structure (called a *lattice*) throughout. Of course, pure silicon is very difficult to obtain (grow) and in reality never exists. Silicon is called pure when it has less than 1 part per 10^{10} impurities. Electron movement is minimal. Pure silicon is called *intrinsic* material. The model in Figure 3–3b represents an intrinsic silicon crystal.

Doped Semiconductor Material

Figure 3–4 compares two types of dopant, N and P. Doping a material is the addition of other atoms into the lattice structure of the material. This changes that structure. The P-type element (also called an *acceptor atom*) has three electrons in its valence. Figure 3–4a shows a model of silicon with an atom of P-type material (boron) in its crystal lattice structure. The reader will note that the acceptor atom has a space called a *hole* (absence of an electron). Typical donor elements (atoms) are boron, aluminum, and gallium. The holes are called *majority carriers* or positive charged carriers. *Minority carriers* in P-type material are *electrons.* The N-type element (also called a *donor atom*) has five electrons in its valence. Figure 3–4b shows a model of silicon with an atom of N-type material (phosphorus) in its crystal lattice structure. The reader will note that the donor atom has a loosely bound electron. Typical donor elements (atoms) are phosphorus, arsenic, and antimony. The free electrons are called majority carriers or conduction electrons. Minority carriers in N-type material are holes. N-type material is silicon that has been doped to be electrically conductive by means of mobile electrons and therefore is negative. Conversely, if the silicon has been doped for conduction by holes, it is said to be positive or P-type.

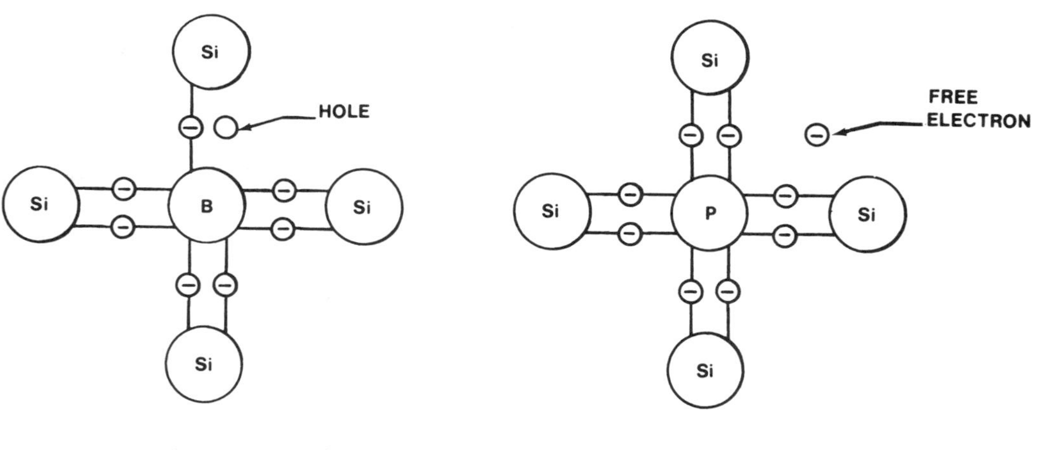

Figure 3–4 Dopants.

The type of dopant (N or P) is extremely important to determine bias polarity in the design of electrical circuits. Impurity dopant is added in the order of 1 or 2 parts per million. The exact doping level is extremely proprietary information. The second important characteristic of the dopant is resistivity. A semiconductor is, as its name implies, semi or partially conductive. The degree of conductivity of the silicon is fixed by the amount of dopant that is added. The resistivity of the end product (semiconductor device) is determined by the type and amount of the dopant material added to the silicon. Typical conductive metals have a resistivity of 10^{-3} ohm per square centimeter. Intrinsic (pure) silicon has a resistivity of 3×10^5 ohms per square centimeter at room temperature. N-type or P-type materials are semiconductors that have been doped to have an excess or a lack of electrons in the valences of its atoms. This semiconductor material is called *extrinsic*.

Energy Model

The valence model provided us with a space relationship. The energy model allows us to study the semiconductor in terms of energy.

In order to interface the valence model with the energy model, let's look at the comparison in Figure 3-5. This illustration compares the electron orbits with energy levels. The first orbit is the first energy level, and so on.

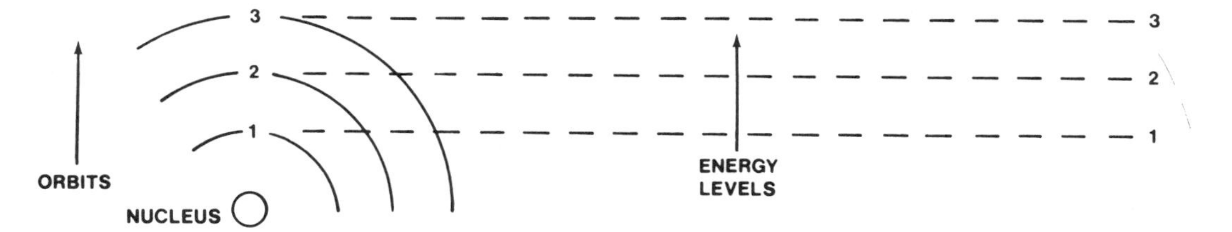

Figure 3-5 Comparison of orbits and energy levels.

It takes a certain amount of energy to move an electron from one orbit to another. These orbits are at distinct energy levels. Electrons cannot exist between these levels. Each energy level is a band of energy. Figure 3-6 illustrates an energy diagram of an intrinsic crystal of silicon. Intrinsic silicon falls in group IV of the periodic table of elements because it has four electrons in its valence. The significant energy bands in this intrinsic silicon material are the conduction band and the valence band. Valence electrons must be removed by some force and drawn across the forbidden gap. Electrons cannot exist in the gap.

There are three ways in which electrons can be drawn across the forbidden gap: the application of heat, the application of an electric field, and the application of light. To remove electrons from the valence band, it takes 1.1 electron volts of energy at room temperature. The forbidden gap decreases with an increase in temperature and increases as temperature decreases.

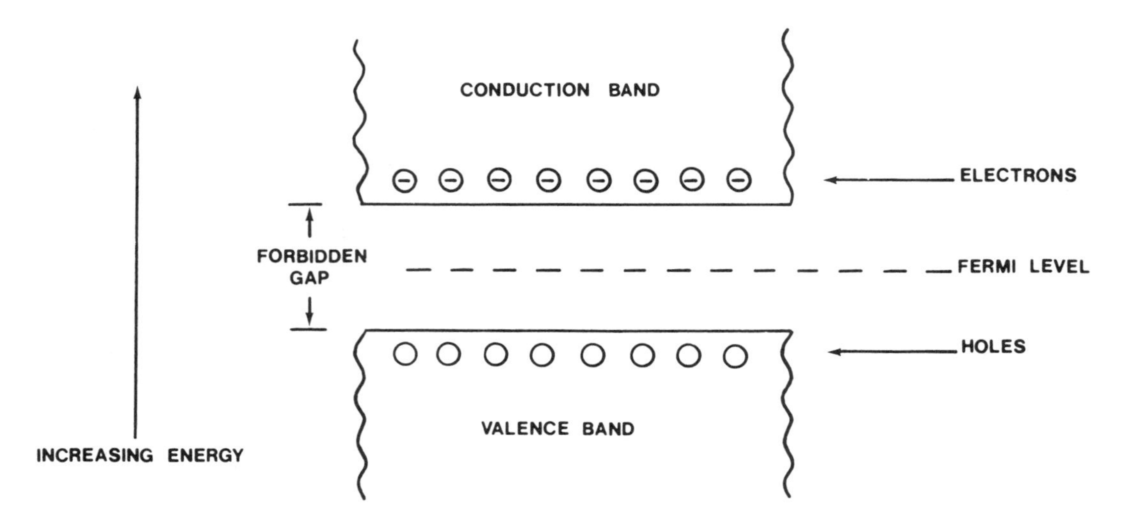

Figure 3–6 Energy-level schematic of intrinsic silicon.

As you may see, when an electron leaves the valence band it leaves a positive charge called a *hole*. The hole may appear to move just as the electrons move to other atoms. Holes draw electrons from neighboring atoms, leaving another hole, and so on. It is imperative that a state of electrical equilibrium exist in an intrinsic crystal. Therefore, the number of electrons in the conduction band will equal the number of holes in the valence band. The *Fermi level* is a level in the forbidden (energy) gap where there is an equal opportunity for an electron to exist with a change in energy above the Fermi level as there is for a hole to exist with a change in energy below it. In short, the Fermi level is centered between the conduction and valence bands in intrinsic silicon. With impurities added the Fermi level will move toward one or the other band.

Energy Levels in Extrinsic Silicon

It has probably occurred to the reader that an intrinsic semiconductor is not an extremely useful substance. It can be used for several purposes: for instance, as a resistance. However, the best use of semiconductor material is always achieved when dopants are used to create desired parameters and special characteristics.

Under a previous heading we discussed the valence model of extrinsic semiconductors. Let's now consider the energy levels of this doped material.

In Figure 3–7 an energy-level schematic of a donor (N-type) material is shown. Donor material falls in group V of the periodic table of elements because it has five electrons in its valence. Just as with intrinsic silicon, this material has a conduction band, a valence band, a forbidden gap and a Fermi level. Since only four of these valence electrons fit into the crystal lattice structure of the silicon, a free electron is introduced for each dopant atom. At absolute zero, the electrons will fill a posi-

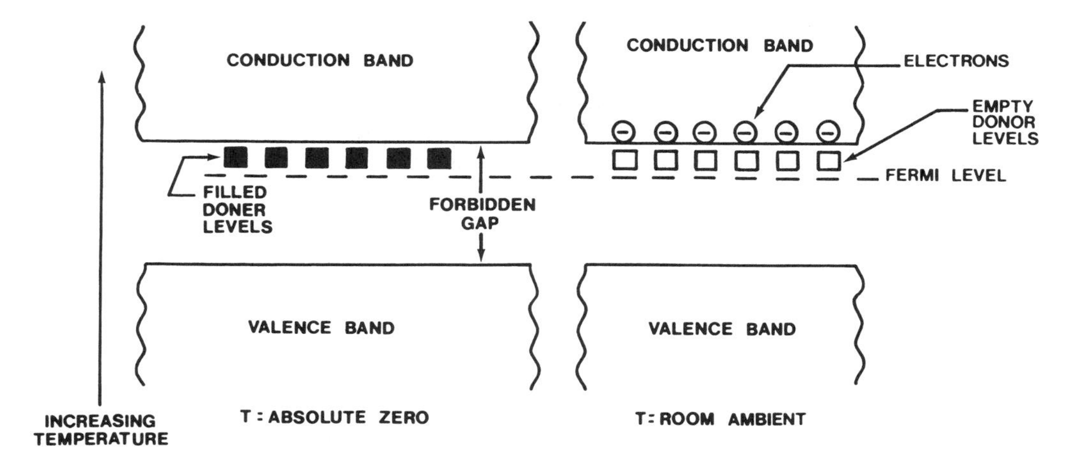

Figure 3–7 Energy-level schematic of N-type (donor) material.

tion at the lower side of the conduction band. As temperature increases toward room ambient, the electrons move into the conduction band where they are available for conduction. Because of the introduction of electrons in the conduction band, the Fermi level will be closer to the conduction band. Note that temperature did not effect the Fermi level. This crystal is at equilibrium.

In Figure 3–8, an energy-level schematic of an acceptor (P-type) material is shown. Acceptor materials fall in group III of the periodic table of elements because they have three electrons in their valence. Just as with intrinsic silicon, this material has a conduction band, a valence band, a forbidden gap, and a Fermi level. Since only three of these valence electrons fit into the crystal lattice structure of the silicon,

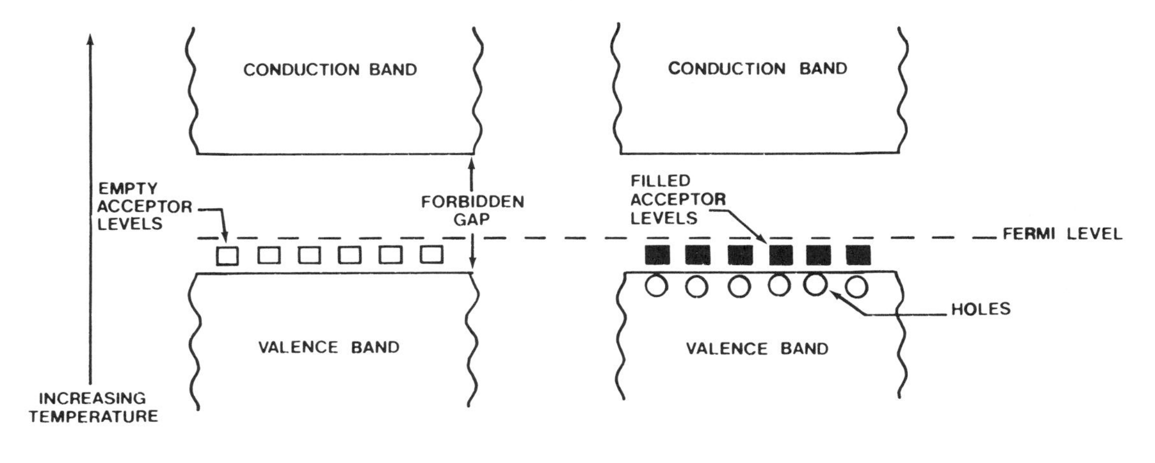

Figure 3–8 Energy-level schematic of P-type (acceptor) material.

a hole is introduced for each dopant atom. At absolute zero, the acceptor level holes exist just above the valence band. As temperature increases toward room ambient, electrons move into the empty acceptor level holes and leave holes in the valence band where they are free for conduction. Because of the introduction of holes in the valence band, the Fermi level will be closer to the valence band. Note that temperature did not effect the Fermi level. This crystal is at equilibrium.

Energy Levels in an Unbiased PN Junction

In Figure 3–9, a PN junction is shown. The junction was formed by diffusion. Diffusion is accomplished by exposing one type of semiconductor material to gas which consists of the other type of dopant. The gas is controlled by temperature and time. In any event, the N-type and P-type materials are back to back with one another. In the N-type material, electrons are called majority carriers and holes are minority carriers. In the P-type material, holes are the majority carriers and electrons are the minority carriers. When the junctions are formed, electrons from the N-type material are attracted by the holes from the P region. These electrons diffuse across the junction and recombine with some of the holes. This action leaves a small positive charge in the N region. In a similar pattern, holes from the P region are attracted by electrons from the N region. The holes diffuse across the junction and recombine with some of the electrons. This action leaves a small negative charge in the P region. An electric field is generated by these charges which repels further electron–hole recombination. The interface between the two material types becomes depleted of carriers, either electrons or holes. The area is called a *depletion region*. The field created in this depletion region is an actual difference of potential called a *barrier*. In doped

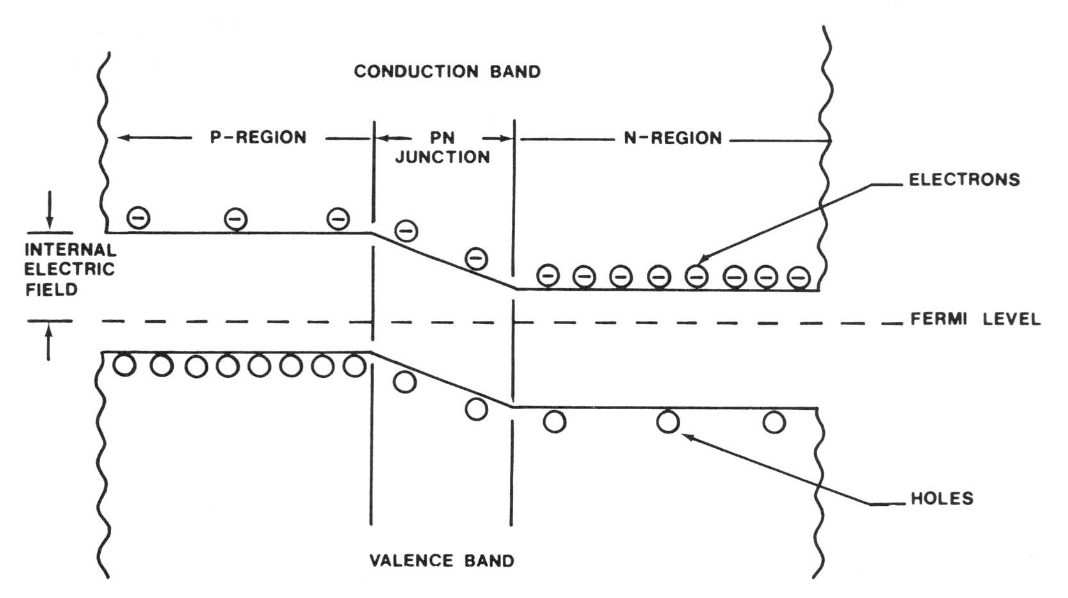

Figure 3–9 Energy levels in an unbiased PN junction.

silicon the barrier potential is around 0.65 volt. In germanium the potential is around 0.3 volt at 25 °C. This voltage varies with temperature. A curve is provided with PN devices which show the variation.

Energy Levels in a Biased PN Junction

In Figure 3–10, energy levels for a PN junction are illustrated. If a forward bias is applied to the junction the external electric field is in opposition to the barrier voltage (internal electric field). Electrons from the N-type material are attracted to the positive terminal. Holes from the P-type material are attracted to the negative terminal. An electric current may flow through the junction. In Figure 3–10 you see that the external electric field has shifted the Fermi level.

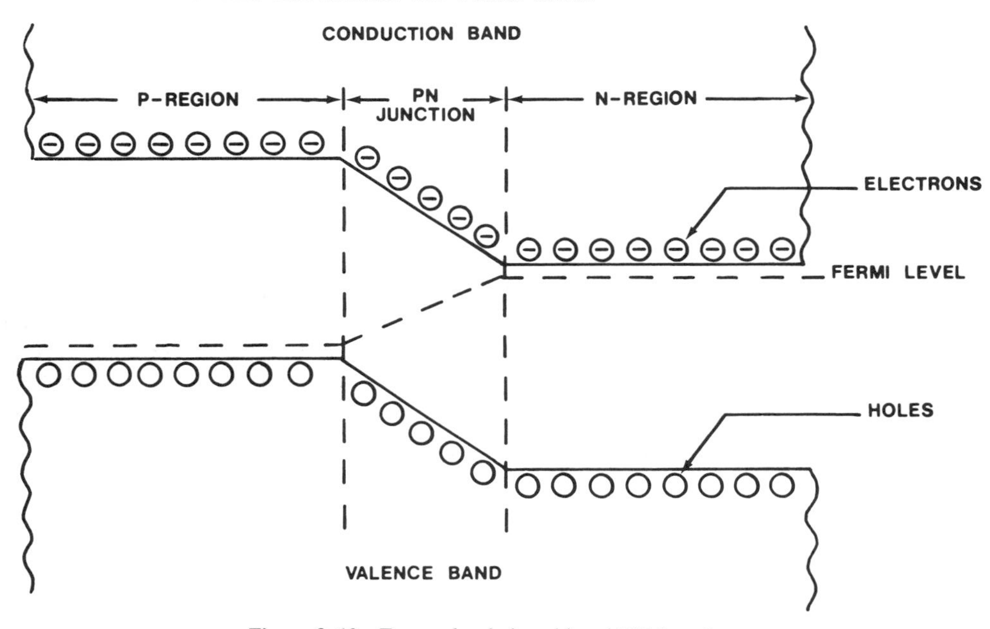

Figure 3–10 Energy levels in a biased PN junction.

SELF-CHECK QUESTIONS

Self-check questions allow students to evaluate how well they have learned what they are studying.

1. Define an atom and name its three components.
2. What determines the atomic weight of an atom?
3. What are *electron shells?*
4. What distinguishes an insulator from a conductor?
5. Name three ways that electrons can jump the energy gap.
6. If atoms are moved to a higher energy state, are they absorbing or emitting radiation?

Four

Electroluminescence

Probably the most useful of all the optoelectronic effects is electroluminescence. This effect involves the application of electrical energy, in the form of current flow, to a semiconductor material causing it to emit light. This light, in the form of photons, is used in applications such as light-emitting diodes (LEDs) and in LED combinations to provide digital and alphanumeric displays. Another application of electroluminescence is the injection laser diode (ILD) used to launch light waves in optical fibers.

OBJECTIVES

After studying this chapter and completing the self-check questions at the end of the chapter, the reader will be able to:

1. Define the term *electroluminescence.*
2. Compare several electroluminescent materials by their operation wavelengths.
3. Explain the function of electroluminescent device performance curves.
   a. Spectral response
   b. Operating time/power
   c. Forward current/output power
   d. Diode current/voltage
4. Interpret LED specifications.

5. Present some of the applications of electroluminescent devices.
   a. Current flow indications
   b. Inverter logic
   c. Open collector logic
   d. Switch closure indication
6. Compare the cross sections of the red and orange LEDs and the material of which they are made.
7. Choose the correct LED by selection of basic factors of LED usefulness.
8. State some of the typical characteristics of LED performance curves.
9. Name uses of flashing LEDs.
10. Determine methods of mounting LEDs.
11. Become familiar with LED displays.
12. Compare the LEDs direct dc drive and the strobe (multiplexer) drive.
13. Describe the monolithic display as to function.
14. Describe the intelligent display as to function.
15. Compare the cross sections of the homostructure and heterostructure ILD and the material of which they are made.
16. Describe the structure of a scattered-light LCD display.

GENERAL

Electroluminescence is a photoelectric effect in which electrical energy is converted to light energy. This is, of course, the opposite phenomenon to photodetection, where light energy is converted to electrical energy.

Figure 4–1 represents the fundamental operation of an electroluminescent device, the light-emitting diode (LED). Voltage is applied to a circuit by a battery or some other direct-current power source. The circuit contains a load resistance and an electroluminescent device (LED). The load resistance limits the amount of current that can be drawn in the circuit. The LED converts the electron flow to radiant energy. The LED is forward-biased, that is, the battery polarity is applied positive on the P material, negative on the N material. Electrons are caused to flow through a load resistance into negatively (N) charged material. Holes (absence of electrons) are created by electrons drawn from the positively (P) charged material to the positive pole of the battery. An energy gap is developed between the negatively charged material and the positively charged material at the junction. It is at this junction that electrons combine with holes to create photons. The electron–hole combinations cause the LED device to radiate energy. This radiated energy is seen through a glass or plastic envelope similar to a small light bulb.

Another electroluminescent device is the injection laser diode (ILD). The ILD is a heterostructure device. Electrons are injected into a junction region where there

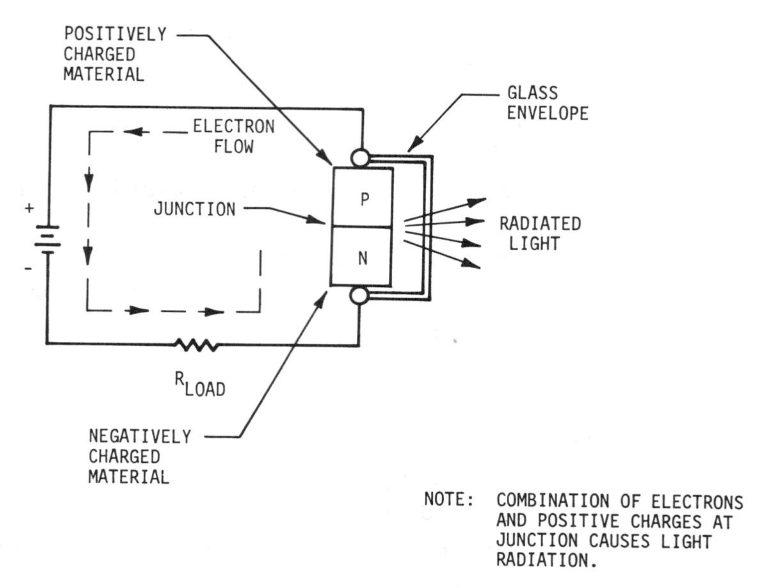

Figure 4–1 Electroluminescence (*Mike Anaya*).

are deficiencies of free electrons in lattice structures where electron–hole combination takes place. Photons are emitted. Mirror ends reflect the photons back into the active region where more electron–hole combinations take place. The mirror ends serve as a feedback mechanism, making the ILD a semiconductor laser. Control of the brightness is made by adjusting the current flow.

Electroluminescent Materials

The materials that are used for manufacturing solid-state electroluminescent devices are called *compound semiconductors.* This material is chosen by its band-gap energy and its radiation wavelength. There are two common electroluminescent devices, the light-emitting diode (LED), often called a solid-state lamp (SSL), and the injection laser diode (ILD). The basic materials of which electroluminescent devices (LEDs) are made are gallium arsenide (GaAs), gallium phosphide (GaP), and gallium arsenide phosphide (GaAsP). Refer to table 4–1.

Gallium arsenide (GaAs) has an energy gap of around 1.4 electron volts at a wavelength of 900 nanometers. Gallium phosphide (GaP) has an energy gap of around 2.2 electron volts at a wavelength of 560 nanometers. Gallium arsenide phosphide (GaAsP) has an energy gap of around 1.9 electron volts at a wavelength of 650 nanometers. These are estimated values and you will find that each manufacturer's LEDs will vary somewhat in wavelength and energy gap.

The first material, gallium arsenide (GaAs), does not emit visible radiation. The spectral wavelength range of GaAs material falls in the infrared range. Although the basic material GaAs emits radiation at 900 nanometers, it may be doped with other materials to provide a wider range of wavelengths. Infrared wavelengths cannot be seen, but are generally included in the classification of LED. Some manufacturers call them infrared LEDs or simply IR LEDs. Others call them infrared emitters.

Gallium phosphide (GaP) emits a green light from 520 to 570 nanometers with its peak around 560 nanometers. It can also radiate a red light between 630 and 790 nanometers with a peak around 690 nanometers. The gallium arsenide phosphide (GaAsP) emits light over an orange-red range depending on the amount of GaP in the material. The GaAsP emits red light between 640 and 700 nanometers with its peak around 650 nanometers. With the correct amount of GaP in the material mix, a yellow light is emitted with a peak around 610-nanometer wavelength.

New compounds of material are coming into operation. For instance, the gallium aluminum arsenide (GaA1As) has been available for some time. It emits light over a red range from 650 to 700 nanometers with a peak around 670 nanometers.

The efficiency of the LED is very dependent on the emitted wavelength, with a drastic fall-off in efficiency as the wavelength gets shorter. The basic material that the injection laser diode (ILD) is constructed from is aluminum gallium arsenide (A1GaAs). The A1GaAs emits infrared light from around 838 to 844 nanometers with a peak near 840 nanometers. The efficiency of an injection laser diode (ILD) lies in its ability to operate in a narrow bandwidth with high intensity.

Direct-Gap versus Indirect-Gap Materials

Electroluminescent devices are used as light radiators because they are direct-gap materials. Other semiconductor materials are indirect-gap materials. *Direct gap* and *indirect gap* refer to the transition of electrons from the conduction band to the valence band. Electron transition takes place when electrons move through a PN junction from N material to P material and combine with holes. When the electrons move from the high-energy state of the conduction band to low-energy state of the valence band, photons of energy are released.

In the direct-gap concept the electrons and holes have equal momentum. Electrons in the conduction band and holes in the valence band may directly recombine. In the indirect-gap concept electrons and holes do not have the same momentum. Electrons and holes are captured or trapped by secondary donor and acceptor levels so that direct transition cannot happen. The electrons and holes do combine and some radiation may occur, followed by transition to the valence band. Direct-gap transition is in one step from the conduction band to the valence band. Indirect-gap transition is in two steps, from the conduction band through an entrapment area to the valence band. The GaAs and the GaAsP diodes are direct-gap radiators. The GaP diode is an indirect-gap radiator.

PERFORMANCE CURVES

Figure 4–2 is a set of typical performance curves for an electroluminescent device. This particular set of curves is for an Alcatel Model T–820 Stripe Geometry GaA1As double-heterojunction edge-emitting LED. Other electroluminescent devices have similar curves.

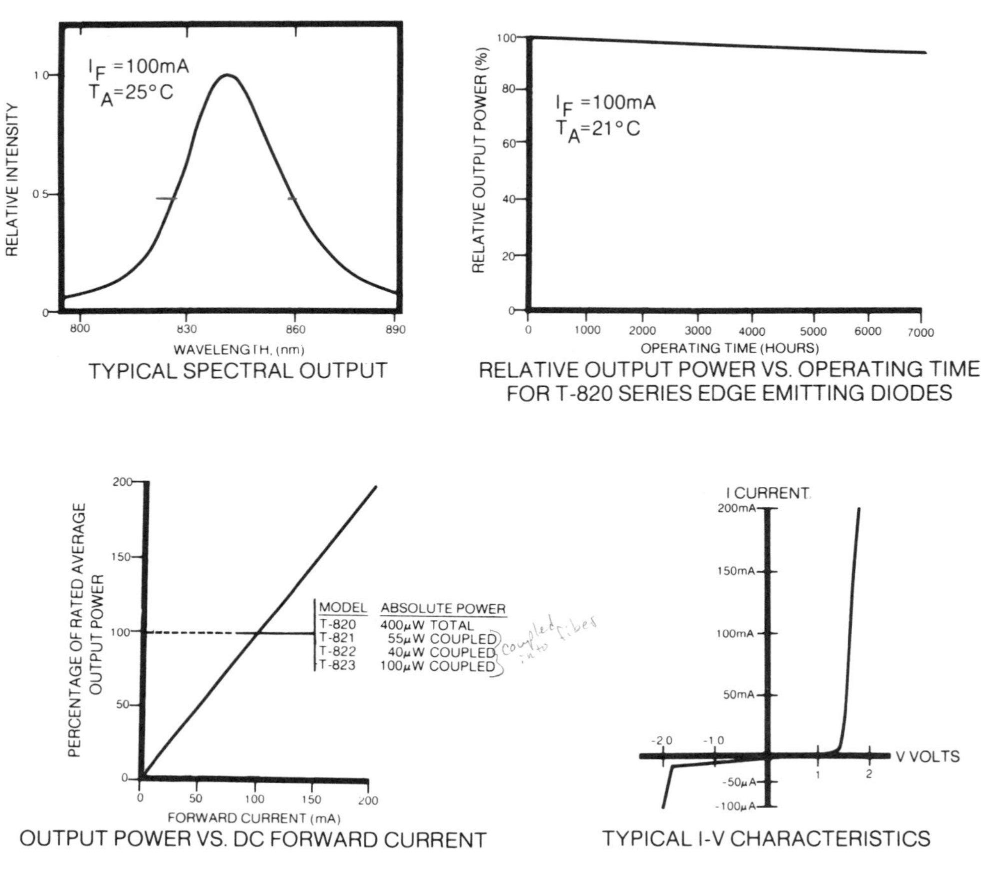

Figure 4–2 Performance curves for electroluminescent devices (*courtesy of Alcatel Cable Systems Group*).

Spectral Response Curve

The first curve in the illustration is the spectral response curve on the top left side. The curve plots wavelength in nanometers against relative intensity. Note that the peak relative intensity occurs at 840 nanometers. The spectral width is around 40 nanometers. The spread is from 825 to 865 nanometers with maximum intensity between 830 and 860 nanometers. The wavelength range of operation is in the infrared. Some of the typical wavelengths versus relative intensity values for other electroluminescent materials are listed below.

Operating Time–Power Curve

The curve in the top right side of the illustration plots operating time in hours against relative output power. The reader will note that the number of hours operating at

TABLE 4-1 TYPICAL WAVELENGTHS VERSUS MATERIAL

Material	Color	Wavelength peak (nm)
GaP	High red	700
GaP	Green	560
GaAs	Infrared	900
GaAs	Infrared	940
GaAsP	Infrared	940
GaAsP	Green	540

the limits of the curve is around 7000 hours. The percent of power loss in that amount of time is only several percent.

Forward Current–Output Power Percent Curve

The lower left curve in the figure plots forward current in milliamperes (mA) against percentage of rated average output power. You will note that this is an extremely linear curve that depicts a direct change in power for increments of current.

Diode Current–Voltage Curve

The curve on the lower right side of the illustration is a familiar curve used for all diodes. It plots forward and reverse voltage against current. As voltage increases in the forward-bias direction (right on the curve) current does not flow until around 1.3 volts is reached. In reverse-bias, current flows slightly until a reverse-bias voltage reaches around −1.9 volts. At that point and further, current will increase rapidly in the reverse direction until the diode is destroyed.

Other Performance Curves

There are other performance curves. Case temperature versus normalized output is one and temperature changes due to forward voltage as a function of forward current is another.

SPECIFICATIONS OF ELECTROLUMINESCENT DEVICES

Table 4–2 is a standard set of specifications for an electroluminescent device. Other electroluminescent devices have similar specifications. Case temperature is a standard as are forward and reverse voltage and continuous and pulsed current. This particular device is used with fiber optics; therefore, the optical power output is specified for different optical fiber use. A final useful specification for most radiating devices is the size of the emitting area.

TABLE 4-2 SPECIFICATIONS FOR TYPICAL LED DEVICE (CASE TEMPERATURE = 25°C)

Maximum ratings			
Continuous forward current	100 mA		
Pulsed forward current (10 kHz, 50% duty cycle)	200 mA		
Reverse voltage	2.0 V		
Operating and storage temperature range	−40 to 75°C		
	Nominal	**Units**	**Conditions**
Peak emission			
Wavelength	840	nm	I_F = 100 mA
Spectral width (full-width half max.)	40	nm	I_F = 100 mA
Forward voltage	1.5	V	I_F = 100 mA
Thermal impedance (chip to external heat sink)	20	°C/W	
Optical power output			
Unencapsulated	400	μW	I_F = 100 mA
Coupled	55	μW	I_F = 100 mA
Coupled	40	μW	I_F = 100 mA
Coupled	100	μW	I_F = 100 mA
Rise/fall times (10 to 90%)	15	ns	
Emitting area	1 × 20	μm	

Courtesy of Alcatel Cable Systems Group.

ELECTROLUMINESCENT DEVICE APPLICATIONS

Visual indications are by far the most versatile and usable applications for the electroluminescent device. In Figure 4–3a, an LED is utilized in a simple dc circuit. The V_{dc} energizes the circuit. Forward bias is applied to the LED as a current flows through the load. The LED indicates that current is flowing through the load. In Figure 4–3b, the LED is placed in series with a current-limiting resistance and an inverter output. When the inverter output is low (ground), the LED is forward biased and comes on. The current-limiting resistance limits current to the LED.

In Figure 4–3c, the LED is placed in series with a current-limiting resistance and in shunt with an inverter output. When the inverter output is high (positive), the LED is forward-biased and comes on. The current-limiting resistance limits current to the LED. When the inverter output is low, the LED is shorted and does not light. In Figure 4–3d, the LED is placed in the collector circuit of a transistor lamp driver circuit with a load resistance. Switch S_1 represents some action such as a door opening or a liquid at a specified level. The switch closes and forward biases the transistor base–emitter junction. This action causes the transistor to saturate and in turn the LED comes on.

The electroluminescent device is especially useful in opto-isolator applications (see Figure 4–4). In Figure 4–4a, an NPN transistor is used as a driver for an LED that is part of an opto-isolator. Logic T^2L HI input supplies forward bias to the transistor. The transistor turns on to saturation, supplying a ground for the LED in the opto-isolator. When current flows through the LED, it radiates a beam of light. The phototransistor detects the light on its base, causing output current (I_o) to flow.

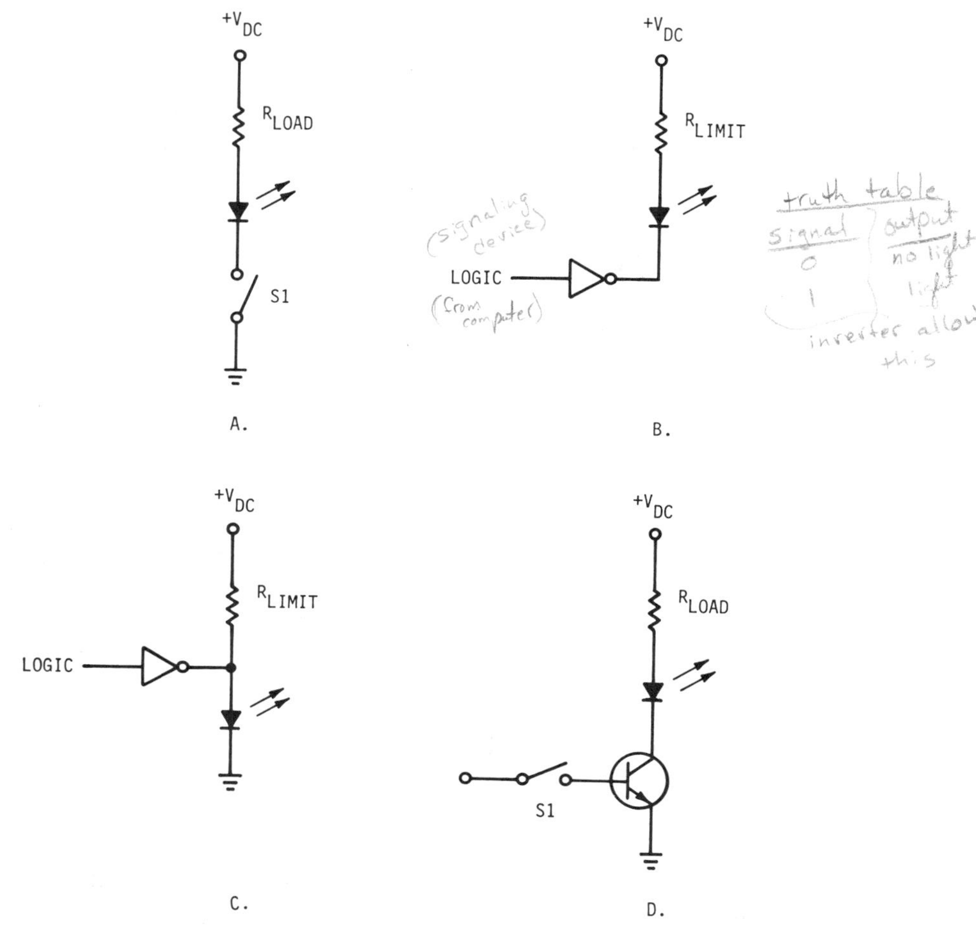

Figure 4–3 Electroluminescent applications (*Mike Anaya*).

In Figure 4–4b, a PNP transistor is used as a driver for an LED that is part of an opto-isolator. Logic T^2L LO input supplies forward bias to the transistor. The transistor turns on to saturation, supplying a positive voltage for the LED in the opto-isolator. When current flows through the LED, it radiates a beam of light. The photo transistor detects the light on its base, causing output current (I_o) to flow.

In Figure 4–4c, a buffer NAND gate is used as a driver for an LED within an opto-isolator. When the NAND gate has a low output, a ground is supplied for the LED. Current flows through the LED. The LED radiates a beam of light. The photo-transistor detects the light on its base, causing output current to flow.

In Figure 4–5 an electroluminescent device is applied to fiber optic light communication. An input square-wave signal is driven into a transmitter module. The transmitter module converts the square wave to a current flow for the diode. The diode is an injection laser diode (ILD). When current flows through the diode, the

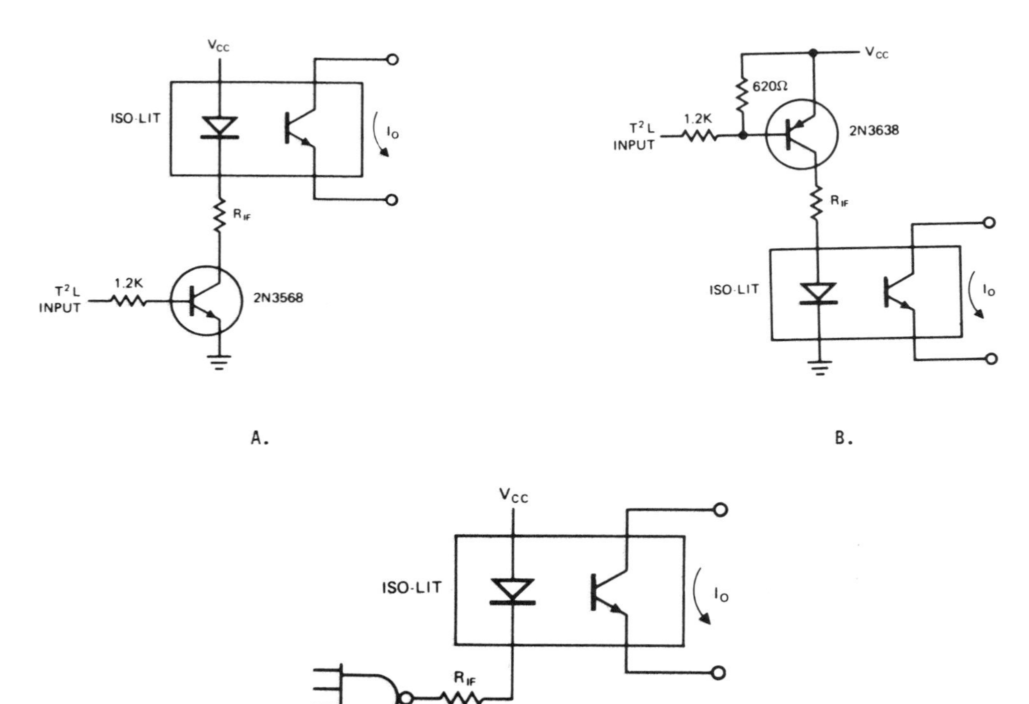

Figure 4–4 Electroluminescent devices used as opto-isolators (*courtesy of Siemens Optoelectronics Division*).

diode radiates a high-intensity light that is launched into the optical cable through the source-to-fiber connector. The light wave is transmitted by reflection to a detector on the receiver side. The detector is probably a photodetector diode. The light wave is coupled from the optical cable by a fiber-to-detector connector. The detector converts the light energy to a current flow. The receiver module converts the current to a replica of the original square-wave input signal.

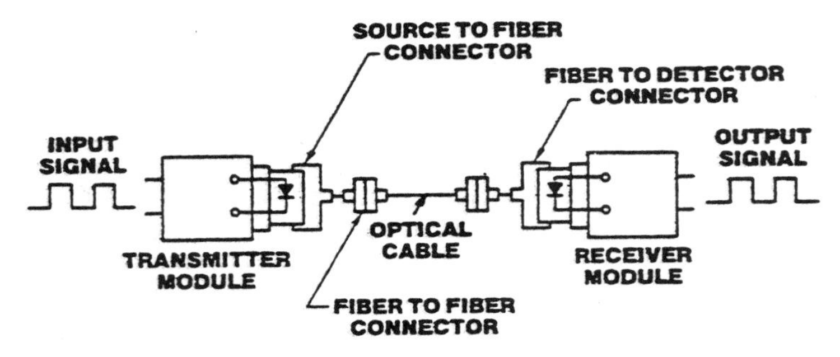

Figure 4–5 Electroluminescent device used with fiber optics (*courtesy of ITT Cannon Electric*).

LIGHT-EMITTING DIODES

The LED represents the best of the electroluminescent devices. Electroluminescence, you may recall, is the emission of light from a solid-state device by application of current flow through the device.

Cross Section of an LED

In Figure 4–6, a standard red LED is illustrated. In Figure 4–7, a bright orange LED is shown. You will note that in general these both have similar cross sections. Both have substrates on which expitaxial layers are grown. Each has a gold (Au) ball bond for current application.

In the cross sections of Figures 4–6 and 4–7, you will note that the layers of GaAsP are grown in substrates of GaAs and GaP, respectively. The subtrates are grown in ingots as in other solid-state devices. A lattice structure which is pure and free from imperfections is produced under a controlled environment. The epitaxial layers are gradually changed to maintain a perfect crystalline structure. Note that the red LED epitaxial layers are doped with tellurium (Te), while the orange LED epitaxial layers are doped with sulfur (S). The wafer is then coated with a layer of silicon nitride (SiN) which acts as a barrier to the diffused P layer. The P layer diffusant is made of zinc (Zn). It is placed in the opening in the silicon nitride (SiN) to form the PN junction. The outer casting on the red LED in Figure 4–6 is made of aluminum (A1) on the front side, with gold–tin (AuSn) on the backside contact. A gold ball bond is used to establish bias contact. The outer casing on the orange LED in Figure 4–7 is made of gold–beryllium (AuBe) with a backside reflective contact of gold–germanium (AuGe). A gold ball bond is used to establish bias contact. Figure 4–8 is a plan view of a typical standard red LED die (chip).

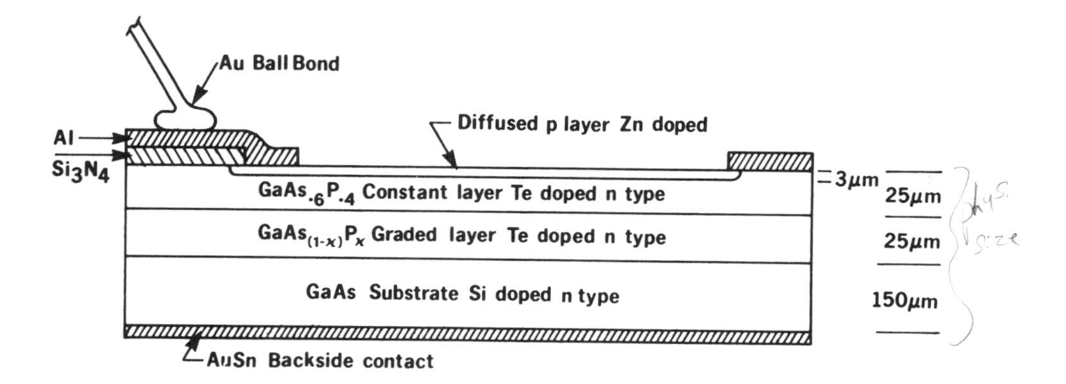

Figure 4–6 Red LED cross section (*courtesy of Siemens Optoelectronics Division*).

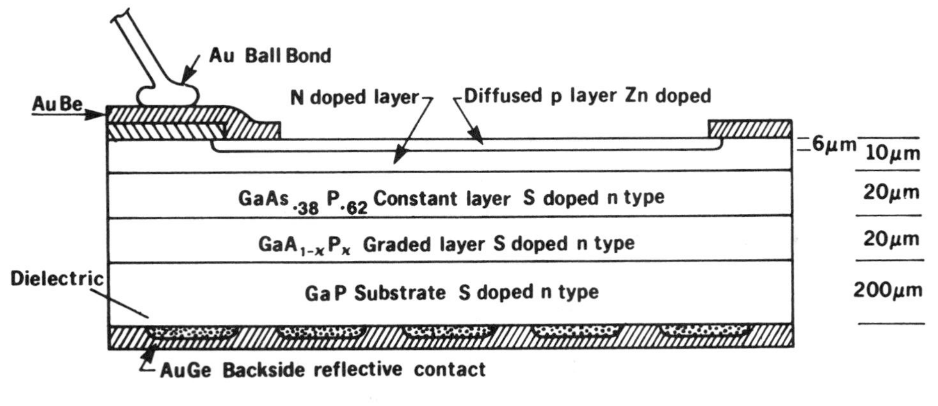

Figure 4–7 Orange LED cross section (*courtesy of Siemens Optoelectronics Division*).

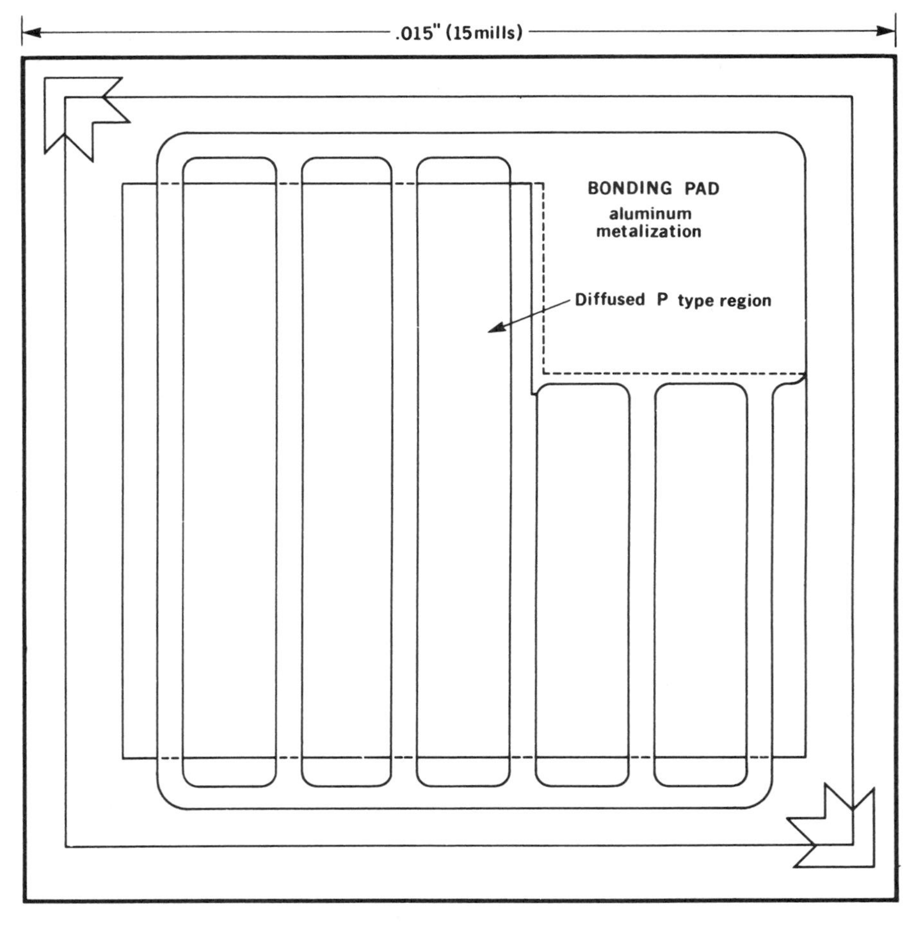

Figure 4–8 Plan view of typical LED die (chip) (*courtesy of Siemens Optoelectronics Division*).

It must be stated that the manufacturing processes involved in the growing of substrate ingots and the further growth of epitaxial layers, impurity diffusions, and thin-film deposition is a complex and highly technical task. Since this is the case, most manufacturers have secret or proprietary procedures because of the high cost of research and development involved.

LED Lamps

In Figure 4–9 are shown a series of typical LED lamps. LED lamps are made so as to be as bright as possible within their chemical capabilities. The LED chip is placed within a plastic structure that provides a maximum full-flooded front-radiating area and wide-angle viewing. They are also made to be easily soldered on a PC board or snapped into a mounting clip.

The choice of an LED for a specific use is obvious in many ways. The choice usually boils down to five basic factors.

1. The color of an LED is selected by the user for a particular use. Generally, the color is in various shades of red and often orange. Amber and green are also available, but the difficulty in locating suitable raw material may preclude the choice.

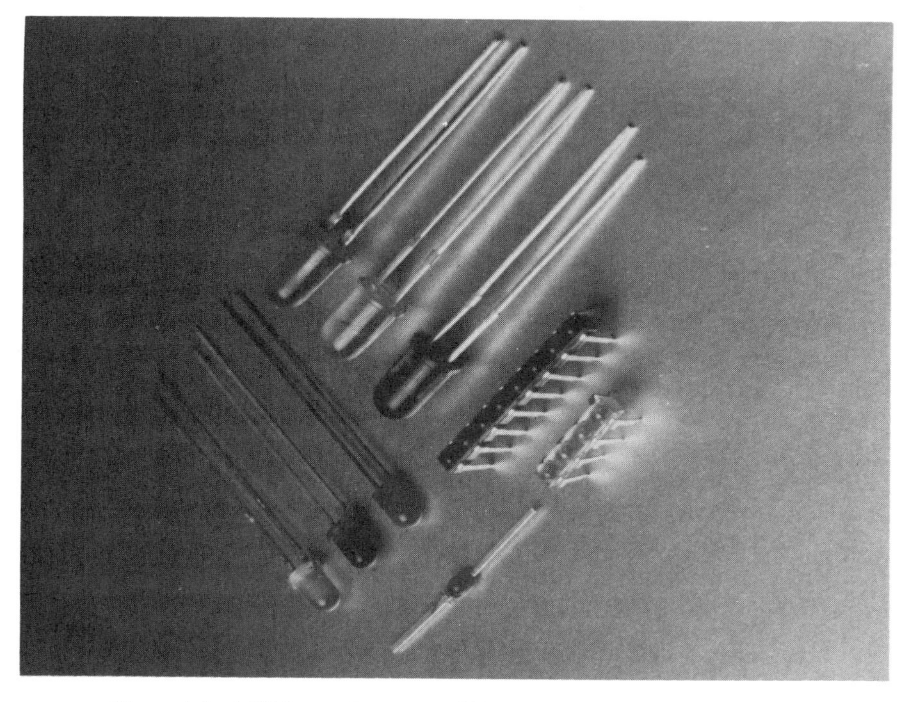

Figure 4–9 LED lamps (*courtesy of Siemens Optoelectronics Division*).

2. The size of the LED depends, of course, on where the device is to be installed. Many combinations of chip and casing sizes are available which range from 5 to 300 mils in diameter. The larger the size, the more visible the device radiance.
3. Most LED chips are Lambertian in angular distribution of light. That is, the luminance of the LED can be seen equally well from all directions. With suitable design the angular light pattern may be changed from a very broad pattern to a very narrow pattern. By placing the chip at the focus of the lens system, a narrow high-intensity beam is obtained. By using diffusing lens material a large area source is obtained but luminance is reduced.
4. The luminous intensity of the device will govern the visibility under background contrast conditions when viewed at normal distances. A typical LED lamp has a luminous intensity of 1 millicandela. Luminous intensity is flux per solid angle.
5. When it is not possible to provide a dark contrasting background, or when the source (LED) is viewed at very close distances, the luminance becomes important. Luminance is the intensity per unit area of a source. Values are from 100 to 5000 foot-lamberts.

The electronics person may find the use of photometry and radiometry a confused mass of strange units and values. Chapter 10 strives to organize these units.

LED Characteristic Curves

The characteristic curves in Figure 4–10 are representative of a typical red LED lamp. These particular curves are for high bright gallium arsenide phosphide (GaAsP). The curves represent the following characteristics:

1. *Relative luminous intensity versus angle:* relative intensity in percent versus the number of degrees off angle in percent
2. *Luminous intensity versus ambient temperature:* percent of relative luminance versus the ambient temperature in Celsius (°C)
3. *Spectral distribution:* relative intensity versus wavelength λ in nanometers
4. *Forward current versus forward voltage:* forward current in milliamperes versus forward voltage in volts of direct current
5. *Luminous intensity versus forward current:* luminous intensity in millicandela versus forward current in milliamperes

These are typical of LED characteristics and are self-explanatory.

Infrared LEDs

A typical infrared LED is made from gallium arsenide (GaAs). This is an excellent material to make the diode from because most of the material's energy is given up in the form of radiant energy. With forward bias the diode emits a narrow wave-

RELATIVE LUMINOUS INTENSITY VS ANGLE

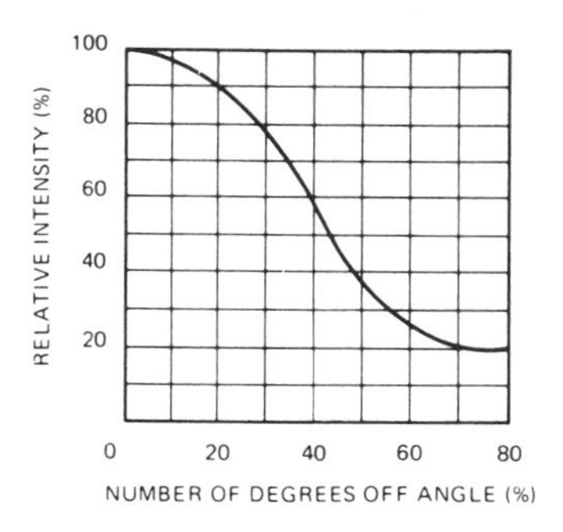

LUMINOUS INTENSITY VS AMBIENT TEMPERATURE

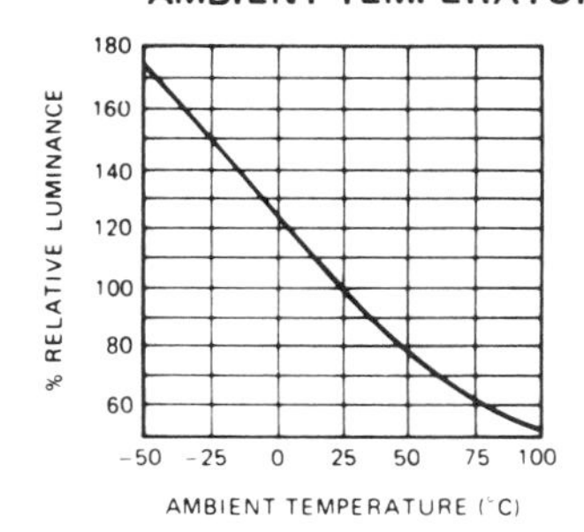

SPECTRAL DISTRIBUTION

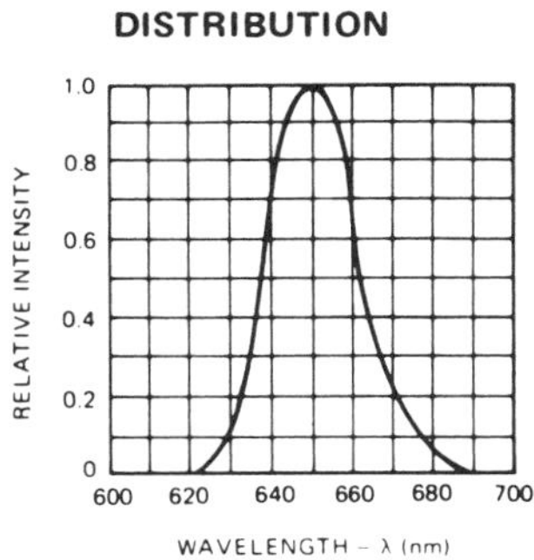

FORWARD CURRENT VS FORWARD VOLTAGE

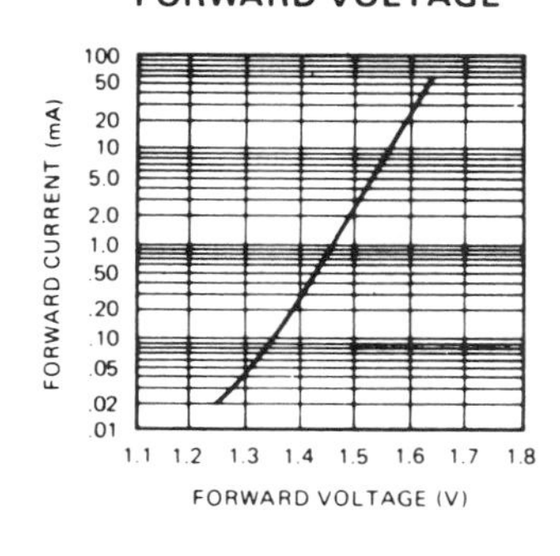

LUMINOUS INTENSITY VS FORWARD CURRENT

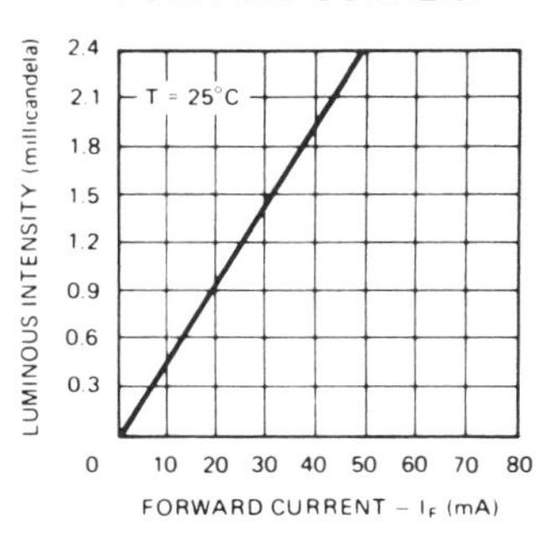

Figure 4–10 Typical LED characteristic curves (*courtesy of Siemens Optoelectronics Division*).

length of around 900 nanometers. Typical specifications include thermal ratings, radiant intensity curves, and forward current and voltage ratings.

Operating LEDs on AC Power

Often it is desirable to place an LED in an alternating-current (ac) circuit (see Figure 4–11). There are basically three applications typical of the LED operation in ac circuits. In the first application (Figure 4–11a) an LED is placed in series with a rectifier diode and a series current-limiting resistance. High power is dissipated in the resistor.

In the second application (Figure 4–11b), a rectifier diode is placed in shunt with an LED. A capacitor and current-limiting resistor are in series with the parallel diodes. For current in either direction the voltage drop across the LED or rectifier is a negligible part of the applied ac volts. Capacitor current is almost equal to the ac supply voltage divided by the capacitor reactance ($I_{cap} = V_{ac}/X_c$). The resistor in series is used to limit turn-on transient currents. Power dissipation is very low and is limited to the LED and rectifier losses, resistance losses, and capacitive reactance losses.

In the third application (Figure 4–11c), three LEDs are operated from the same source. The LEDs are simply placed in series in the circuit with a shunt rectifier diode. Current is essentially the same as for one LED.

Mounting LEDs

The LED comes in standard sizes. The most prevalent sizes used in electronics are the T1 and T1 3/4 sizes. These numbers relate to the diameter of the lamp in eighths of an inch. There are, of course, other shapes and sizes, such as the subminiature

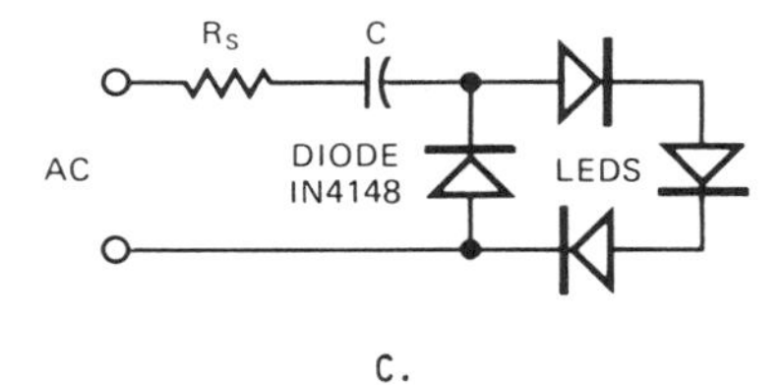

Figure 4–11 LED applications using ac power (*courtesy of Siemens Optoelectronics Division*).

and the rectangular. Most lamps have plastic lenses, but others may be made with hermetically sealed glass. Some have spherical lenses. As you may realize, mounting techniques are dependent on the shape, size, and use of the LED.

There are many ways to mount an LED. Some of these are by soldering directly into PC boards, plugging in preassembled sockets, panel mounting, or mounting with clips. Lead bending is allowed, but bending must follow manufacturers' recommended bend radii. The leads should be clamped next to the base during bending to relieve stresses. Mechanical side loads or other force should not be placed on the LED during installation. Figure 4–12 illustrates functional mounting of LEDs.

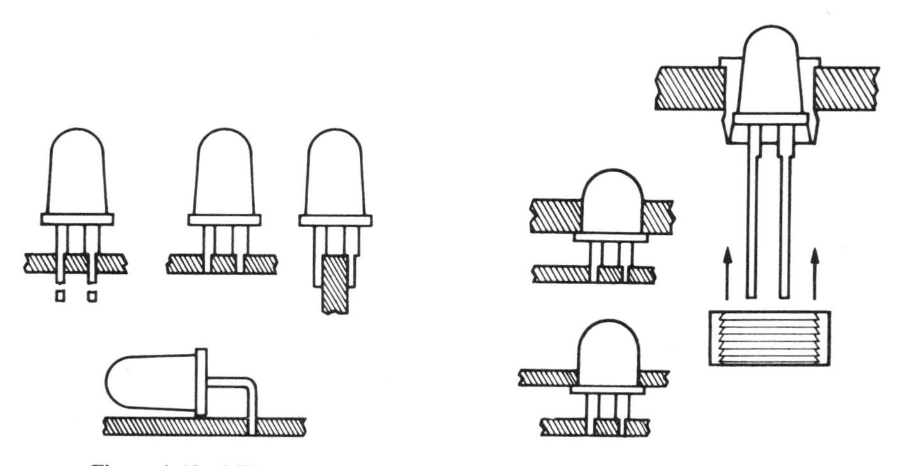

Figure 4–12 LED mounting (*courtesy of Siemens Optoelectronics Division*).

Most LEDs are subject to thermal stress. A typical LED chip is mounted on a substrate or lead frame with a wire bond from the top of the chip to a metallized trace on the substrate. Temperature changes cause these various metals to expand and contract at different rates. High temperatures cause reduced lifetime rather than immediate failures. Internal LED junction temperature depends on ambient temperature, power applied to the LED, and thermal resistance from LED chip to ambient.

LED DISPLAYS

Electronic displays include incandescent, liquid crystals, and the light-emitting diode, among others. This section is dedicated to the LED display. Probably the most useful form of the LED is in digital displays. The LED digital display is reliable and inexpensive. It is usually driven from a low-level dc voltage but has comparatively high current draw. Some disadvantages are that under high-intensity ambient light they tend to look faded. As you would understand, large displays are expensive, since they would require large quantities of GaAsP material. Manufactures such as Siemens have developed light-pipe construction to enhance the reflective light. This allows the display to be seen for distances up to 10 feet. The process involves the placement of a reflecting light pipe over the LED chip. The LED emits light that is reflected off the light-pipe cavity walls and emerges as a larger lighting surface. Light pipes are often filled with glass particles to enhance reflection.

LED displays come in a variety of sizes and colors. The most popular heights are 0.3 and 0.6 inch. Character sizes usually run up to 1.0 inch. Displays come in one-, two-, three-, and four-digit. However, they are not limited to number. Colors are variations of red, yellow, and green.

Seven-Segment Displays

Figure 4–13 is an illustration of typical red reflector displays. The displays are called seven-segment because the numbers on the display are made up of seven diffused segments. There are, of course, other types of display fonts such as the nine- and 16-segment fonts and the dot-matrix fonts. Each font has special uses such as numeric, hexidecimal, alphabetic, alphanumeric, and other characters. It is beyond the scope of this book to cover all these in detail. Therefore, we shall limit our discussion to the seven-segment display.

Figure 4–13 shows numerical displays with a decimal point on left, right or both sides of the number. These are representiatives of the seven-segment displays available. Figure 4–14 shows characteristic curves for optoelectronic devices.

Figure 4–13 Red reflector displays (*courtesy of Siemens Optoelectronics Division*).

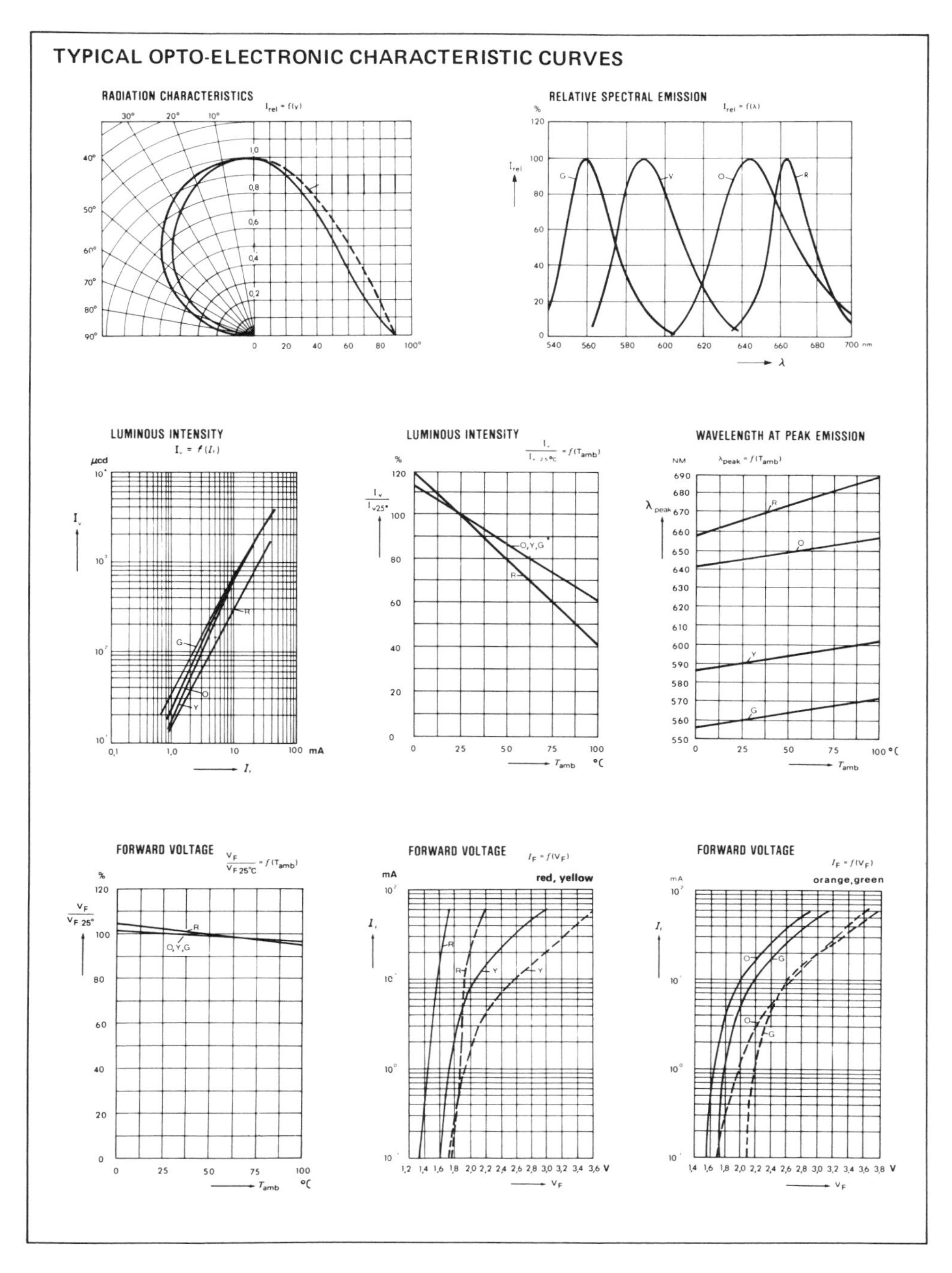

Figure 4–14 Typical optoelectronic characteristic curves (*courtesy of Siemens Optoelectronics Division*).

DL-7750/7751/7760R
DLO=7650/7651/7653O

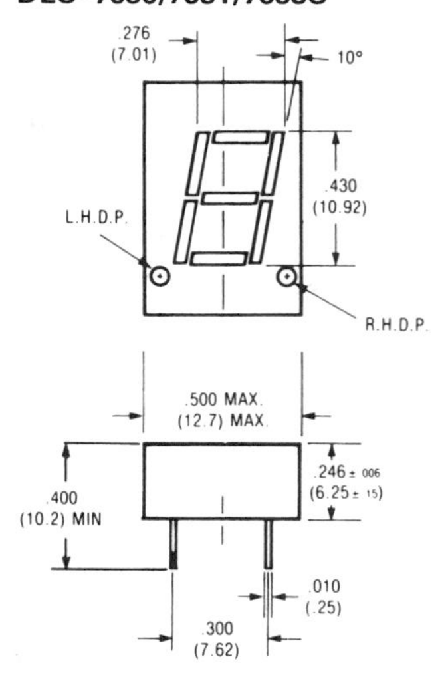

DL-7650O/DL-7660Y
DL-7670G/DL-7750R

Pin	Function
1	Cathode -a
2	Cathode -f
3	Anode
4	No Pin
5	No Pin
6	Cathode -d.p.
7	Cathode -e
8	Cathode -d
9	No Conn.
10	Cathode -c
11	Cathode -g
12	No Pin
13	Cathode -b
14	Anode

TYPICAL DRIVE CIRCUITRY

DL-707, DL-707R

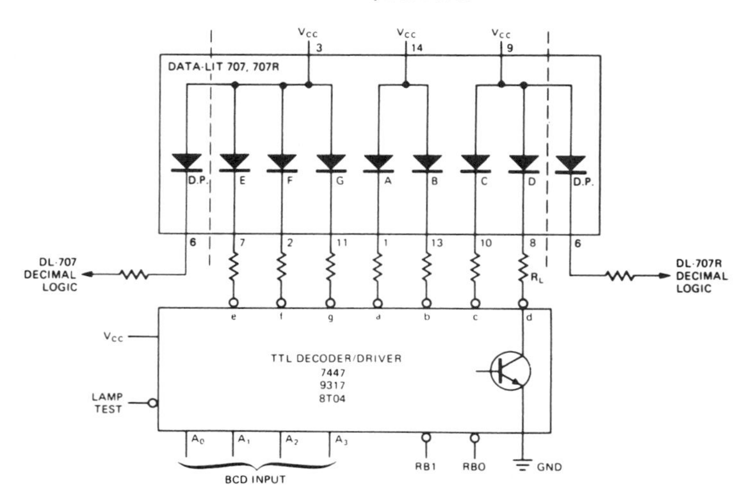

Figure 4–15 Direct dc single-diode LED-drive circuitry (*courtesy of Siemens Optoelectronics Division*).

LED Driver Circuitry

There are two methdos of driving LED displays. These are the direct dc drive through a decoder/driver of a single LED display and the multiplexing (strobing) of a multiple number of LED digits.

Direct dc drive. Figures 4–15 and 4–16 are representative of the direct dc drive through a decoder/driver. In Figure 4–15 the pin configuration of a 0.43-inch seven-segment common-anode red numerical display. This is a simple method of driving the segments. The digit has its own decoder/driver. Outputs from the decoder control current flow to each LED through current-limiting resistors. The display and the decoder/driver are TTL matched. Decimal logic is provided along with input and output blanking functions (RBI and RBO).

PIN CONFIGURATIONS

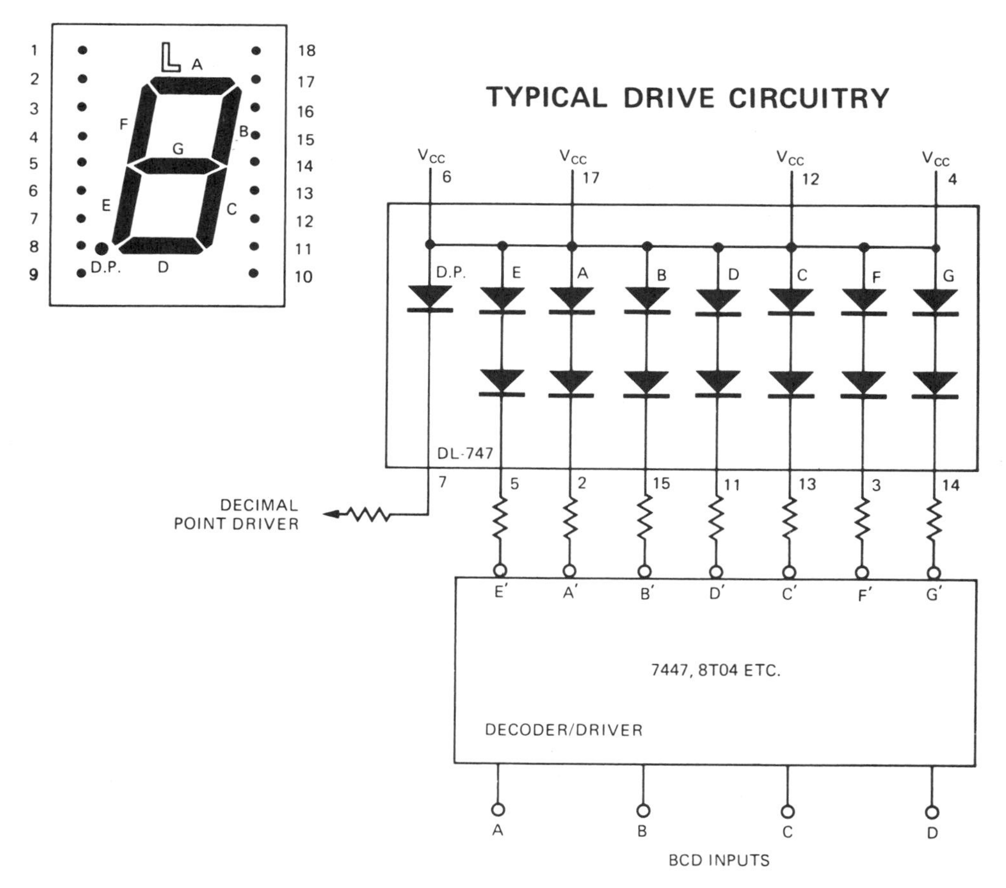

Figure 4–16 Direct dc double-diode LED-drive circuitry (*courtesy of Siemens Optoelectronics Division*).

In Figure 4–16 the typical drive circuitry of a 0.6-inch seven-segment series LED display is shown. While the two diodes in series may cover a larger viewing area, the source current from a drop in supply voltage V_{cc} may not illuminate the LEDs sufficiently. As in the standard dc driver circuit with a single LED, a decimal-point driver is supplied.

Multiplexing LED Displays. The second method of driving seven-segment displays is by multiplexing (strobing). Driving a single-digit display with its own decoder/driver is the simplest method, but does not carry the same efficiency as the multiplexed display. The costs are also a factor, for a large reduction in parts and wiring means cost-effectiveness. Furthermore, the reduction in current requirements may even mean a smaller power supply.

Figure 4–17 is the drive circuitry for a multiplexed eight-digit LED display. BCD data are input to a decoder/driver. This particular driver is a TTL logic device as are the rest of the components. The corresponding cathodes of each digit are bused together. The buses each have their individual current-limiting resistors. The display data are presented to the decoder driver together with an enable signal to the appropriate anode driver. Each digit anode is driven by a switch capable of passing the full current to each display's segments. A simple switch is the PNP high-current or amplifier transistor. In operation the anode switches are operated one at a time in the desired sequence while the appropriate digital data are presented to the decoder/driver. The digit-select multiplexing logic enables each anode driver in a time-sharing process.

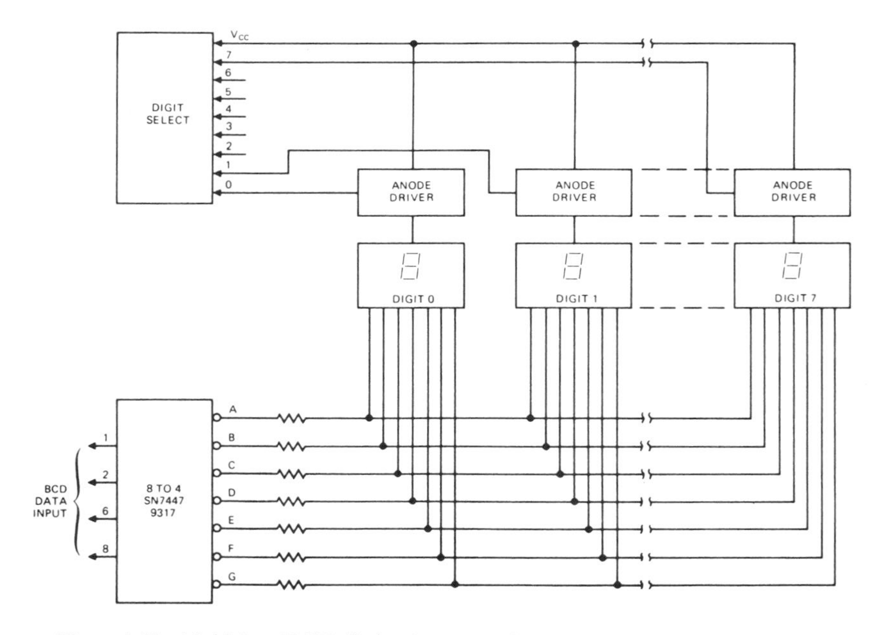

Figure 4–17 Multiplexed LED display (*courtesy of Siemens Optoelectronics Division*).

A number of factors must be taken into account when deciding on the design of a multiplexed display. Besides the optical output, thermal considerations are very important. Most quarter-inch-size LED characters are rated at 30 milliamperes dc maximum per segment. Under pulsed operation, high currents can be used provided several thermal considerations are taken into account. The average power dissipation must not exceed the maximum rated power. The power pulse width must be short enough to prevent the junction from overheating during the pulse. This implies that the pulse width must get shorter as the amplitude increases.

Present experience indicates that for pulses of 10 microseconds the amplitude should be limited 100 milliamperes maximum. Shorter pulses of higher amplitude may be used, but the circuit problems become severe if the pulse width is very short.

MONOLITHIC DISPLAYS

The monolithic (dual-in-line package) display is different from other displays in that all the segments made of light-emitting diodes are diffused on a single chip (see Figure 4–18. The chip is usually made from GaAsP as a substrate of N-doped material, while

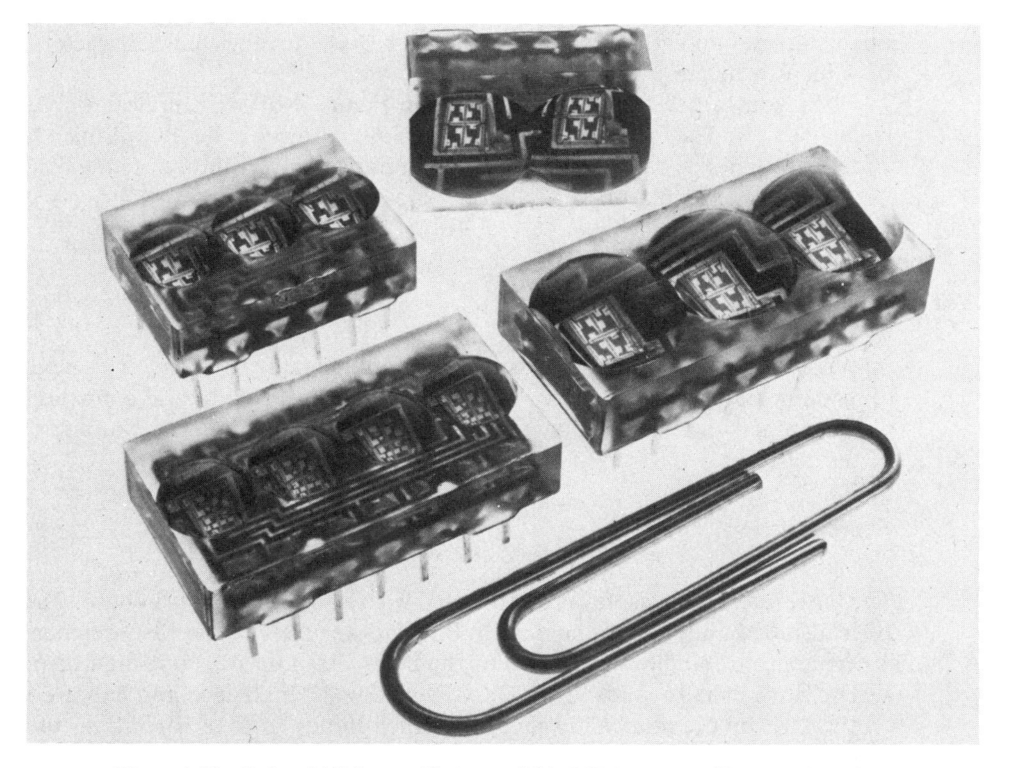

Figure 4–18 Red multidigit magnified monolithic DIP (*courtesy of Siemens Optoelectronics Division*).

the junction is P+ diffused. The operating wavelength of the monolithic DIP is around 655 nanometers. Since the price of GaAsP is quite high, the monolithic chip size is kept small. An external lens is used to magnify the size of characters. The actual heights of the character are much smaller than the apparent size of the digits.

The lens is placed on the chip by two methods, immersion and nonimmersion. The immersion method molds the lens directly to the chip in cast epoxy. An aluminum contact for each digit segment is wire bonded to an anode contact. The chips are attached to a ceramic or lead frame where the common cathode electrical connection is formed.

The nonimmersion lens differs from the immersion lens in that there is a layer of air between the chip and the lens. Nonimmersion lenses are constructed by attaching the chip to a circuit board. Board size is dependent on the number of display digits. Aluminum contacts for each LED segment are wire bonded to the anode traces on the printed circuit board. Precision-molded lenses are aligned and attached to the printed-circuit board. A second lens is usually attached to add height to the characters.

Figure 4–19 is a pinout-and-dimension illustration of typical red monolithic DIP displays. Figure 4–19a represents a two-digit cast lens display. It is common-cathode multiplexed. As you can see, it has characters that are 0.15 inch in height and 0.11 inch wide. Figure 4–19b shows three- and four-digit encapsulated lens displays. These are also common-cathode multiplexed. Both these displays have characters that are 0.11 inch in height and 0.08 inch wide.

As you can see in the photograph in Figure 4–18, the monolithic DIP is extremely small. They are also low cost and lower power. The monolithic DIP has a wide viewing angle that allows for maximum outdoor visibility. These displays are used for hand calculators and digital watches.

INTELLIGENT DISPLAY DEVICES

Intelligent displays are compact modular components that convert computer data to readable numbers and letters. They combine CMOS ICs, hybrid construction, and plastic immersion optic display technology as a single product to simplify the work of the designer.

Construction

The basic intelligent display is constructed with a four-layer hybrid base. The CMOS IC is installed underneath the dual-in-line package (DIP). The LED-segmented chip is diffused on a single chip on top of the DIP. This unit is a four-digit display. The LED chip is usually made from GaAsP. The wide-angle lens and cap are installed on the DIP by the immersion method, which molds the lens directly to the chip in cast epoxy. An aluminum contact for each digit segment is wirebonded to common contacts (see Figure 4–20).

DL-44M

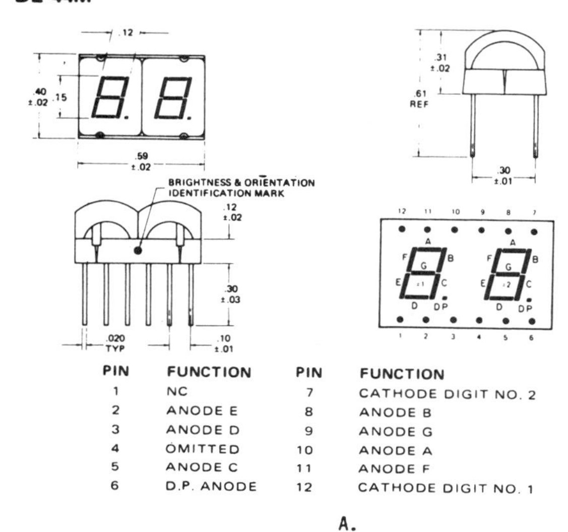

PIN	FUNCTION	PIN	FUNCTION
1	NC	7	CATHODE DIGIT NO. 2
2	ANODE E	8	ANODE B
3	ANODE D	9	ANODE G
4	OMITTED	10	ANODE A
5	ANODE C	11	ANODE F
6	D.P. ANODE	12	CATHODE DIGIT NO. 1

A.

DL-330M

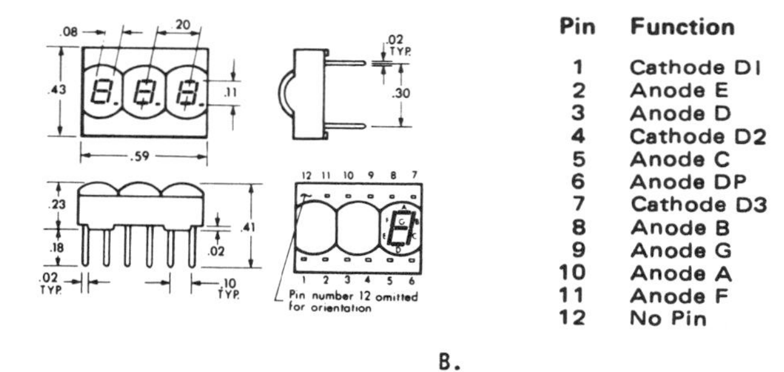

Pin	Function
1	Cathode D1
2	Anode E
3	Anode D
4	Cathode D2
5	Anode C
6	Anode DP
7	Cathode D3
8	Anode B
9	Anode G
10	Anode A
11	Anode F
12	No Pin

B.

DL-340M

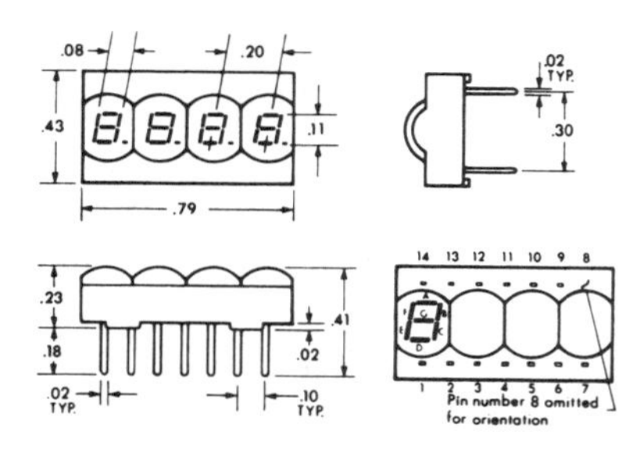

Pin	Function
1	No Connection
2	Anode E
3	Anode D
4	Anode C
5	Anode D.P.
6	Anode G
7	Cathode 4
8	No Pin
9	Anode B
10	Cathode 3
11	Anode F
12	Cathode 2
13	Anode A
14	Cathode 1

C.

Figure 4–19 Pinout and dimension outline of red multidigit monolithic DIPs (*courtesy of Siemens Optoelectronics Division*).

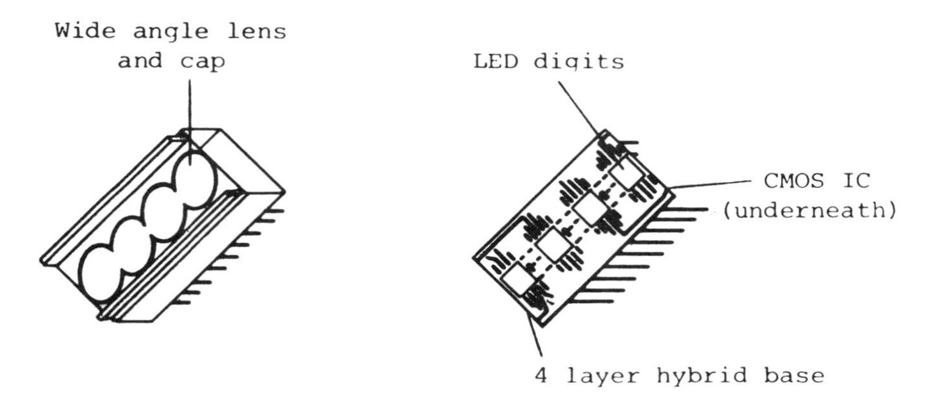

Figure 4–20 Construction of the intelligent display (*courtesy of Siemens Optoelectronics Division*).

Electrical Description

Under normal usage, a display would have traditional problems of hookup. These are segment decoding, driving, and multiplexing. The intelligent display eliminates these design problems. In May 1977, Siemens introduced the first of the intelligent displays. As a system, the intelligent display is a single device that takes alphanumeric signals from a microprocessor data bus, stores them, then translates and displays the signals as numbers and letters.

Figure 4–21 is an internal block diagram of the Siemens DL–1416 alphanumeric intelligence display. The unit consists of a monolithic LED display and a single integrated-circuit chip. The display is a four-character, 16-segment alphanumeric monolithic LED, mounted on a 1.2 × 1.0 inch PC card. The LED is magnified to a height of 160 mils. Behind the four-digit display is a complementary-metal-oxide semiconductor (CMOS). The CMOS chip is large-scale integration (LSI). It contains the necessary logic to translate 7-bit data from a microprocessor into any of 64 alphanumerical characters. Functional components within the CMOS LSI chip are 16 segment drivers, four digit drivers, 64-character ROM, four-word, 7-bit RAM, an internal oscillator for multiplexing, a multiplex counter/decoder, a cursor RAM, a write address decoder, and level shifters for the inputs.

Figure 4–22 is the pin layout of the alphanumeric intelligent display. The inputs to the chip are as follows:

$\overline{CE}$: *chip enable* (active low). This determines which device in an array will actually execute the loading of data. When the chip enable is in the high state, all inputs are inhibited.

A_0, A_1: *digit address.* The address to the display determines the digit in which the data will be written. Address order is right to left for positive-true address.

D_0–D_6: *data lines.* The seven data input lines are designed to accept the 64-character code set.

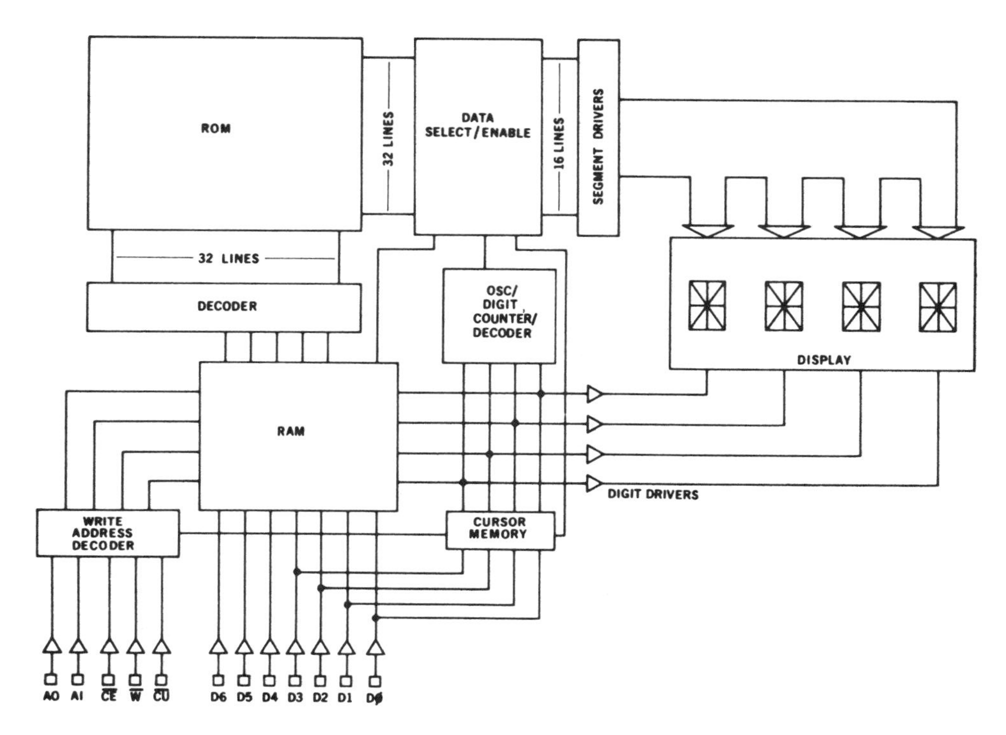

Figure 4–21 Internal schematic of an alphanumeric intelligent display (*courtesy of Siemens Optoelectronics Division*).

$\overline{\text{W}}$: *write* (active low). Data to be written into the display must be present before the leading edge of write. The data and address must be stable until after the trailing edge.

$\overline{\text{CU}}$: *cursor* (active low). When the CU is held low, the display enables the user to write or remove a cursor in any digit position. The cursor function lights

Pin		Function	Pin		Function
1	D5	Data Input	11	A1	Digit Select
2	D4	Data Input	12	Unused	
3	DØ	Data Input	13	Unused	
4	D1	Data Input	14	Unused	
5	D2	Data Input	15	Unused	
6	D3	Data Input	16	Unused	
7	CE	Chip Enable	17	Unused	
8	W	Write	18	V+	
9	CU	Cursor Input	19	V−	
10	AØ	Digit Select	20	D6	Data Input

20 19 18 17 16 15 14 13 12 11
DIGIT 3 DIGIT 2 DIGIT 1 DIGIT 0
TOP VIEW
1 2 3 4 5 6 7 8 9 10

Figure 4–22 Alphanumeric intelligent display PIN layout (*courtesy of Siemens Optoelectronics Division*).

CHARACTER SET

D6 D5 D4 D3 \ D2 D1 D0	L L L	L L H	L H L	L H H	H L L	H L H	H H L	H H H
L H L L		!	"	#	$	%	&	'
L H L H	(	)	*	+	,	-	.	/
L H H L	0	1	2	3	4	5	6	7
L H H H	8	9	:	;	<	=	>	?
H L L L	@	A	B	C	D	E	F	G
H L L H	H	I	J	K	L	M	N	O
H L H L	P	Q	R	S	T	U	V	W
H L H H	X	Y	Z	[	\	]	^	_

(Column headings: D0 = L H L H L H L H; D1 = L L H H L L H H; D2 = L L L L H H H H.)

NOTE: All undefined data codes that are loaded or occur on power-up will cause a blank display state.

LOADING DATA

$\overline{CE}$	$\overline{CU}$	$\overline{W}$	ADDRESS A_1	ADDRESS A_0	D6	D5	D4	D3	D2	D1	D0	DIGIT 3	DIGIT 2	DIGIT 1	DIGIT 0
H	X	X	X	X	X	X	X	X	X	X	X	NO CHANGE	NO CHANGE	NO CHANGE	NO CHANGE
L	H	L	L	L	H	L	L	L	L	L	H	NO CHANGE	NO CHANGE	NO CHANGE	A
L	H	L	L	H	H	L	L	L	L	H	L	NO CHANGE	NO CHANGE	B	A
L	H	L	H	L	H	L	L	L	L	H	H	NO CHANGE	C	B	A
L	H	L	H	H	H	L	L	L	H	L	L	D	C	B	A
L	H	L	L	L	H	L	L	L	H	L	H	D	C	B	E
L	H	L	H	L	H	L	L	H	L	H	H	D	K	B	E
L	H	L	–	–	–	–	–	–	–	–	–	SEE CHARACTER SET			

X = DON'T CARE

LOADING CURSOR

$\overline{CE}$	$\overline{CU}$	$\overline{W}$	ADDRESS A_1	ADDRESS A_0	D6	D5	D4	D3	D2	D1	D0	DIGIT 3	DIGIT 2	DIGIT 1	DIGIT 0
H	X	X	X	X	X	X	X	X	X	X	X	D	K	B	E
L	L	L	X	X	X	X	X	L	L	L	H	D	K	B	■
L	L	L	X	X	X	X	X	L	L	L	L	D	K	B	E
L	L	L	X	X	X	X	X	L	L	H	L	D	K	■	E
L	L	L	X	X	X	X	X	L	H	L	L	D	■	B	E
L	L	L	X	X	X	X	X	H	L	L	L	■	K	B	E
L	L	L	X	X	X	X	X	H	H	H	H	■	■	■	■
L	L	L	X	X	X	X	X	L	L	L	L	D	K	B	E

X = DON'T CARE

Figure 4-23 Character set for Siemens DL-1416 intelligent display (*courtesy of Siemens Optoelectronics Division*).

all 16 segments in the selected digits without erasing the data. After the cursor is removed, the digit will again display the previously written character.

V+: *positive supply*. TTL compatible +5 volts.

V−: *negative supply. Ground.*

Figure 4–23 illustrates the 64-character set along with loading data and loading cursor. Data entry is asynchronous and random access. Each digit will continue to display the character last written until replaced by another. The cursor function causes all 16 segments of a character to turn on. The cursor is not a character and upon removal leaves the character unchanged. The cursor is loaded as follows: $\overline{CE}$ and $\overline{CU}$ are held low. A write ($\overline{W}$) pulse will now store the cursor in all digit positions for which the first four data lines (D_0, D_1, D_2, D_3) are held high. If previously stored, the cursors can only be removed if their respective data lines are held low while $\overline{CE}$, $\overline{CU}$ are low and write ($\overline{W}$) occurs. A "lamp test" function is realized by simply storing a cursor in all four digits of a display. All undefined data characters stored in memory will override a cursor display.

Data are loaded as follows: The chip enable ($\overline{CE}$) held low will enable data loading. The desired data code (D_0–D_6) and selected digit address (A_0, A_1) should be held stable while write ($\overline{W}$) is low for storing new data. Data entry may be asynchronous and random order.

A Simple System Hookup

A simple 16-digit display system can be set up in an extremely easy manner. Data lines, address lines, and digit select lines along with ±dc power are connected. The significant part of this simple hookup is that only the decoder and 4 four-digit displays are required along with microprocessor inputs. Simplicity in system connection is the intelligent displays greatest asset (see Figure 4–24).

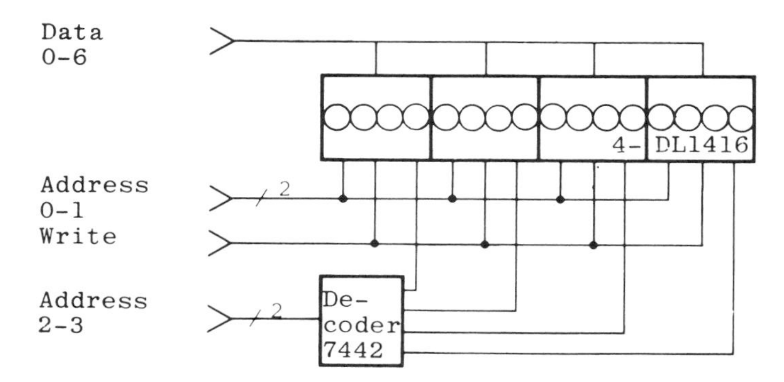

Figure 4–24 Simple system hookup (*courtesy of Siemens Optoelectronics Division*).

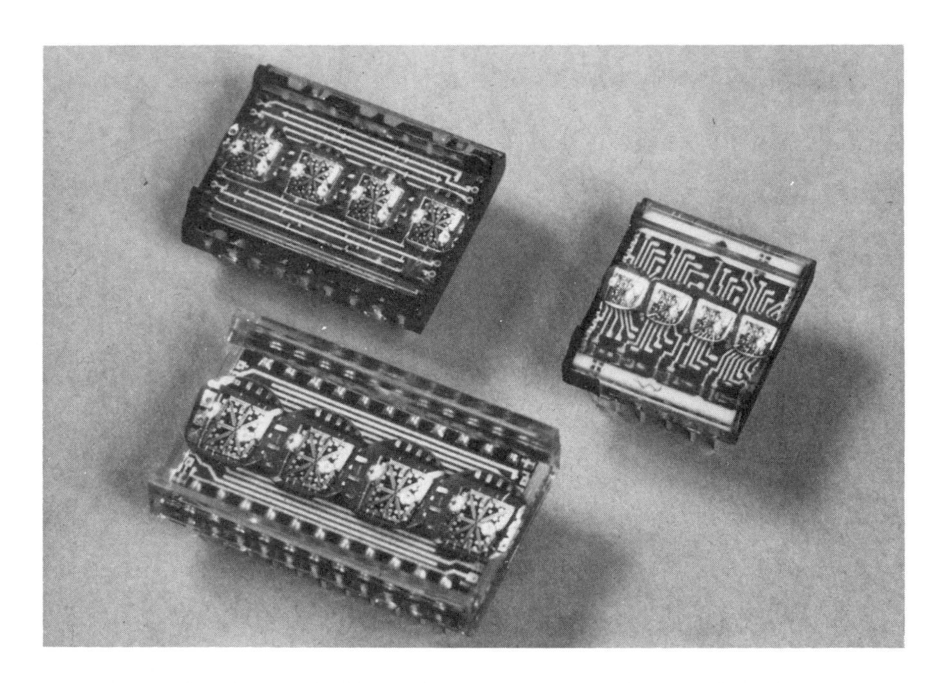

Figure 4-25 Siemens intelligent displays (*courtesy of Siemens Optoelectronics Division*).

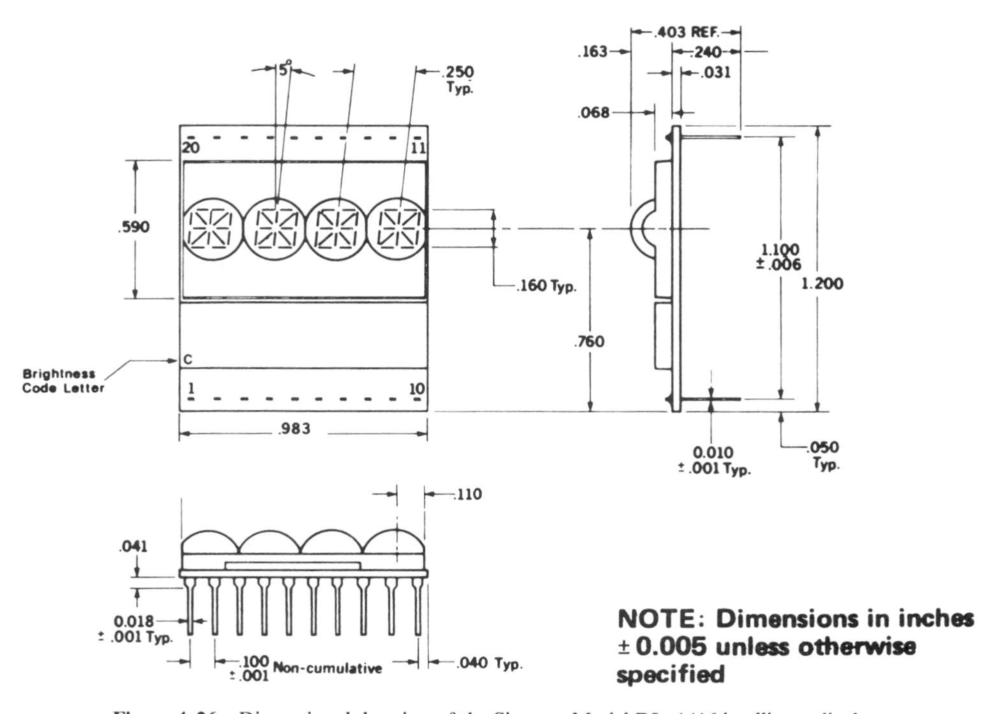

Figure 4-26 Dimensional drawing of the Siemens Model DL-1416 intelligent display (*courtesy of Siemens Optoelectronics Division*).

Siemens Intelligent Displays

Figure 4–25 illustrates three types of Siemens intelligent displays. Figure 4–26 provides a dimensional drawing for one of the displays, the model DL–1416. Optical specifications for model DL–1416 include luminous intensity, viewing angle, digit size, spectral peak wavelength, and spectral line half-width.

Siemens Programmable Display

State of the art in displays are the four-character dot matrix, programmable display from Siemens Optoelectronics Division. This device has a built-in CMOS integrated circuit which contains control and display, an ASCII decoder, a multiplexer, a read/write memory, a driver, and bus interface circuitry. The device is software driven. Software-controlled features include programmable highlight attribute (blinking and nonblinking), asynchronous memory clear function, lamp test, display blank func-

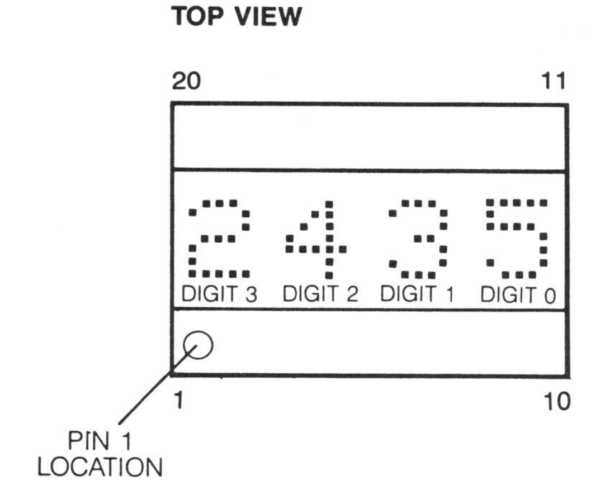

PIN ASSIGNMENTS

PD 2435, PD 2437 PINOUT					
Pin		**Function**	**Pin**		**Function**
1	$\overline{RD}$	READ	11	$\overline{WR}$	WRITE
2	CLK I/O	CLOCK I/O	12	D7	DATA MSB
3	CLKSEL	CLOCK SELECT	13	D6	DATA
4	$\overline{RST}$	RESET	14	D5	DATA
5	CE1	CHIP ENABLE	15	D4	DATA
6	$\overline{CE0}$	CHIP ENABLE	16	D3	DATA
7	A2	ADDRESS MSB	17	D2	DATA
8	A1	ADDRESS	18	D1	DATA
9	A0	ADDRESS LSB	19	D0	DATA LSB
10	GND		20	V_{CC}	

Figure 4–27 Pinout for Siemens programmable display (*courtesy of Siemens Optoelectronics Division*).

Figure 4–28 Siemens programmable display (*courtesy of Siemens Optoelectronics Division*).

tion, single- or multiple-character blinking function, and programmable brightness levels. The display has a 96-character ASCII format with an 8-bit bidirectional data BUS. The clock is internal or external. Characters are bright green or high-efficiency red in color. The device is stackable, has a 5 × 7 dot matrix format, and can be read from 8 feet. See Figure 4–27 for pinout and Figure 4–28 for a photograph of the device.

THE 10 ELEMENT LINEAR DISPLAYS

The 10-element linear display are 10 individual linear bar displays, designed to display information in easily recognizable bar graph form. They may be stacked end to end for expandable display lengths. An interlock ensures that each bargraph will align correctly with the next one. The bars are matched LEDs. They are contained in a standard dual-in-line package. Colors come in red, green, and yellow. See Figure 4–29 for pinout and Figure 4–30 for a photograph of the display. The 10-element display is used with solid-state meters, position indicators, and instrumentation.

RBG-4820 OBG-4830 YBG-4840 and GBG-4850

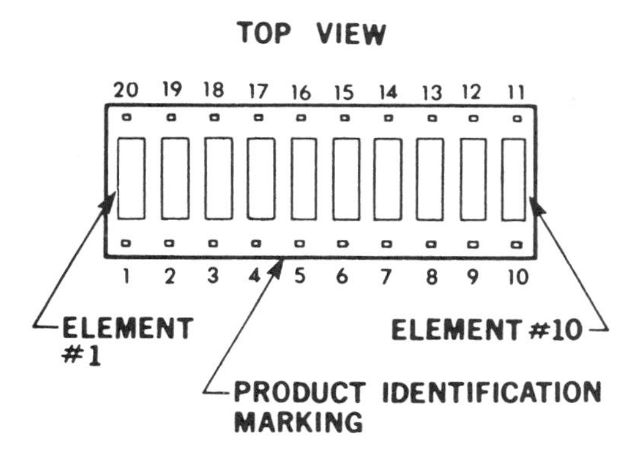

PIN	FUNCTION	PIN	FUNCTION
1	ANODE 1	11	CATHODE 10
2	ANODE 2	12	CATHODE 9
3	ANODE 3	13	CATHODE 8
4	ANODE 4	14	CATHODE 7
5	ANODE 5	15	CATHODE 6
6	ANODE 6	16	CATHODE 5
7	ANODE 7	17	CATHODE 4
8	ANODE 8	18	CATHODE 3
9	ANODE 9	19	CATHODE 2
10	ANODE 10	20	CATHODE 1

Figure 4–29 Pinout for 10-element linear display (*courtesy of Siemens Optoelectronics Division*).

Figure 4–30 Miscellaneous linear arrays and light bars (*courtesy of Siemens Optoelectronics Division*).

THE 101-ELEMENT LINEAR ARRAY

The 101-element linear array is an instrument-quality red LED array. It provides an analog representation of digital data as an expanding bar or as a position indicator when used as a moving dot. The LEDs are connected 10 LEDs to a group, with 10 groups. A single element is brought out separately. The array is addressed by 22 single-in-line pins extending from the back of the circuit board. Colors are available in orange, red, yellow, and green. See Figure 4–31 for pinout and Figure 4–30 for a photograph of the array. The 101-element linear display is specifically designed for multiplexed operation.

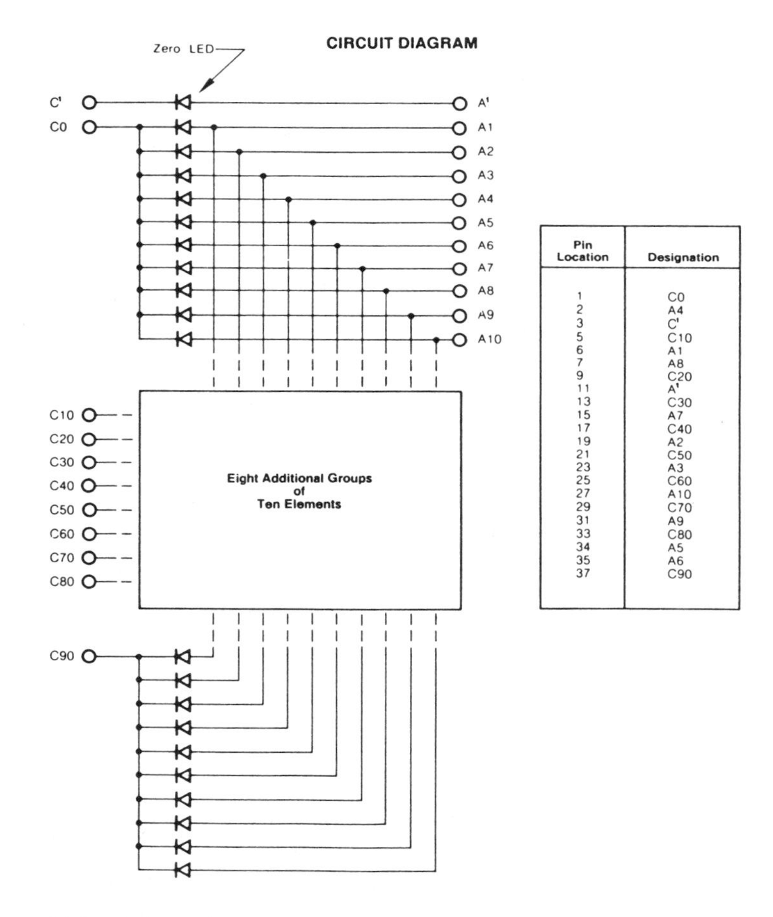

Pin Location	Designation
1	C0
2	A4
3	C'
5	C10
6	A1
7	A8
9	C20
11	A'
13	C30
15	A7
17	C40
19	A2
21	C50
23	A3
25	C60
27	A10
29	C70
31	A9
33	C80
34	A5
35	A6
37	C90

Note A particular element is selected by the common cathode number and the anode number
For example element 56 is lighted by addressed C50 and A6

Figure 4–31 Pinout for 101-element linear display (*courtesy of Siemens Optoelectronics Division*).

LIGHT BARS

Light bars are square displays. They are configured in a dual-in-line package with a mechanical barrier to isolate two rectangular light-emitting areas. Four LED chips are used for uniform light emitting. The light bar may be stacked in X-Y configuration. Colors available are red, yellow, and green. See Figure 4–32 for pinout and Figure 4–30 for a photograph of the light bar. Light bars are used with message annunciators, position/status indicators, and bar graphs.

Package Dimensions in Inches (mm)

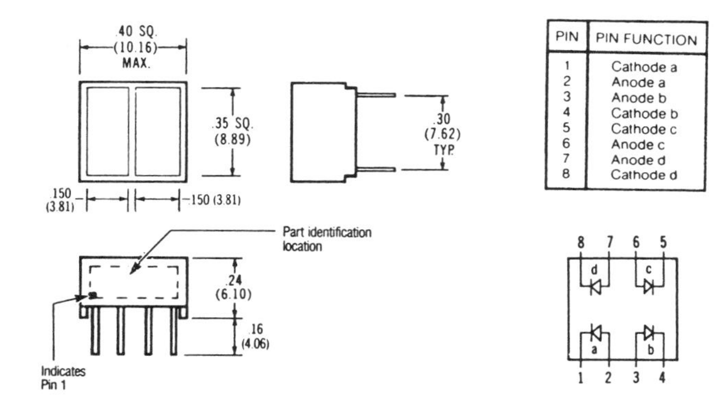

PIN	PIN FUNCTION
1	Cathode a
2	Anode a
3	Anode b
4	Cathode b
5	Cathode c
6	Anode c
7	Anode d
8	Cathode d

Figure 4–32 Pinout for light bars (*courtesy of Siemens Optoelectronics Division*).

INJECTION LASER DIODE (ILD)

The semiconductor injection laser diode (ILD) is extremely well suited for use within the fiber-optic industry. The prime reasons for their applicability are their inherent ruggedness, extreme efficiency, and small size. They can be pumped and modulated by injection current. Figure 4–33 is an Alcatel Model T912 double-heterojunction

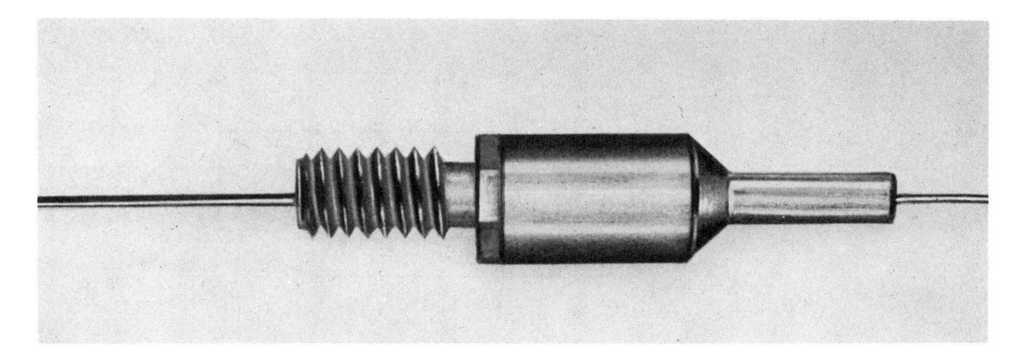

Figure 4–33 The T-912 AIGaAs injection laser diode (*courtesy Alcatel Cable Systems group*).

AlGaAs injection laser typical of the present-day state-of-the-art ILD's on the market. The term heterojunction requires some explanation. Early in the development of semiconductors, lasing action by stimulated recombination of carriers injected across a PN junction was predicted. In 1962, lasing action was achieved in a crystal of gallium arsenide (GaAs) (see Figure 4-34). It was accomplished by three independent research teams simultaneously. The firms were GE, IBM, and Lincoln Laboratory. The unit was called a homostructure PN-junction laser. The word homostructure is used because the injection laser diode was made from one material, gallium arsenide (GaAs). The lasers are called junction laser diodes, junction lasers, and injection laser diodes because electrons are injected into the junction region. The gallium arsenide (GaAs) is lightly doped with suitable impurities to form a PN junction just as silicon is doped when manufacturing solid-state diodes for use as rectifiers. The ends of the GaAs crystal are polished to a mirrorlike finish. Light is generated by injecting current (electrons) into the P-type region where there are deficiencies of free electrons in its lattice structure. The N-type region has an excess of free electrons. When the electrons and holes recombine within the junction region (active, region), photons are emitted as radiant energy. The mirror ends tend to reflect the photons back into the active (more recombinations) and serve as a feedback mechanism. The sides of the device are optically diffused. Brightness is controlled by adjusting the current flow. As noted, the ILD is placed in a circuit in reverse of a standard diode because current (electrons) are injected into the P-type material.

The homostructure diode quickly fell into obscurity for it was found that the ILD could be improved considerably with the use of the heterostructure. In the heterojunction, the active region is bounded by wider band-gap regions. Heterojunctions fall into three categories, the single heterostructure (SH), the double heterostructure (DH), and the separate confinement heterostructure (SCH). The single heterostructure has only one heterojunction so that injected carriers are confined by the junction at only one boundary of the active area. The double heterostructure (DH) laser carriers and waveguide have boundaries on both sides of the active region. In the separate confinement heterostructure (SCH), the carriers are confined in a region within the waveguide.

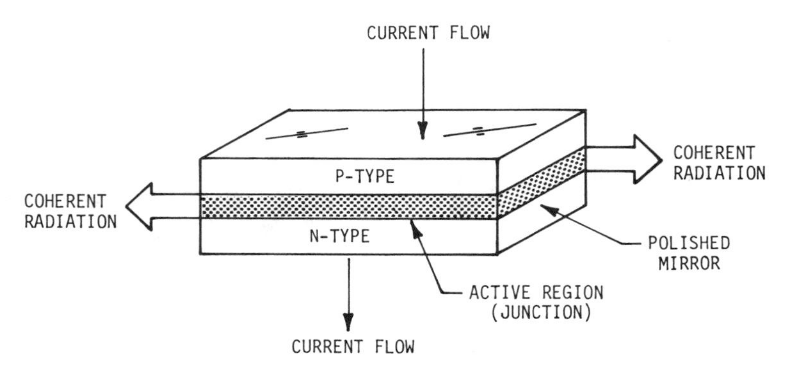

Figure 4-34 A homostructure PN junction laser (*Mike Anaya*).

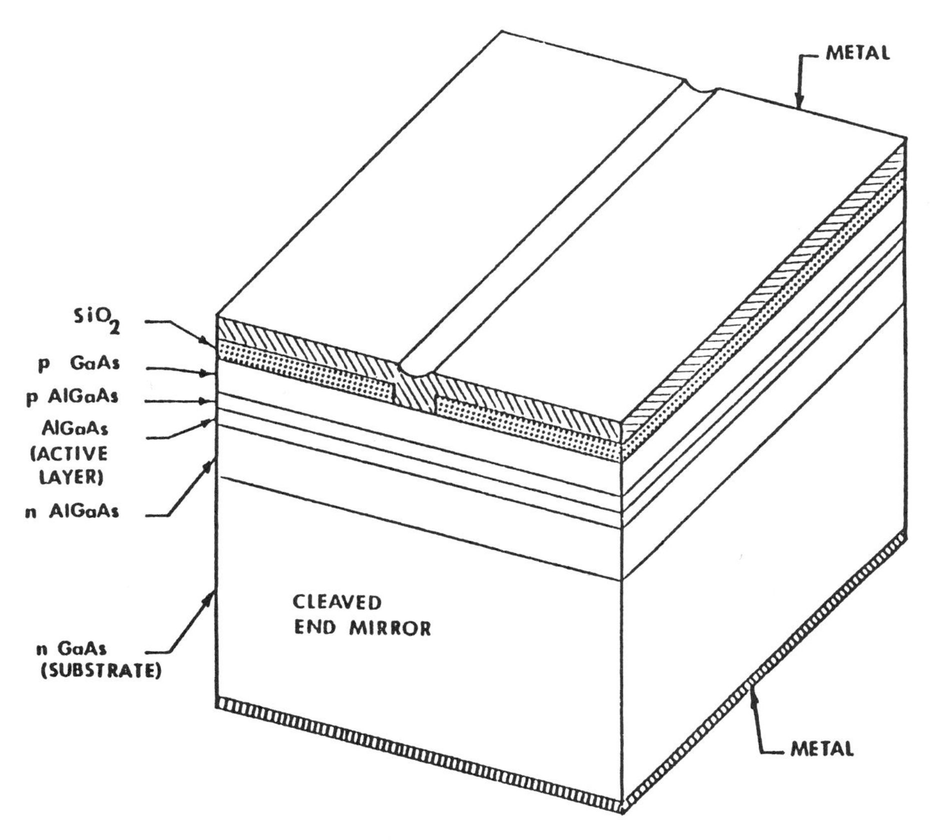

Figure 4–35 A heterostructure PN junction laser (*courtesy of Alcatel Cable Systems group*).

A cross section of the T912 double heterojunction injection laser diode is illustrated in Figure 4–35. Note that the active area is aluminum gallium arsenide (A1GaAs) while one boundary to the active region is N-type A1GaAs and the opposite boundary to the active region is P-type A1GaAs. On the top of the structure is a layer of P-type gallium arsenide (GaAs), while on the base is an N-type gallium arsenide (GaAs) substrate. The T-912 device is of a stripe geometrical design, that is, a layer of silicon dioxide (S_iO_2 is placed between the junctions and a copper. The purpose of the S_iO_2 is to provide a narrow stripe rather than a wide structure that may eliminate noise.

The output characteristics of the T-910 are compatible with PIN and APD detectors and photocathodes. They are ideally suited for use as light sources in high-performance, long-distance, high-bandwidth fiber optic communication systems. They may also be used as target designators, intrusion alarms, and other applications.

A dimensional outline of the T-912 is illustrated along with some of its characteristic curves. The edge-emitting chip is mounted P-type side down on a copper sink, which also provides an on-axis alignment with the coaxial stud header (see Figure 4–36).

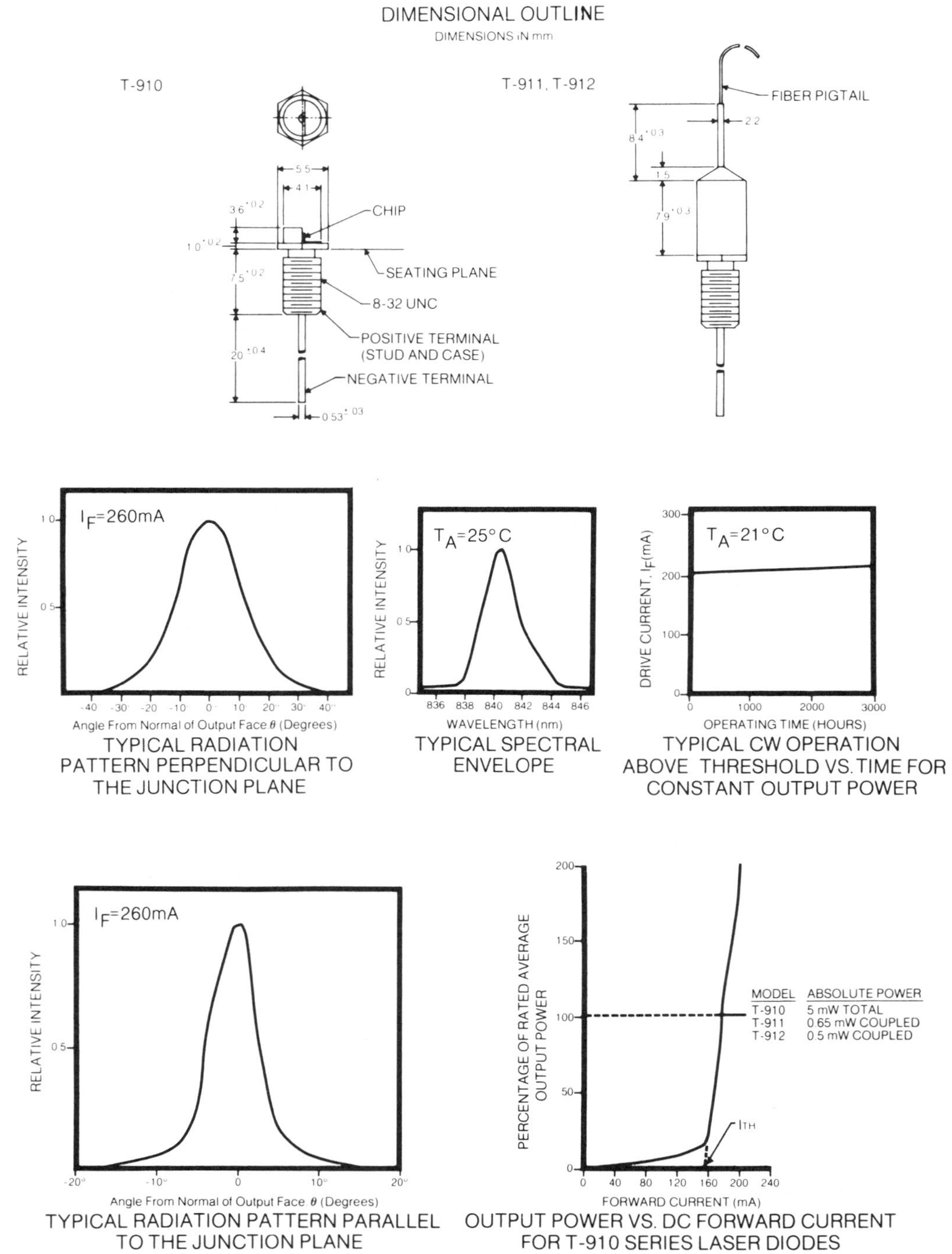

DIMENSIONAL OUTLINE

DIMENSIONS IN mm

T-910

T-911, T-912

CHIP

SEATING PLANE

8-32 UNC

POSITIVE TERMINAL (STUD AND CASE)

NEGATIVE TERMINAL

FIBER PIGTAIL

TYPICAL RADIATION PATTERN PERPENDICULAR TO THE JUNCTION PLANE

TYPICAL SPECTRAL ENVELOPE

TYPICAL CW OPERATION ABOVE THRESHOLD VS. TIME FOR CONSTANT OUTPUT POWER

TYPICAL RADIATION PATTERN PARALLEL TO THE JUNCTION PLANE

OUTPUT POWER VS. DC FORWARD CURRENT FOR T-910 SERIES LASER DIODES

MODEL	ABSOLUTE POWER
T-910	5 mW TOTAL
T-911	0.65 mW COUPLED
T-912	0.5 mW COUPLED

Figure 4–36 Characteristic curves of an AlGaAs injection laser diode (ILD) (*courtesy of Alcatel Cable Systems group*).

LIQUID-CRYSTAL DISPLAYS

Liquid-crystal displays (LCDs) are now used as very low power replacements for LED displays, particularly in portable, battery-powered applications. Since they do not give off light like the rest of the devices in this chapter, they really should not be here. However, they are so similar to LEDs in application that they are included.

Liquid-crystal material contains long, cigar-shaped molecules in a "mesophase" liquid—a material state between liquid and solid. The liquid has a property that makes it useful for displays—it will either reflect light or pass it through, depending on whether a small voltage is present or not. An LCD cell contains a thin layer of this liquid sandwiched between two thin sheets of glass. Very often the bottom sheet is mirrored to reflect light. Transparent electrodes of any desired shape are deposited on the inside of each sheet.

The simplest type of display is the scattered-light type shown in Figure 4–37. With no voltage applied the molecules align themselves and provide very little opposition to light passing through. If the bottom layer of glass is mirrored, the display will appear silvery.

With a few volts (3 to 20) applied, a small current flows through the liquid and bumps the molecules into random patterns. Incoming light cannot pass through the cell and is scattered instead. The cell appears dark under the electrode with the applied voltage. For adequate viewing the ratio of bright to dark (the contrast ratio) should be better than 2:1. Depending on the ambient light conditions, these cells can be superb or marginal.

The cell will work with either an ac or a dc voltage. However, over time, a dc voltage will cause the cell to polarize and cease to operate. Ac voltages with equal positive and negative swings are used to offset this.

A newer display type provides higher contrast ratios and is used on the relatively large screens of portable computers. The liquid used is a "twisted-nematic" type where the crystals align themselves normally in a slow twist. As light passes through the layer of liquid its polarization will be rotated (see Figure 4–38).

To make a display with this liquid, polarizers are added to the front and back of the display with one oriented at 90° to the other. With no voltage applied, light passes through from front to back and then back to the front again because the liquid adds the extra polarization twist needed. With voltage applied the liquid is disturbed and the extra rotation is not available. Although a mirrored back surface can be used with either display, better effects (higher contrasts) are achieved if some battery power can be spared to backlight the display.

For calculators, only a numeric display is needed and so the transparent electrodes are deposited in the traditional 7-segment pattern used with LEDs. Computers must also display alphanumeric and graphic symbols and so a dot matrix is needed—typically, 5 × 7 dots to form a character. The dots are formed at the intersection of horizontal and vertical grid lines. Each of the dots on one horizontal line are scanned in sequence. Only one dot will be activated at any one time and the whole process

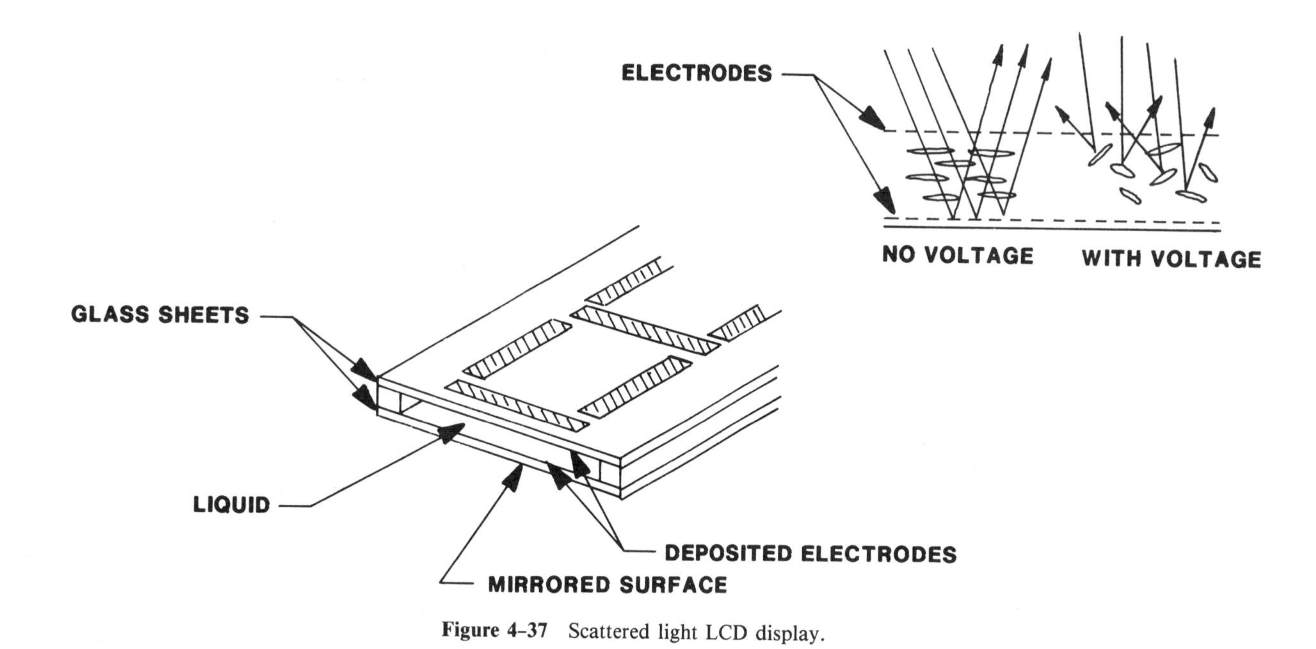

Figure 4–37 Scattered light LCD display.

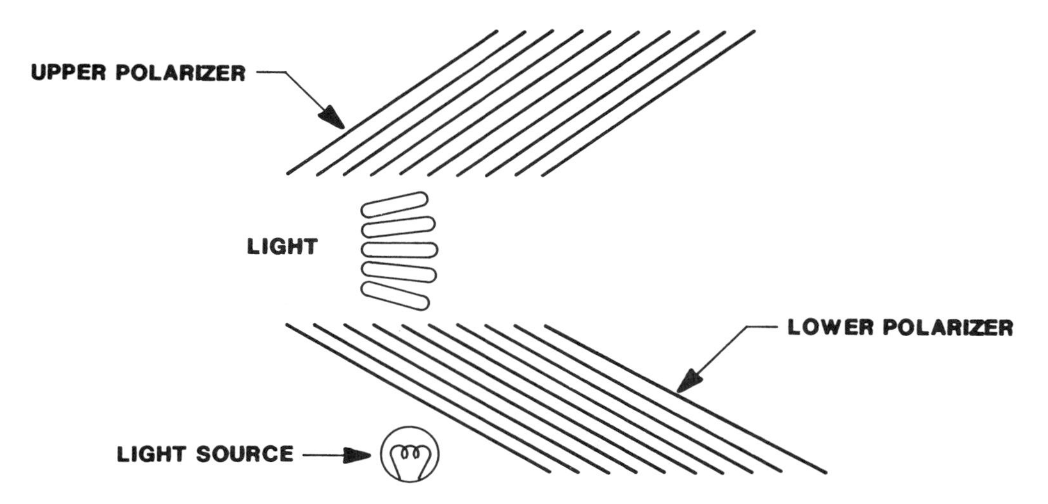

Figure 4–38 Twisted-nematic LCD display with backlighting.

is repeated 60 times each second (see Figure 4–39 and 4–40). A pixel will be dark only when there is a difference between the row and column voltages.

LCDs suffer from one problem: The switching time from on to off is a little slow, around 100 milliseconds, and gets worse as temperature drops. This causes a smearing effect when information moves on the display. However, for most calculator/computer applications, this has not been a serious problem.

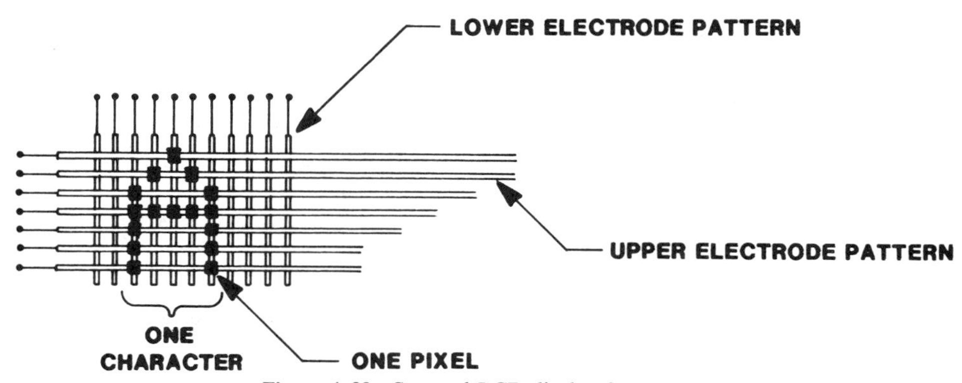

Figure 4–39 Scanned LCD display for computers.

SELF-CHECK QUESTIONS

Self-check questions allow students to evaluate how well they have learned what they are studying.

1. Define the term *electroluminescence*.

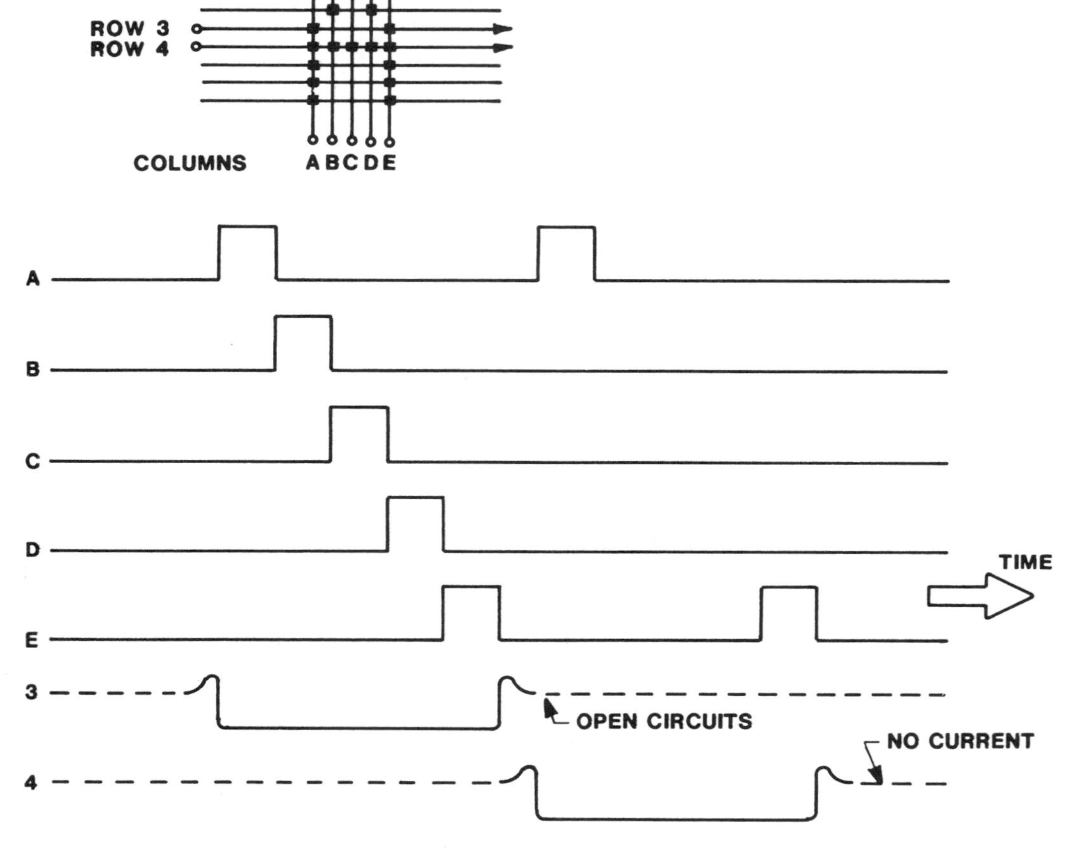

Figure 4–40 Row and column scanning.

2. What is probably the fundamental electroluminescent device? What is the other recognized device?
3. Explain the basic operation of an LED.
4. What are some of the materials used to create electroluminescent devices?
5. What is meant by direct-gap material? Indirect gap?
6. Define the spectral response curve.
7. Define the operating time–power curve.
8. Define the forward current–output power percent curve.
9. Define the diode current voltage curve.
10. What are some of the specifications used with electroluminescent devices?
11. Name some of the circuit applications of the LED.
12. What are two advantages of liquid-crystal displays (LCDs) over LED displays?
13. What are two disadvantages of liquid-crystal displays (LCDs)?

Five

Photodetection

The human eye detects visible radiation in wavelengths between 3900 to 7500 angstroms. Other detectors are capable of detecting electromagnetic radiation from radio waves to x-rays. The wavelength detection that we are considering includes ultraviolet, visual, and infrared wavelengths, that is, wavelengths from 0.005 to 4000 microns. There are three basic forms of photodetection. They involve photoemission, photoconduction, and photovoltaic action. All quantum detectors respond directly to the action of incident light waves. The first, photoemission, involves incident light that frees electrons from a detector's surface. This usually occurs in a vacuum tube. We shall describe this action in the next paragraph. With photoconduction the incident light on a photosensitive material causes the photodetector to alter its conduction. The third, photovoltaic action, generates a voltage when light strikes the sensitive material in the photodetector. In application, solid-state detectors operate in either the photoconductive mode or the photovoltaic mode.

OBJECTIVES

After studying this chapter and completing the self-check questions at the end of the chapter, the reader will be able to:

1. Describe the three forms of photodetection.
   a. Photoemission
   b. Photoconduction
   c. Photovoltaics
2. State several photodetector types.

3. Expound on the attributes of a good photodetector.
4. Explain the function of photodiode detector performance curves to include the following.
   a. Spectral response
   b. Spectral response versus temperature
   c. Source resistance versus temperature
   d. Capacitance versus bias voltage
   e. Dark current versus temperature
5. Explain the function of phototransistor detector performance curves.
   a. Angular response
   b. Collector dark current versus temperature
   c. Relative response
   d. Fall time versus collector current
   e. Turn-off delay times for current
   f. Turn-on delay times
6. Interpret photodiode specifications.
7. Interpret phototransistor specifications.
8. Examine detector circuit applications.
9. Compare the operation of the planar diffused photodiode and the Schottky barrier photodiode.
10. Describe the PIN diode and its operation.
11. Analyze the spectral response curve for photodetectors.
12. Examine typical photodiodes and their applications.
13. Examine typical phototransistors.
14. Describe the vidicon camera tube and its operation.
15. Describe the solid-state camera array and its operation.
16. Describe the opto-isolator and its operation.
17. Analyze the silicon solar cell and its characteristics.
    a. The photoelectric effect
    b. Solar cell equivalent circuit
    c. Power in an ideal solar cell
    d. Spectral response
    e. Cell efficiency
    f. Solar cell material
18. Compare the several types of solar cell and their functions.
    a. Homojunction cell
    b. Heterojunction cell
    c. Schottky barrier cell
    d. Polycrystalline cell
    e. Insulator cell
    f. Cadmium sulfide cell

g. Gallium arsenide cell
h. Amorphous silicon cell
i. Electrochemical cell

GENERAL

Photoemission

Photoemissive action usually takes place in a vacuum tube. The action is due to radiant energy striking a sensitive material causing an emission of electrons. The effect of light on certain materials causes electrons to be displaced from the surface. The emitted (radiated) electrons are called *photoelectrons.* In Figure 5-1, light is applied to the cathode of a basic phototube. The light causes photoelectrons to be drawn through a load by the positive anode. Thus the light determines the current flow and the voltage drop across the load resistance. Current flow is also controlled by the potential applied to the anode.

When light is applied to the surface of the cathode, it illuminates the cathode. Illumination is expressed in lumens or footcandles. One lumen is the amount of luminous flux of 1 square foot of a spherical surface, 1 foot from a 1-candle source. A footcandle is the intensity of illumination at all points on the illuminated surface 1 foot from a-candle source. (Stated in comparison, intensity of illumination is 1 footcandle when the amount of luminous flux is 1 lumen per square foot.)

Alkali metals have the best ability to emit electrons because of light. Some of these metals are cesium, lithium, and rubidium. Thin coats of the alkali metal are placed on the cathode to provide a radiation-sensitive coating. Some phototube devices are the gas-filled phototube, the multiplier phototube, and the vacuum phototube. Phototubes are used in applications such as relay control and film/sound reproduction. While the phototube itself is rarely used in modern applications, two of its descen-

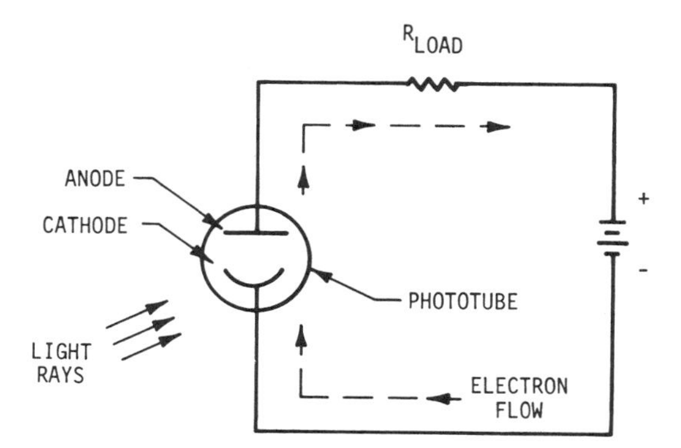

Figure 5-1 Photoemission (*Mike Anaya*).

dents are heavily utilized: the vacuum-tube *vidicon* and the solid-state *charge-coupled device* (CCD) array. Both are discussed later in this chapter.

Photoconduction

Operation through photoconduction involves a change in resistance of the photosensitive material. A wafer of photoconductive material is placed underneath a glass window to protect it from exposure. The photoconductive material is tied to a load resistance and a power source. The clear glass window allows light radiation to strike the photoconductive material freeing valence electrons. The resistance of the photoconductive material decreases, causing current through the load to increase. The resistance of the photoconductive material may change from several million ohms to several hundred depending on the current demand of the device (see Figure 5–2).

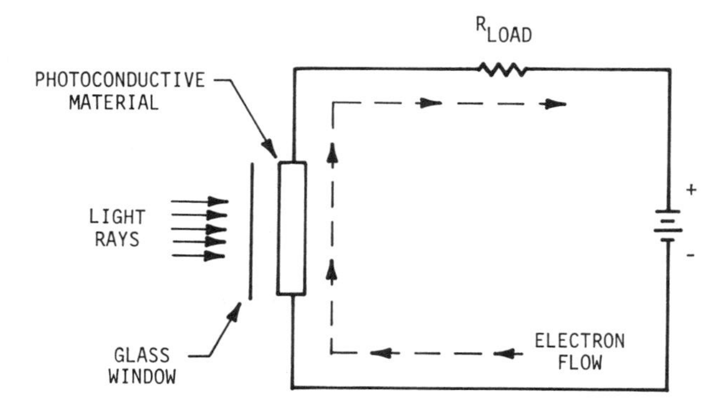

Figure 5–2 Photoconduction (*Mike Anaya*).

Photovoltaic Cell Operation

Operation of the photovoltaic cell involves the use of dissimilar metals to generate an electromotive force in response to radiated light. In Figure 5–3, a light-sensitive material is placed beneath a thin layer of transparent metal and next to a dissimilar metal. The light-sensitive material is exposed to radiation through the thin transparent metal which acts as a filter. When exposed, free electrons are removed from the light-sensitive material, causing electrons to flow to the dissimilar metal. This creates current flow and a difference of potential between the two terminals connected to the load.

Photodetector Types

The most basic form of detection is the photoresistor. The photoresistor is a small slice of photoconductive material whose resistance increases or decreases as light energy is applied. Electrons are released by the light and flow toward a positive power sup-

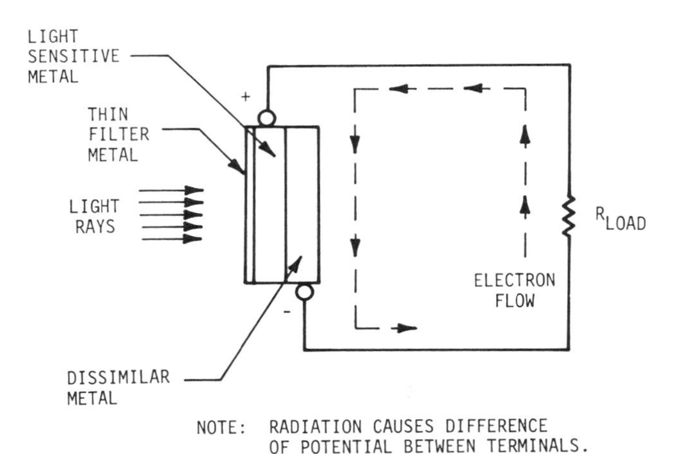

Figure 5-3 Photovoltaic cell operation (*Mike Anaya*).

ply. The basic assumption in the photoresistor is to control electrical energy from the power supply. The photoresistive material is nonreflective. There are no junctions in a bulk photoresistor, therefore lead polarity does not matter.

A second detector is the single-junction photodiode. A photodiode is the optical version of the standard diode. It is constructed of a PN junction. Photons of light energy are absorbed into the device. Hole-electron pairs are generated. The pairs are combined at different depths within the diode depending on the energy level of the photon. A wide, thin surface area is used to ensure maximum absorption. Current flow is dependent on the amount of radiation absorbed. Photodiodes operate in the photoconductive mode with reverse dc bias applied. They operate in the photovoltaic mode without bias.

The solar cell is a photodiode that is heavily doped. The depletion area is extremely thin. The cell is coated to avoid reflection. Hole-electron pairs diffuse to the depletion area of the diode where they are drawn out as useful current. Output current is dependent on input radiation. Solar cells are not biased and are photovoltaic in operation.

Phototransistors are two-junction devices that have a large base area. The base region of the phototransistor absorbs the photons of energy and generates hole-electron pairs in the large base-collector region. The collector, being reverse-biased, draws the holes toward the base and the electrons toward the collector. The forward-biased base-emitter junction causes holes to flow from base to emitter and electrons from emitter to base. Forward bias causes the phototransistor to operate just as the conventional transistor operates. The basic function, then, is that the light energy induces the transistor to operate. Collector bias is provided by a power supply. Base bias is provided by the incident light. This in turn controls collector current.

Other detection devices include photo field-effect transistors, the photothyristor, and opto-isolators. Photo field-effect transistors are constructed in the same manner as the standard-junction FET. A lens is used to focus light on the gate. The gate, being light-sensitive, excites electrons from the gate into a conduction band. Electron movements causes current flow and a voltage drop. The voltage drop induces the FET load current through its drain resistance and thus modifies the drain to source voltage. Photoposition detection devices provide an output that is proportional to the direction of the input beam. The photothyristor is a four-layer PNPN device. Photons of energy create hole-electron pairs in a thin junction. These pairs are drawn across all three junctions producing a current from anode to cathode. The opto-isolator combines a light source with a detector within a single package. Current is applied to a source LED (light-emitting diode). The LED emits photons that are detected by a detector. The detector can be in the form of a photoresistor, photodiode, phototransistor, or photothyristor.

Each of the devices and detection methods that we have discussed has applications in the infrared and the ultraviolet wavelengths with detector-material changes. Typical detection devices are described and illustrated later in this chapter.

Photodetection Materials

A detector material is specially selected so that when it is exposed to light rays it will absorb the light energy. If the detector responds to the light energy (flux), and not the wavelength, the detector is said to be nonselective. If the detector responds by varying detection at different wavelengths, it is selective. Selectivity or responsivity is the detector's response per unit of light. Wavelength responsivity is called *spectral response.* Frequency response is the speed by which the detector responds to changes in radiation amplitude. Fluctuations in output current and/or voltage are referred to as *noise.* Noise is usually caused by current that flows in the detector regardless of whether light is applied. Current, such as this, is termed *dark current,* because it flows even without radiation. A common specification for a detector is the signal-to-noise ratio. The ratio is the signal current divided by the noise current.

Photoresistor resistor materials are called *bulk photoconductors.* Bulk photoconductors are usually made from cadmium sulfide (CdS), cadmium selenide (CdSe), lead sulfide (PbS), and silicon (Si). The type of material is dependent on the application. Bulk photoconductors do not have a junction, they are simply made from one material. This material should respond to a broad range of wavelengths from ultraviolet through infrared. The ultimate purpose of the bulk photoconductor is to convert light intensity to current flow, therefore the material selected should free electrons easily with the application of light. Light should not reflect but be transmitted easily into the detector. To be efficient the bulk (resistive) material should be extremely sensitive to light radiation and be as easy to apply as a resistor. The primary problem with photoresistive material is that a problem arises with temperature stability for faster-acting materials.

The photodiode is usually made from silicon or germanium much the same as standard diodes. There are basically two processes in which the photodiode material is formed. These are the planar process and the Schottky barrier process. The planar process uses silicon as the solid and gas as dopants. By diffusion, the dopant gases penetrate the solid surface of the silicon. Diffusions may be made on diffusions, therefore several layers of dopants may be diffused on one device. The Schottky barrier photodiode operates in much the same manner as the planar diffuse. Their construction differs in that the Schottky barrier diode has its active region metallized to an N-type bulk silicon material. The Schottky barrier photodiode has some advantages over the conventional photodiode in that it is easier to manufacture and operates at wavelengths of less than 500 nanometers. The conventional photodiodes operate well at wavelengths over 500 nanometers. The reason for the wavelength difference is that short wavelengths are absorbed near the surface of the diodes while long wavelengths may penetrate deeper into the structure. Again, as in bulk resistors, temperature is a continuous problem. The PIN diode (another photodiode) has heavily doped N and P regions with a very lightly doped I (intrinsic) region between. The resistance of the I region is from 10 ohms per cubic centimeter, while the N and P regions are less than 1 ohm per cubic centimeter. The intrinsic layer provides a large depletion area that allows the wavelengths to penetrate deeper into the diode. This large depletion area complements faster speeds, lower noise, and greater efficiency at longer wavelengths.

The phototransistor is made from silicon or germanium and has much the same characteristics as the standard or conventional transistor. Phototransistors are constructed in much the same manner as photodiodes and standard transistors using the planar diffused method. The base area of the phototransistor is usually made large so as to provide an area which incident light can penetrate and generate electron-hole pairs. Phototransistors are subject to the typical problem of all transistors, temperature variations. These problems may be solved with biasing techniques and thermal-stability resistors.

Infrared detectors must use thermal detectors because their wavelengths are longer. Lead sulfide (PbS) responds to infrared wavelengths of 2 to 4 microns. Indium antimonide (InSb) is sensitive to infrared wavelengths up to 5.6 microns. Silicon (Si) and germanium (Ge) are only sensitive to the near-infrared wavelengths.

Ultraviolet detectors utilize materials such as telluride, cesium, or a fluorescent coating of sodium salicylate as cathodes within phototubes. Bulk semiconductors made of cadmium selenide (CdSe) may respond to ultraviolet in the range of 300 to 400 nanometers.

PERFORMANCE CURVES FOR PHOTODIODE DETECTORS

Figure 5–4 is a set of curves for the United Detector Technology, Inc. Models PIN-LSC/5D and PIN-LSC/30D diode detectors.

Spectral response. The first curve in the illustration is the spectral response curve on the top left side. The curve plots wavelength in nanometers against respon-

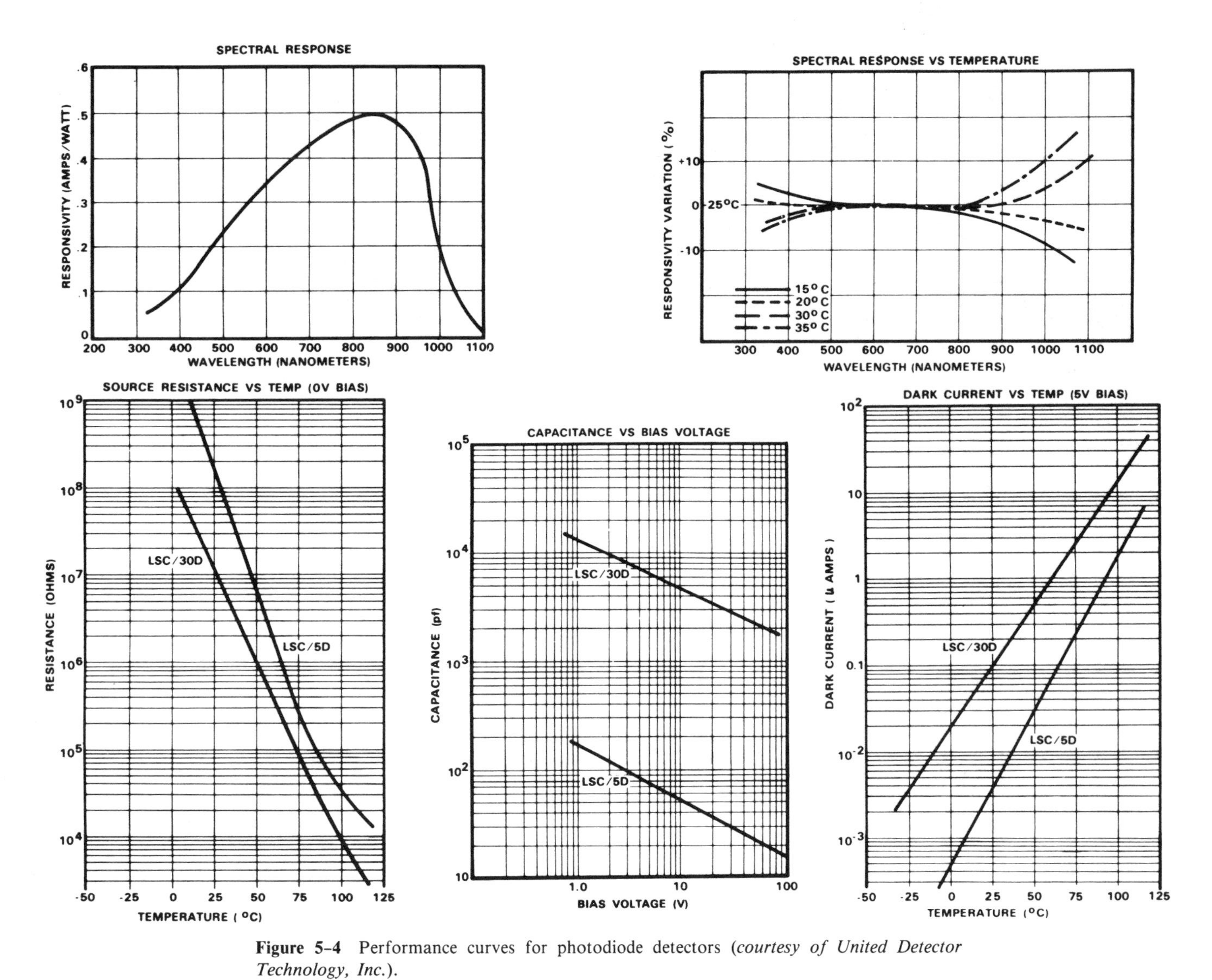

Figure 5-4 Performance curves for photodiode detectors (*courtesy of United Detector Technology, Inc.*).

sivity. Note that the peak responsivity occurs around 850 nanometers. The spectral range is between 350 and 1100 nanometers.

Spectral response versus temperature. This curve is on the top right in the illustration. There are four curves involved here at four separate temperatures in Celsius. The curves plot wavelength in nanometers against percent of responsivity variation. Note that at all temperatures zero variation occurs between 500 and 800 nanometers with fall-off both positive and negative at other wavelengths.

Source resistance versus temperature. The curves for two photodiode detectors on the bottom left of the illustration plot resistance in ohms against temperature in degrees Celsius. The curves are logarithmic and show an operating range at temperatures between 0 and +75 °C.

Capacitance versus bias voltage. The curve on the lower center of the illustration plots bias voltage in volts against capacitance in picofarads. The two curves are logarithmic. They show varying capacitance at maximum bias levels.

Dark current versus temperature. These curves are on the lower right side of the illustration. They plot temperature in Celsius against dark current in microamperes. The two curves are logarithmic. They show varying dark current at varying temperatures. Dark current you may recall is that current which flows regardless of the devices exposure to radiation.

Other curves. There are other curves for photodiodes worth mentioning. The linearity curve plots incident radiant energy in watts against photocurrent in amperes. The current versus voltage characteristics (I–V) plot bias voltage against photocurrent.

PERFORMANCE CURVES FOR PHOTOTRANSISTORS

Figure 5–5 is a set of performance curves for phototransistors. This particular set is for the Siemens LPT–100 series phototransistor. Other phototransistors have similar curves.

Angular response. The curve on the top left of the illustration plots angle in degrees against relative response in percent. Note that the peak response is at an angle of 0°. The LPT–100 has a spread of around −30 to +30°. The LPT–110 has a wider spread.

Collector dark current versus temperature. The top right curve plots temperature in Celsius against leakage current in nanoamperes.

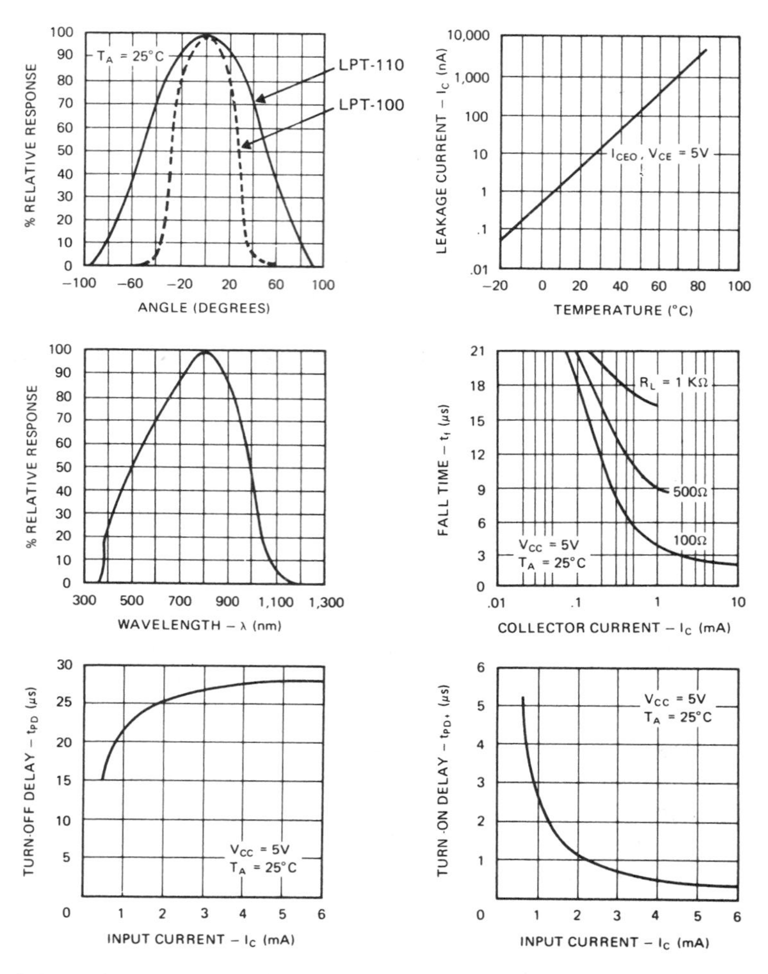

Figure 5–5 Performance curves for phototransistors (*courtesy of Siemens Optoelectronics Division*).

Relative response. The relative response on the center left of the illustration is probably the curve most frequently provided by manufacturers for photoelectric devices. This curve as you can see peaks at 800 nanometers with a spread of around 500 to 1000 nanometers.

Fall time versus collector current. The curve on the center right is a logarithmic curve that plots collector current in milliamperes against fall time in microseconds, with several loads (R_L). Fall time is defined as the time required for the collector-emitter current (I_{CE}) to decrease from 90 to 10% of peak value. Rise time is defined as the time required for collector-emitter current to increase from 10 to 90% of peak value.

Turn-off delay times for current. The curve in the lower left of the illustration plots input current in milliamperes against turn-off delay in microseconds as current increases to 2 milliamperes, then tends to level off.

Turn-on delay times. The lower-right curve plots input current in milliamperes against turn-on delay in microseconds. The curve drops dramatically from 5 to 1 microseconds as current increases to 2 milliamperes, then tends to level off.

Other curves. There are other characteristic curves for phototransistors worth mentioning. These are photocurrent characteristics, collector currrent versus collector voltage and collector-base characteristics. The latter two are also standard transistor curves.

SPECIFICATIONS OF PHOTODIODE DETECTORS

Table 5–1 is a rather standard set of specifications for a photodiode detector. This particular set establishes maximum and minimum values for the United Detector Technology Inc. models PIN–LSC/5D and PIN–LSC/30D. Other photodiodes have similar

TABLE 5–1 SPECIFICATIONS FOR PIN–LSC/5D AND PIN–LSC/30D PHOTODIODES

Parameters and (units)	PIN–LSC/5D			PIN–LSC/30D		
	Min.	Typ.	Max.	Min.	Typ.	Max.
Recommended mode or operation		Photovoltaic			Photovoltaic	
Spectral range @ 5% of peak (nm)		350–1100			350–1100	
Responsivity at peak λ (A/W)	—	0.55	—	—	0.55	—
Uniformity of responsivity (with 1 mm spot dia)	—	±2%	±7%	—	±2%	±7%
Position sensitivity at peak λ (A/W/cm)	—	2.2	—	—	0.36	—
Position: central 25%	—	0.2%	0.6%	—	0.2%	0.6%
Linearity: central 75%	—	1.0%	5.0%	—	1.0%	5.0%
Null point drift (μm/°C)	—	2.0	10.0	—	3.0	10.0
NEP @ peak λ kHz ($W/Hz^{1/2}$)	—	2×10^{-13}	—	—	5×10^{-13}	—
Dark current (μA) @ 5 V bias	—	0.003	0.05	—	0.1	5.0
Breakdown voltage @ 100 μA (V)	50	200	—	5	—	—
Source resistance (MΩ)	10	100	—	1	10	—
Series resistance (kΩ)	—	10	20	—	2	10
Capacitance (pF)						
@ 0 V bias	—	190	380	—	14,700	29,000
@ 5 V bias	—	77.5	155	—	6200	9300
@ 10 V bias	—	55	110	—	—	—
@ 50 V bias	—	24	48	—	—	—
Rise time 10–90% (μs) @ 5 V bias	—	0.5	—	—	7.0	—
Fall time 90–10% (μs) @ 5 V bias	—	0.5	—	—	7.0	—
Recommended maximum frequency usage (kHz) @ 5 V bias	—	—	10	—	—	2
Output for 10% linearity (mA)						
@ 0 V bias	0.01	0.02	—	0.05	0.1	—
@ 5 V bias	0.1	0.2	—	0.5	1.0	—
Approximate saturation level (mW) @ 5 V bias	—	2	—	—	10	—

Courtesy of United Detector Technology, Inc.

specifications. Spectral range, responsivity, dark current, and breakdown voltages are standard specifications provided for all photodiodes. Rise and fall times are equally standard parameters.

SPECIFICATIONS OF PHOTOTRANSISTOR DETECTORS

Table 5-2 is a standard set of specifications for a phototransistor detector. This particular set is for the Siemens LPT-100 series transistor. Other phototransistors have similar specifications. Photocurrent, acceptance angles, collector-emitter reverse voltage, and collector-emitter cutoff current are standard specifications provided for all phototransistors. Temperature, power dissipation, and maximum current-voltage are also standard specifications.

TABLE 5-2 SPECIFICATIONS FOR SIEMENS LPT-100 PHOTOTRANSISTOR

Photo at V_{CE} = 5 V H = 5 mW/CM^2 (mA)				*Acceptance angle I = 0.51 on axis ± (deg)	Collector-emitter reverse volt V_{CEO} max. V	Collector-emitter cutoff current V_{CE} = 5 V I_{CEO} (nA)		Package dimensions	
Tungsten			GaAs Min.						
Min.	Typ.	Max.				Typ.	Max.	Page	DWG
0.2	1.4		0.6						
1.0		3.0	0.6	30°				32	34
1.3		2.6	0.6						
0.2	0.88		0.6		30	2.0	100		
0.6		1.8	0.6	50°				32	35
0.8		1.6	0.6						

Maximum temperatures/humidity	
Storage temperature	−55 to +100 °C
Operating junction temperature	−55 to +85 °C
Relative humidity at temperature	98% at +65 °C
Maximum power dissipation	
Total dissipation at +25 °C case temperature	200 mW
Total dissipation at +25 °C ambient temperature	100 mW
Maximum voltages	
BV_{CBO} collector-to-base voltage	50 V
LV_{CEO} collector-to-emitter sustaining voltage	30 V
Maximum current	
I_C collector current	100 mA

Courtesy of Siemens Optoelectronics Division.

DETECTOR APPLICATIONS

Detectors are extremely versatile in that they may be used with a great variety of electronic devices to perform an equally great variety of tasks. In fiber optics, an electroluminescent device is applied to fiber optic light communication. An input square-wave signal is driven into a transmitter module. The transmitter module converts the square wave to a current flow for the diode. The diode is an injection laser diode (ILD). When current flows through the diode, the diode radiates a high-intensity light that is launched into the optical cable through the source-to-fiber connector. The light wave is transmitted by reflection to a detector on the receiver side. The detector is probably a photodetector diode such as the PIN diode. The light wave is coupled from the optical cable by a fiber-to-detector connector. The detector converts the light energy to a current flow. The receiver module converts the current to a replica of the original square-wave input signal.

In Figure 5–6a a bulk photoresistor is placed on the input leg to an operational amplifier. The light intensity varies the resistance, which varies the input to the amplifier. R_F is a feedback resistor, while R_1 sets the operating level of the operational amplifier. Figure 5–6b illustrates a PN-junction solar cell in the input leg of an operational amplifier. As in the photoresistor operation in Figure 5–6a the intensity of the light causes an input change to the operational amplifier.

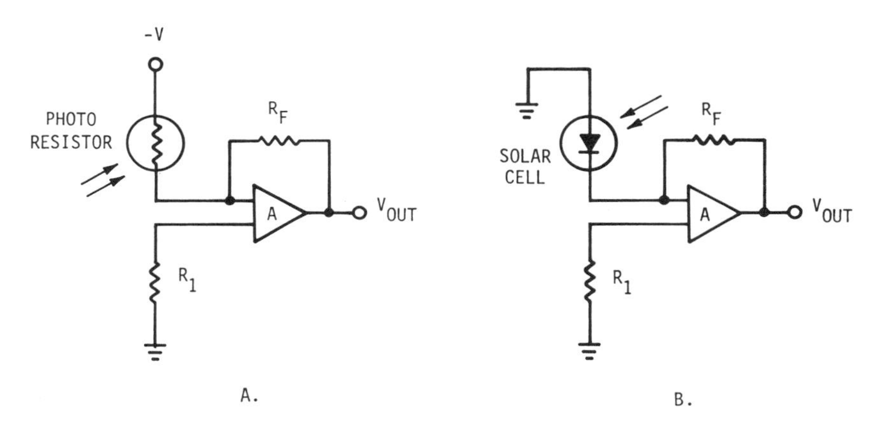

Figure 5–6 Photoresistor and solar-cell detectors (*Mike Anaya*).

In Figure 5–7a a photodetector transistor is used to drive a relay coil. Light beams applied to the base provide forward bias for the transistor. The transistor turns on, energizing the coil of the relay. Relay contacts close and remain closed until the light beam is removed from the transistor base. In Figure 5–7b a phototransistor is used to drive current through a load. Light beams applied to the base forward bias the transistor. The transistor turns on, causing current to flow through a load resistance. In Figure 5–7c a phototransistor is used as a square-wave switch. Light is applied to the base of the transistor in pulses. The transistor turns on and off as the pulses are applied and removed. The output is taken off a resistor in the emitter circuit of the transistor. The output represents the on/off time of the light pulses. Because of the special light energy input, there will be no difference in the ''gains'' of circuits b and c, only a difference in polarity.

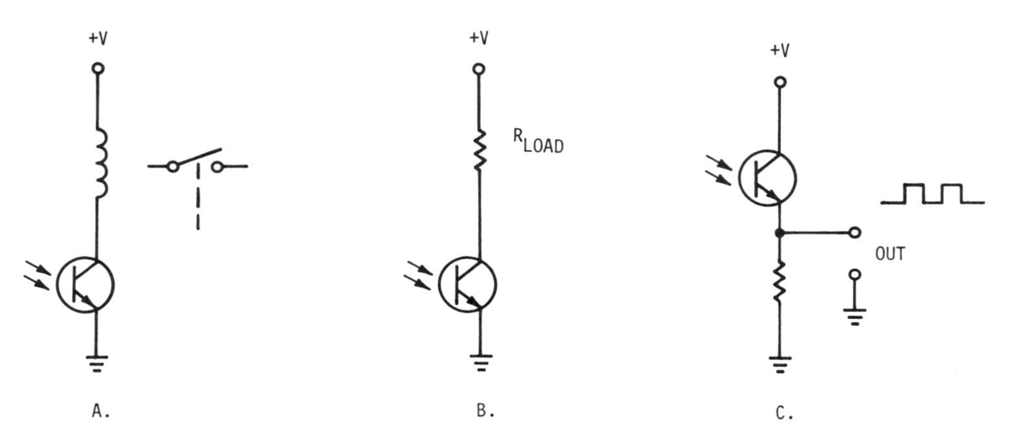

Figure 5-7 Phototransistor applications (*Mike Anaya*).

CONVENTIONAL PN JUNCTION PHOTODIODE OPERATION

Figure 5-8 is a cross-section drawing of a conventional photodiode. There have been many processes by which PN junctions have been formed. The technique that has risen to be the most popular and the one utilized most by industry is the planar process. This process is utilized by integrated-circuit manufacturers. The process uses silicon as the solid and gases as the dopants. By diffusion, the dopant gases penetrate the solid surface of the silicon. Diffusions may be made on diffusions, therefore several layers of dopants may be diffused on one device. This serves to provide manufacturing versatility.

The basic PN junction used in production of the photodiode is the planar diffused. In the illustration, N-type bulk silicon is diffused on one side by N+ dopant

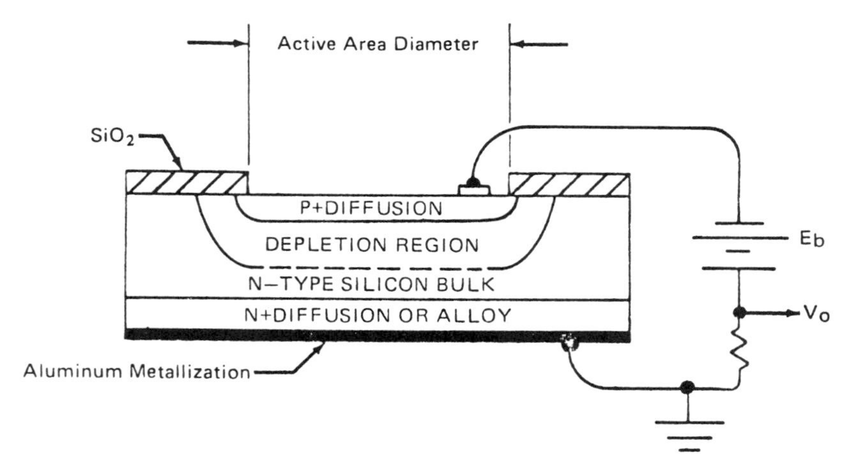

Figure 5-8 Cross section, planar diffused photodiode (*courtesy of United Detector Technology, Inc.*).

and on the opposite side by P+ dopant. A depletion region between the N and P exists free of current carriers. It is in this depletion area that the photons should be absorbed. To operate in the photoconductive mode the device must be reverse biased. To operate photovoltaically, no bias is required. An active area exposes the P+ diffusion to light beams. The light beams are absorbed into the semiconductor. When photons of energy are absorbed into the material, electron-hole pairs are are formed. Short, medium, and long wavelengths of photon power are absorbed at different depths within the PN junction. The depth is dependent on the photon wavelength. Short wavelengths, of course, are absorbed near the surface. Long wavelengths may penetrate the entire structure. To be most useful, the wavelengths should be absorbed in the depletion area.

Current is produced by electron–hole pairs being separated and drawn out in directions of more positive or negative sources whichever is the case. If the electron–hole pairs happen outside the depletion area they will usually combine and no current will be produced. The active region (P+ diffusion) should be extremely thin to ensure maximum penetration. As in other reverse-biased diodes, the depletion area can be made larger by increasing the reverse bias.

In Figure 5–9 the Schottky barrier photodiode is illustrated. Operation of the Schottky photodiode is much the same as the conventional PN junction. The Schottky diode (often called a surface diode) differs in the method by which the P-type material is formed. In the conventional photodiode, the method is diffusion. In the Schottky photodiode, the method is by metallization. In the case illustrated, the active region has a thin gold film metallized to the N-type silicon bulk. The Schottky barrier photodiode has some advantages over the conventional photodiode. It operates well at wavelengths less than 500 nanometers and has much simpler fabrication processes. The Schottky photodiode does not operate well at high temperatures or with high light power.

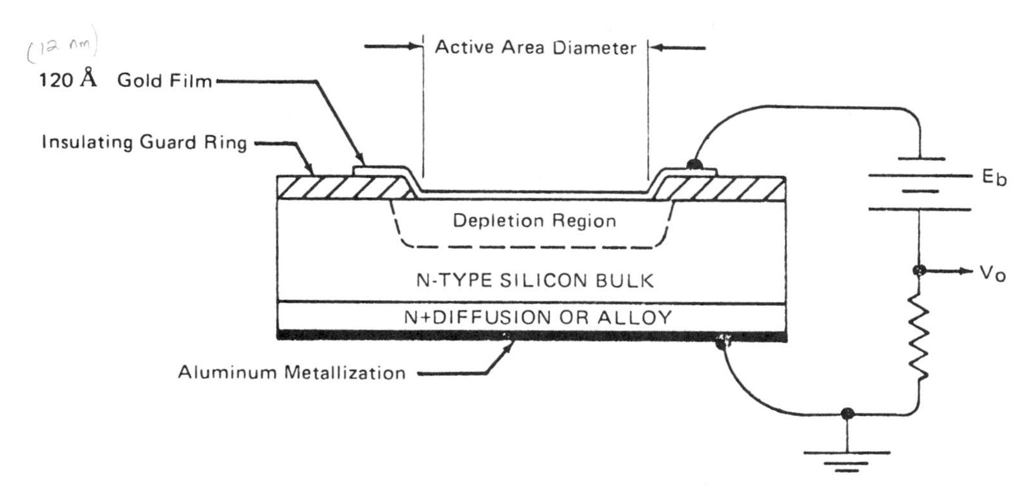

Figure 5–9 Cross section, Schottky barrier photodiode (*courtesy of United Detector Technology, Inc.*).

PIN PHOTODIODES

The PIN diode is so called because of the layer material by which it is constructed. The word PIN is an acronym for P-type, Intrinsic, N-type materials. A PIN photodiode is one in which a heavily doped P region and a heavily doped N region are separated by a lightly doped I region. In Figure 5–8, the N-type silicon bulk would represent the I region. The resistance of the I region can range from 10 ohms per centimeter to 100,000 ohms per centimeter. The P and N regions are less than 1 ohm per centimeter. Since a depletion area can extend further into a nondoped or lightly doped region the PIN photodiode has an extremely large depletion area. This large depletion area provides the PIN photodiode with much faster speeds, lower noise, and greater efficiency at longer wavelengths.

SPECTRAL RESPONSE CURVES

Each manufacturer issues spectral response curves which provide the range of wavelengths in which their devices operate (see Figure 5–10). These curves usually have a shape that corresponds to the CIE photopic response curve. Response is based on the photodiode's ability to absorb photons of light energy. Longer-wavelength photons are so low in energy that they can penetrate the entire photodiode structure without producing an electron pair. Therefore there is a response cutoff as illustrated on the curves. Short wavelengths in the ultraviolet area fall off because electron–hole pairs are combined in the active area of the photodiode without reaching the depletion area. Visible wavelengths have higher response with Schottky barrier than planar diffused photodiodes.

PHOTODIODE APPLICATIONS

Photodiodes are used in photoconductive and photovoltaic applications. The characteristics illustrated in Figure 5–11 are for a planar diffused photodiode. In the photoconductive mode the device must be operated with reverse bias. The problem with reverse bias is that the device develops reverse current, called dark current. The dark current is present with reverse bias when no outside radiation is applied.

In the photovoltaic mode the device operates with zero bias. This makes sense since the photodiode in this mode is the generator producing the load voltage. On the top left side of the illustration, the important spectral response curve is provided. You will note that this curve is similar to the one previously illustrated. This particular curve is for a planar diffused silicon photodiode. The curve has the typical response cutoff at long wavelengths. Short-wavelength cutoff is at ultraviolet wavelength.

A second curve that is meaningful to a photodiode operation is the linearity curve. This curve compares incident radiant energy with photocurrent in amperes. Note that the curve is extremely linear. When the photodiode is exposed to radiant

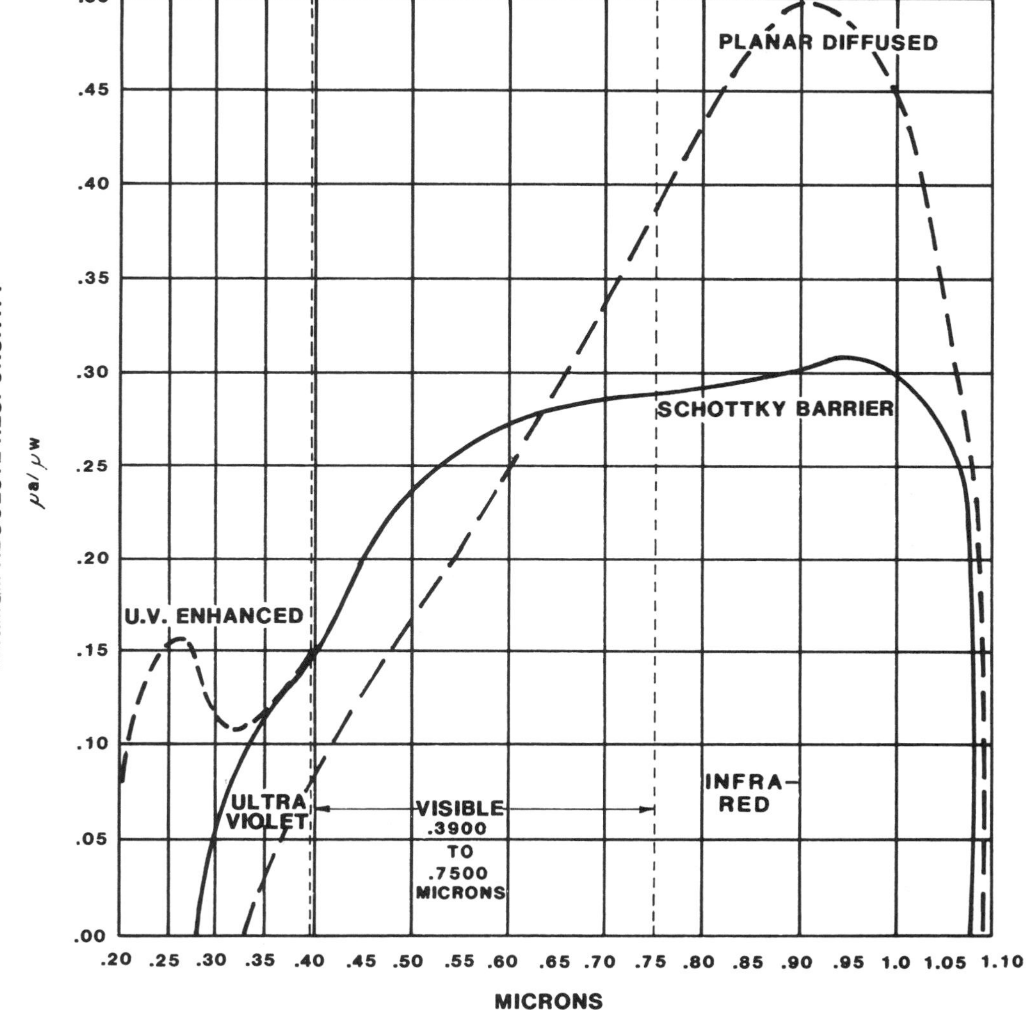

Figure 5–10 Spectral response curves (*courtesy of United Detector Technology, Inc.*).

energy, valence electrons are released. The greater the radiant energy exposure the greater the current flow.

The current–voltage (I–V) characteristic curve is representative of both photovoltaic and photoconductive operation. Note that photovoltaic region of bias voltage is at zero bias with the photodiode generating the load voltage. This is called the *voltage mode*. The photoconductive region requires constant reverse bias. This is called the *current mode*. Current is extracted as a measure of applied radiant energy.

The equivalent circuit in the center of the illustration is representative of all photodiodes. The dark current (I_0) is caused by reverse bias. The signal current (i_s)

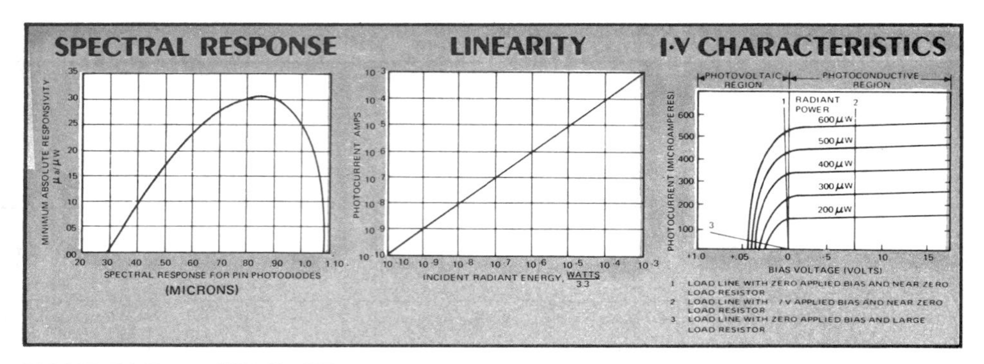

EQUIVALENT CIRCUIT (all models)

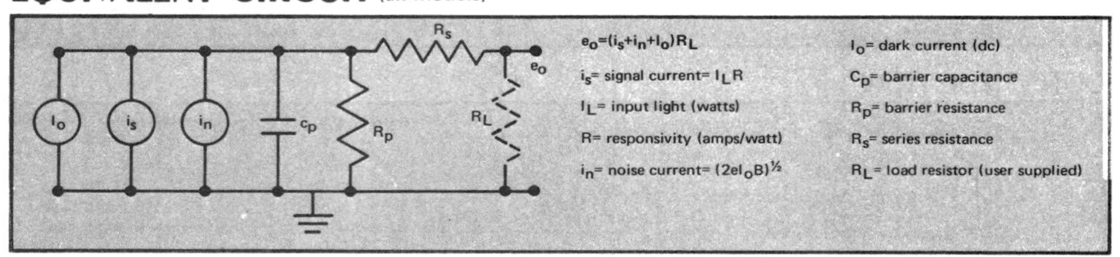

TYPICAL HOOKUPS

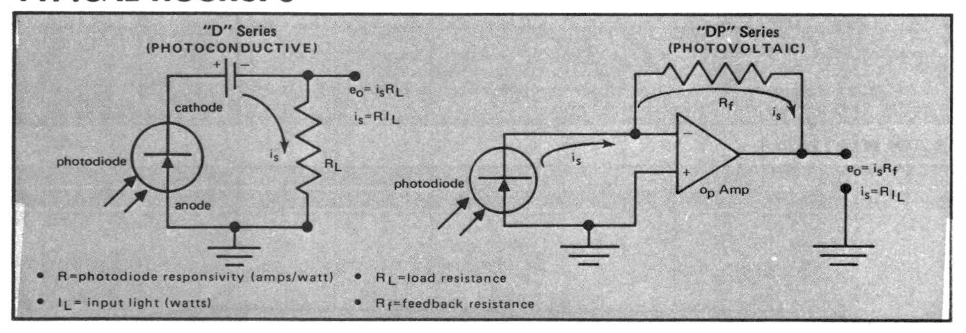

Figure 5-11 Planar diffused PIN diode characteristics (*courtesy of United Detector Technology, Inc.*).

is the current developed by input light. Noise current (i_n) varies with bandwidth, temperature, and dark current. An important ratio for designer use is the signal-to-noise ratio (i_s/i_n). It is obvious that the greater the signal-to-noise ratio the better the design.

Typical hookups are illustrated on the lower part of the illustration. The photoconductive circuit has reverse bias, photovoltaic does not. Signal current flows through the load resistor R_L in the photoconductive circuit. An operational amplifier is used in the photovoltaic circuit to amplify the signal from the photodiode. The input resistance of the operational amplifier is extremely high. The resistance (R_f) is used for feedback.

A planar diffused photodiode is illustrated in Figure 5–12. This particular photodiode is manufactured to operate in the photoconductive mode. That is, the diode must be reverse-biased. The only problem with reverse biasing is that the diode is subjected to the flow of dark current. Dark current you may recall flows when there is no radiant flux on the photodiode. Dark current is an integral part of noise current and must be used in determining signal-to-noise radio. The dark current for this particular device is extremely small (0.002 microampere typically with 10-volt bias).

The United Detector Technology (UDT) spot diodes are bi-cell or quadrant-cell detectors ideally used for a wide range of nulling and centering applications. The devices consist of two or four discrete elements on a single silicon substrate with an active lead from each element. When a light beam is centered on the detector (null or center position is the intersection of active elements), output current from each quadrant is equal. As the light beam moves, current indicates off-center position. The spot diode exhibits stability over time and temperature, high responsitivity, and fast response times per pulse operation. Note that the one photodiode has two discrete elements and the other has four.

The spot diode has broad frequency response, fast rise time, and low barrier capacitance. If these are features that are required for the design, the spot diode must be used in the photoconductive mode regardless of dark current. If dark current is a problem for a design, then the photovoltaic mode should be utilized. The UDT spot diode is applied in lens manufacturing, feedback control systems, guidance systems, laser alignments, machine tool alignment, targeting, and process machinery alignment. It is primarily used in the photoconductive mode of operation and position-sensing detection.

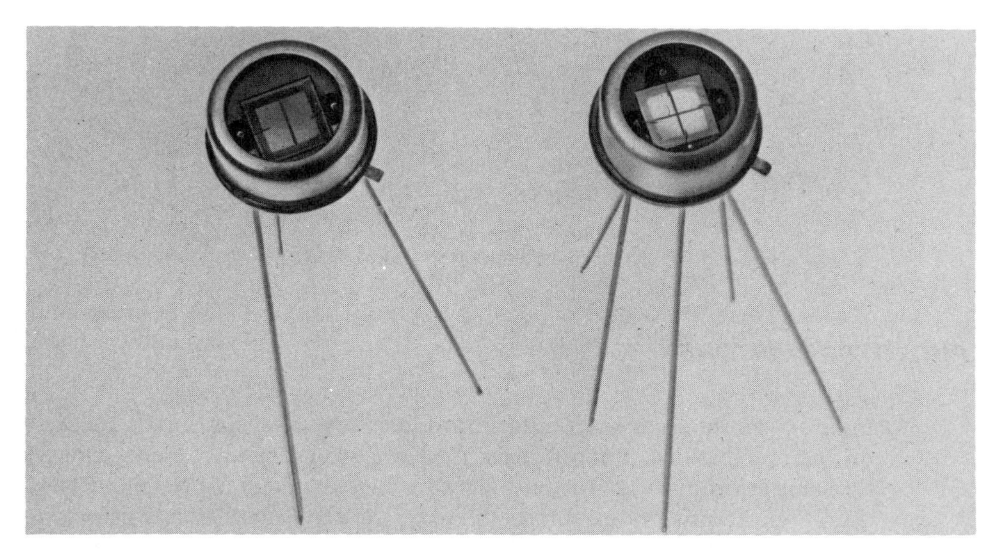

Figure 5–12 Planar diffused photodiode (UDT PIN spot) (*courtesy of United Detector Technology, Inc.*).

Photodiodes that are manufactured to operate in the photovoltaic mode do so with zero bias. This is ideal for the elimination of dark current in applications that demand this requirement. Photodiodes that are manufactured for photovoltaic modes are used in radiometers, photometers, spectrometers, densitometers, colorimeters, particle counters, fluorescent analysis, replacement of phototransistors, point-of-scale scanners, intrusion alarms, opto-couplers, E–O blood analyzers, and videodisc recorders. Large-area photovoltaic photodiodes have application in solar cells, food processing, illumination control, proximity detection, and automatic conveying.

The Schottky barrier photodiode represents the state-of-the-art in large-area, high-sensitivity, fast-response, expanded-spectral-range detectors. The cold-formed Schottky barrier preserves the high resistivity of the intrinsic (I) region of the PIN diode. This serves to provide low barrier capacitance, low dark current, and low noise. The thin gold contact used in metallization allows ultraviolet and blue light to pass easily into the depletion area where electron pairs are more effectively collected. Schottky barrier photodiodes are used in radiometry, photometry, densitometers, colorimeters, laser power meters, laser range finders, ultraviolet flame detection, character recognition, alignment, and other position-sensing detectors. A variety of photodiodes are pictured in Figure 5–13.

Figure 5–13 Other photodiodes (*courtesy of United Detector Technology, Inc.*).

PHOTOTRANSISTORS

Figure 5–14 shows a variety of phototransistors. The shape and size are dependent on the applications. Phototransistors are made from silicon or germanium, much the same as ordinary transistors. They are constructed in the NPN or PNP configuration. Phototransistors are subject to the typical problem of all transistors, temperature variations. These problems may be resolved by biasing techniques and thermal-stability resistors.

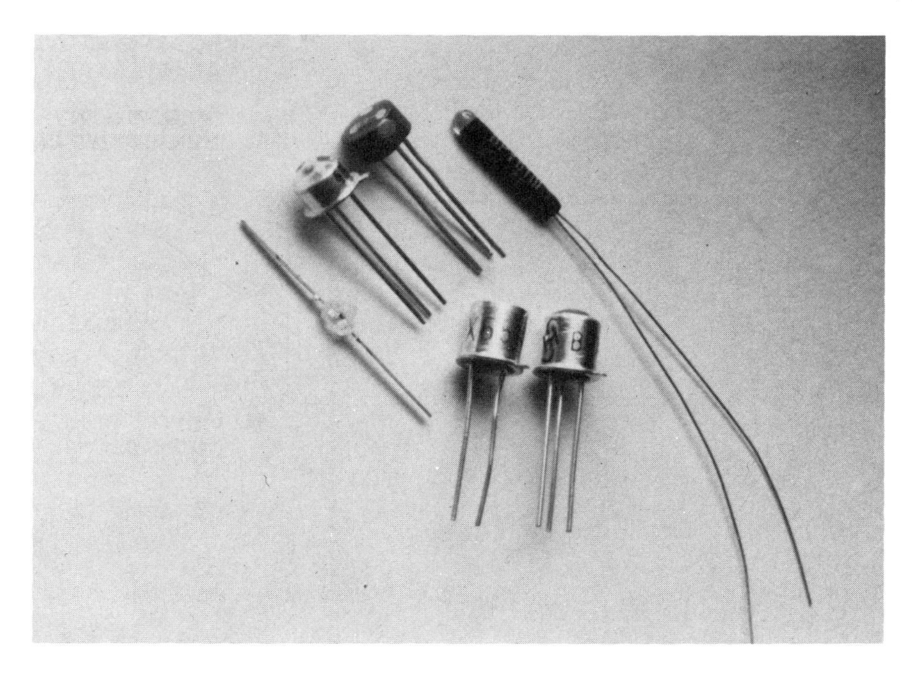

Figure 5–14 Phototransistors (*courtesy of Siemens Optoelectronics Division*).

VIDICON CAMERA TUBE

The vidicon (see Figure 5–15) is a television imaging tube that is still a very popular choice for cameras where extreme low light sensitivity and low power consumption are not major considerations. It is a vacuum tube with a heating element and electron gun at one end and an image plate at the other. The image to be televised is focused, by a set of lenses, on a photosensitive layer that coats the inside of the glass end. The light photons cause electrons to be released from this layer, so a charge image is formed (the image is not in color). The individual charges cannot spread around or decay because the layer has a fairly high resistance.

The charge image is "read" with a scanning electron beam. It originates in the electron gun where electrons are accelerated to a high speed and focused into a narrow beam. The beam is bent vertically and horizontally with a set of deflection coils around the side of the tube and so scans the scene with a raster pattern. The most common scanning will conform to NTSC (National Television System Committee) broadcast standards, which allows 63.5 microseconds for each horizontal scan line and a total of 525 scan lines. The scene is read off a charged photosensitive layer by a scanning electron beam.

As the electron beam hits the target the individual charge areas are discharged back to a neutral condition. The current that flows as a result of this discharge will vary from moment to moment as the electron beam scans different points on the

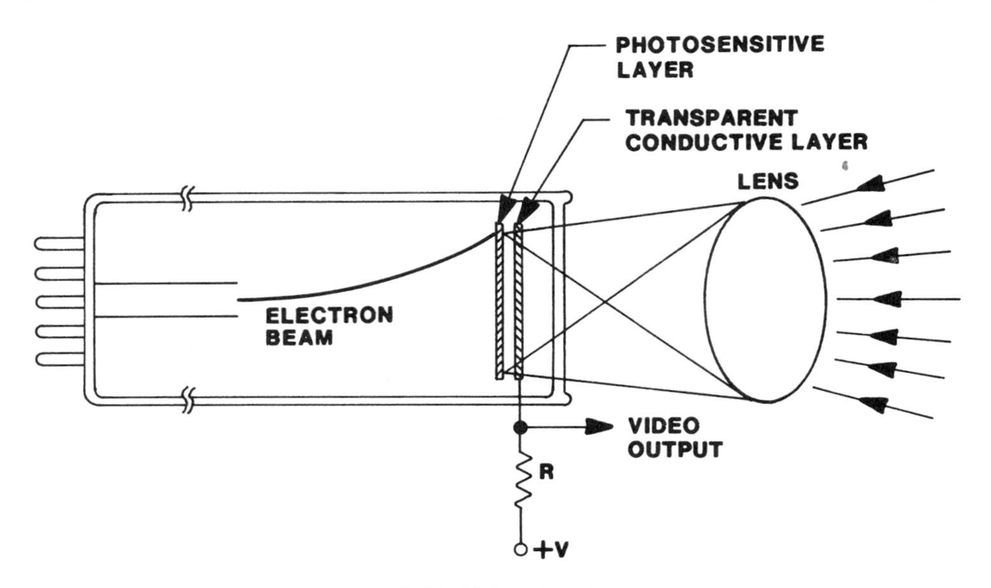

Figure 5-15 Vidicon imaging tube.

target. This current flows through the external resistor and produces the video output voltage ready for further amplification. Vertical and horizontal synchronizing patterns are then added and the resulting composite signal is ready for viewing on a monachrome monitor. The spectral response of a vidicon has the same yellow/green peak sensitivity as the human eye but is slightly broader, reaching farther into reds and violets.

SOLID-STATE CAMERA ARRAY

As with every other area of electronics, solid state has also taken over the job of video imaging. The result is smaller, lighter, requires less power, and is considerably more sensitive in low-light conditions than other imaging systems (see Figure 5-16).

The sensor looks like a normal integrated circuit but with an optical-quality glass window. It consists of a very large array—1000 by 1000 for some applications—of silicon photodiode or metal-oxide-semiconductor (MOS) sensors. The chip also includes its own scanning system so that the light value read by each of the individual sensors can be read in sequence. This eliminates the bulky and heavy deflection coils of the vidicon system. Some solid-state sensors will be described as CCD or charge-coupled devices. This describes one method of reading the charges out from one line within the array. During readout, the sensors pass their charge to their neighbor "on the right" much like an old "bucket brigade fire department."

The spectral response of most solid-state image sensors is much wider than the human eye. The extra sensitivity extends deep into the infrared. As a result, most cameras will require some optical filtering if eye response is to be matched.

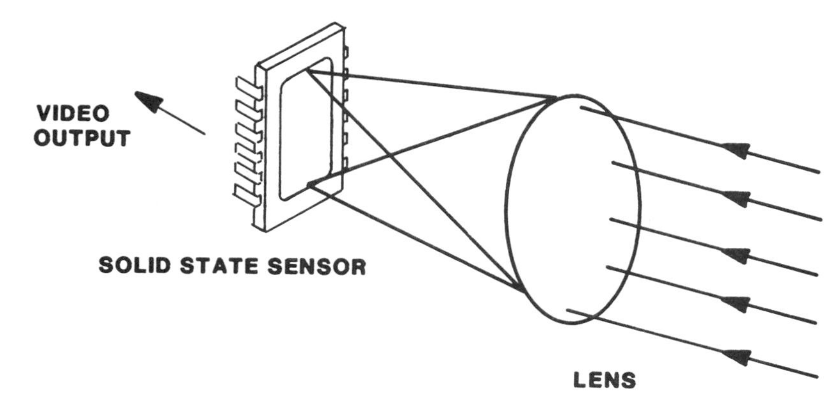

Figure 5–16 Solid state image sensors.

Solid-state sensors are now heavily used for astronomy and satellite camera work. The advantage is their improved sensitivity to low light levels. This is enhanced by leaving the sensors exposed to the low light for long periods of time and not reading the values out many times per second as is the case with commercial television. Note that these image sensors are much simpler than their vacuum-tube equivalents.

OPTO-ISOLATORS

The opto-isolator is a device with a light source coupled to a light sensor. It transmits while maintaining a high degree of isolation between its input and output. Previously, this job was accomplished by relays, isolation transformers, and blocking capacitors. The opto-isolator replaces these devices and adds better reliability and function. There are three basic opto-isolator types (see Figure 5–17). These are the LED-photodiode, the LED-phototransistor, and the LED-photo-Darlington. In each of these cases the LED is the electroluminescent device, whereas the diode and transistors are the detectors. Input leads are supplied from a current source, while output leads may be tied to a variety of electronic circuits, which we discuss later in this section.

With no external optic characteristics, opto-isolators are specified as electrical devices. There are functionally two parameters that define the opto-isolator. These are how well they transfer information from input to output and how efficiently they maintain electrical isolation from input to output. Transfer efficiency is the ratio of output current to input current expressed as a decimal.The current transfer ratio (CTR) varies with temperature and input current.

Current transfer ratio (CTR) is an important factor in opto-isolator use. Typically, if 5 mA of current is fed into the source LED, the detecting diode or transistor will only deliver at a current of 3 or 4 mA. Current transfer ratios, are therefore 0.6 to 0.8 of applied current. An extension to this is the variation of CTR with input current, in other words, the "transfer linearity." This is important for linar applications. Linearities are typically several percent, not very good for high-tech applications.

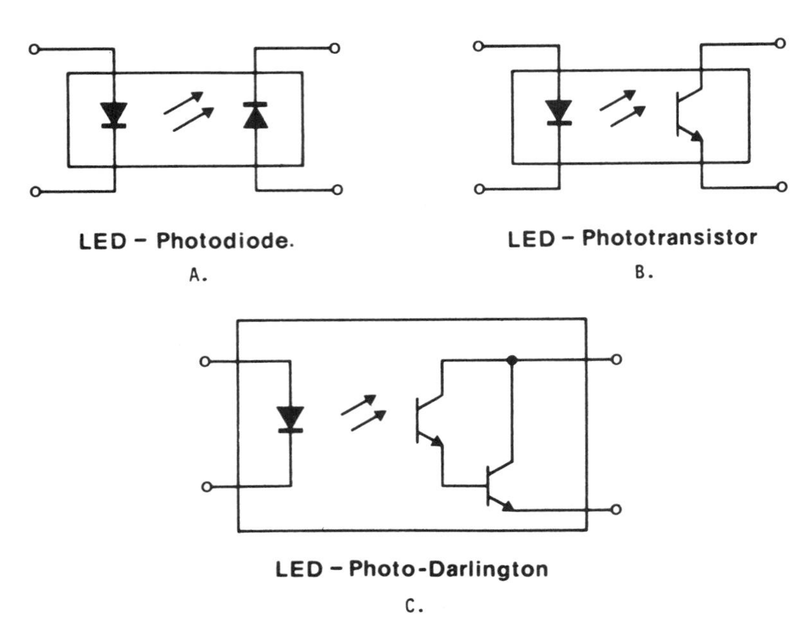

Figure 5–17 Opto-isolator types (*courtesy of Siemens Optoelectronics Division*).

Cross Sections of an Opto-Isolator

A basic opto-isolator consists of a gallium arsenide (GaAs) infrared diode (IR LED) and a silicon phototransistor mounted together in a dual-in-line package. When forward current is passed through the IR LED, it emits infrared radiation at about 900 nanometers wavelength. This radiant energy is transmitted through an optical coupling medium and falls on the base surface of the phototransistor. Phototransistors are designed to have a very large base region and therefore a very large base-collector junction area with a small emitter area. Some incident energy in the form of photons causes the formation of electron–hole pairs in the base region.

Referring to Figure 5–18a, the emitter and the base are grounded. A positive voltage is applied to the collector. This enables the device as a photodiode. The high field across the collector-base junction quickly draws the electrons across into the collector region. The holes drift toward the base terminal and attract electrons from the terminal. It is in this manner that electron current flows between the base and collector causing a voltage drop across the load resistor (R_L). A time constant is developed with the junction capacitance (C_{cb}) and the load resistor (R_L), with a corresponding output voltage rise time. The output current in this configuration is quite small and therefore this configuration is seldom used.

The most common lash-up is the open base connection as in Figure 5–18b. In this configuration, the base is left open and the emitter is grounded. Holes generated

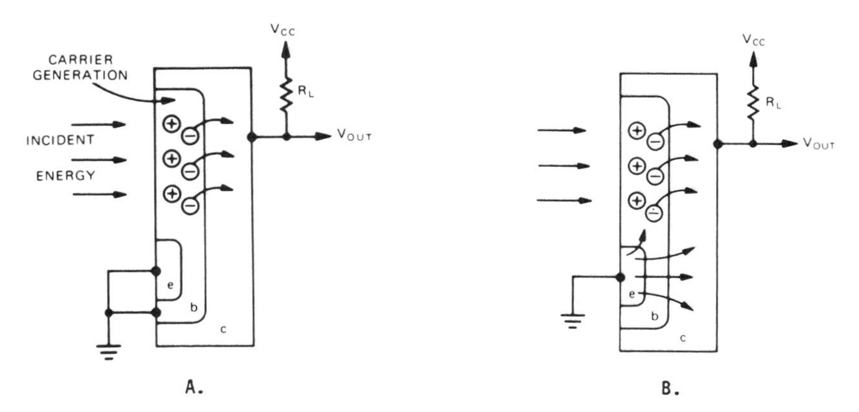

Figure 5–18 Cross section of an opto-isolator (*courtesy of Siemens Optoelectronics Division*).

in the base region cause the base potential to rise, forward biasing the base-emitter junction. Electrons are injected into the base from the emitter to neutralize the excess holes. Because of the close proximity of the collector, the probability of electrons recombining with these holes is remote and most of the injected electrons are immediately drawn into the collector region. Collector current increases as in any forward-bias transistor, depending on the current gain (beta) of the transistor. Current gain for the B configuration is several hundred times greater than in the A configuration. The current gain, however, is beta times as great as in the A configuration, which causes the rise time to increase by a factor of beta.

Increasing Opto-Isolator Speed

The speed at which an opto-isolator responds to a signal depends almost entirely on the detector (sensor). Switching time is a function of the base storage time and the *RC* time constant of the output. The larger the collector-base region, the more sensitive the sensor and the larger the open circuit collector-base capacitance. This results in long storage time and slow switching time. In Figure 5–19a, a base-to-emitter bleed resistor is added. The bleed resistor provides an external path for removal of the stored charge. This decreases the phototransistor's turn-off time, but reduces sensitivity and current transfer ratio (CTR).

In Figure 5–19b, a cascode amplifier is illustrated. In this case the phototransistor is the current source and the effective load resistance R_L is made up of the cascoded isolation amplifier emitter resistance. This emitter resistance may be only a few ohms. The cascoded transistor is chosen with a low open-circuit capacitance.

In Figure 5–19c, an operational amplifier is placed in the output circuit. The phototransistor is used as a current source driving the summing point of a general-purpose operational amplifier. This has the same effect as the cascode circuit. It lowers the effective R_L with resulting higher speeds.

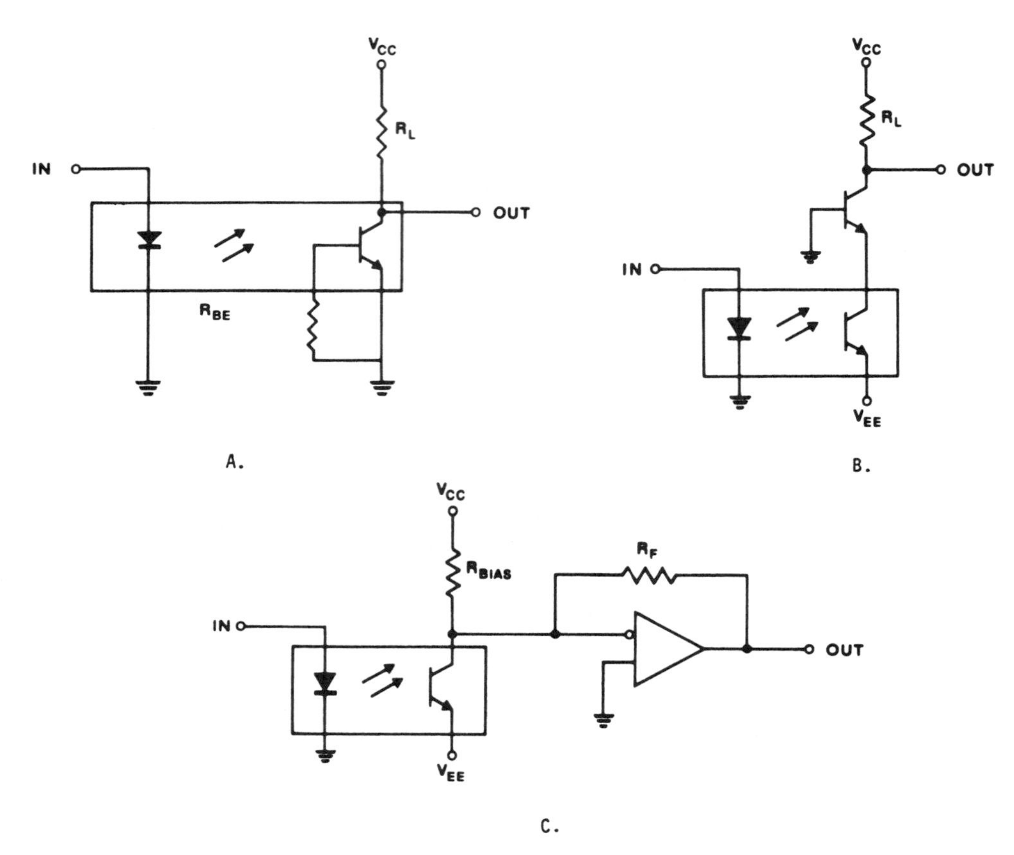

Figure 5–19 Increasing speed of opto-isolator response (*courtesy of Siemens Optoelectronics Division*).

Circuit Applications Using Opto-Isolators

Opto-isolators have a select group of applications. Although most designers are well aware of their distinct capability, there may be a reluctance to utilize them in everyday design. There are three major categories in which the opto-isolator is a standout device because of its unique properties: level translating circuits, remotely actuated switching circuits, and circuits requiring a high degree of electrical isolation. The last category is the most obvious. A prime application of this type would be the electrocardiograph amplifier, which provides the high degree of isolation necessary to ensure patient safety in medical electronics. A second use for an opto-isolator is with an isolated voltage operated potentiometer. The opto-isolator makes the potentiometer totally independent of the control-voltage reference supply. In remotely actuated switching circuits, the opto-isolator is used as a polar switch. The opto-isolator can also be used as class B amplifier, a convenient means of inputting data to a calculator chip, and as a high-voltage series regulator. Let's look at several basic cir-

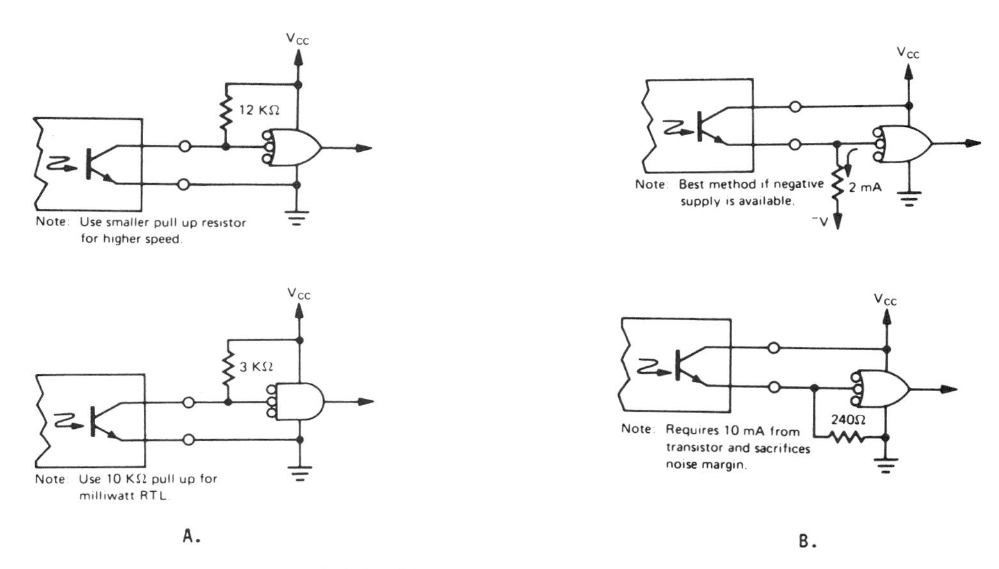

Figure 5–20 TTL logic being driven by opto-isolators (*courtesy of Siemens Optoelectronics Division*).

cuits to see their hookups. One of the simplest uses that can be made of opto-isolators is with transistor-transistor logic (TTL). There are two ways that TTL can be utilized, either driven by or as a driver for the opto-isolator. The matching of the two devices can be handled quite easily within the parameters of the devices.

Figure 5–20 illustrates opto-isolators being used as output driver circuits. In Figure 5–20a two TTL active-low circuits are suggested. In the upper circuit, a gate is driven by an opto-isolator with a 12-kΩ pull-up resistor. This circuit is recommended for DTL and TTL digital interfaces. In the lower circuit, a gate is driven by an opto-isolator with a 3-kΩ pull-up resistor. Figure 5–20b illustrates two TTL active-high circuits. Both circuits utilize OR gates. The upper circuit is recommended if both positive and negative power supplies are available. The lower circuit is suggested if there is no negative supply.

Figure 5–21 represents an opto-isolator used in a logic operation. In the illustration, a pair of opto-isolators are used to perform logical OR functions. This circuit is equally compatible with DTL or TTL logic families. There are two ways in

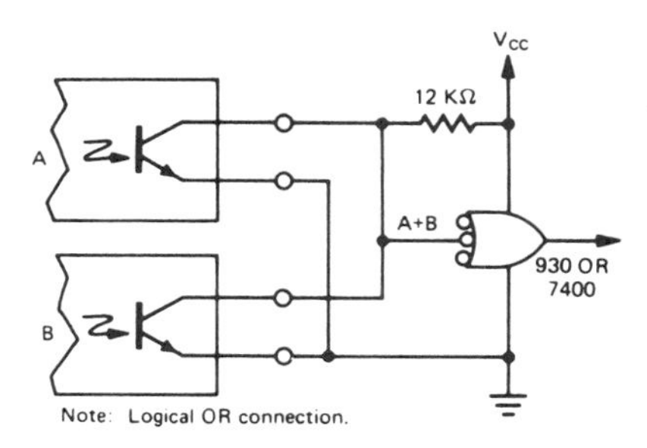

Figure 5–21 Opto-isolator used to perform a logic function (*courtesy of Siemens Optoelectronics Division*).

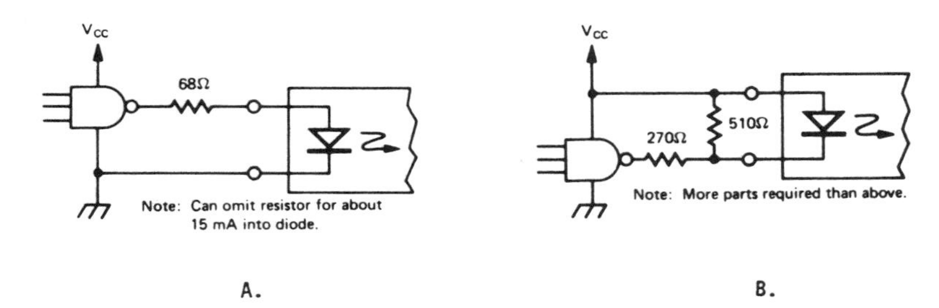

Figure 5-22 TTL logic driving opto-isolators (*courtesy of Siemens Optoelectronics Division*).

which TTL may be used as a driver for the opto-isolator (see Figure 5-22). In Figure 5-22a, the TTL is used in an active level high or a current source mode. The output of the TTL is placed on the anode of the LED. In Figure 5-22b, the TTL is used as an active level low or a current sink mode. The output of the TTL is placed on the cathode of the LED. Care must be taken when using logic gates with opto-isolators that the logic mathematical expression be maintained. A high out of a gate used as a current source would turn the LED on. A high out of the same type gate used as a current sink would turn the LED off.

Figure 5-23 illustrates three examples of linear circuit applications. The first utilizes a differential amplifier. In the a circuit the LED must be forward-biased to

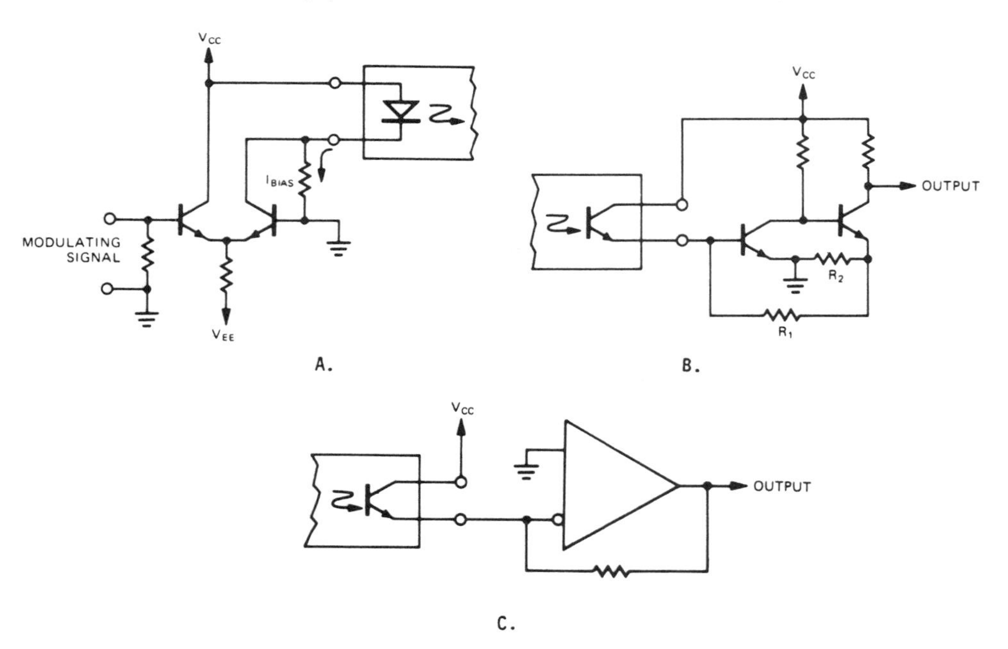

Figure 5-23 Linear circuit applications (*courtesy of Siemens Optoelectronics Division*).

a suitable current flow. Modulating signals may then be impressed on that dc bias. In the b circuit a feedback amplifier is utilized to lower the impedance of the input from the opto-isolator. In the c circuit a high-speed operational amplifier is used to provide lower effective load resistance, therefore higher speeds. To repeat a previous note, the opto-isolator has two basic requirements to live up to, efficient transfer of information from input to output and the maintenance of electrical isolation from input to output.

Typical Opto-Isolators

Typical of the opto-isolators are the Siemens IL-1 and ILQ series (see Figure 5-24). The IL-1 is a 6-pin device while the ILQ is a 16-pin device. The IL-1 houses a single opto-isolator. The ILQ houses four separate opto-isolators. A connection diagram is provided in Figure 5-25.

Figure 5-24 Opto-isolators (*courtesy of Siemens Optoelectronics Division*).

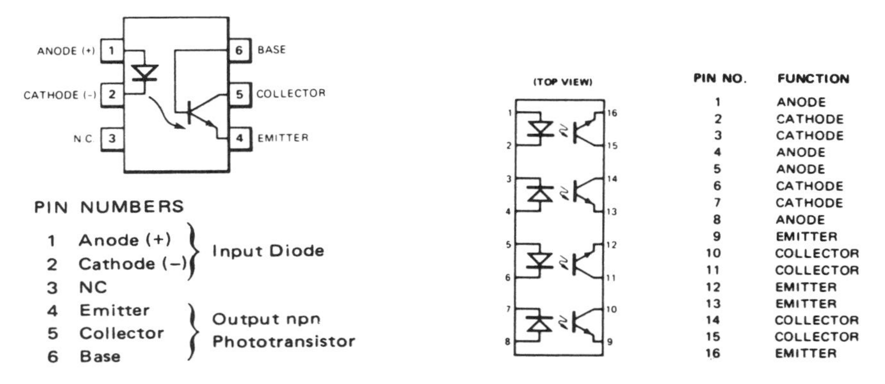

Figure 5-25 PIN connections for Siemens IL opto-isolators (*courtesy of Siemens Optoelectronics Division*).

SILICON SOLAR CELL CHARACTERISTICS

The function of a fundamental solar cell is basically controlled by two phenomena. The first is the increase of electron action when the sun (light) penetrates the PN junction. The second is when photons of light energy bombard atoms and create electron–hole pairs. Short-wavelength photons do not have the energy to generate electron–hole pairs and to elevate electrons from the valence band to the conduction band. These short-wavelength photons are absorbed near the surface of the cell. Silicon is transparent to longer-wavelength photons. These longer wavelength photons penetrate the surface and cross the PN junction into the P material. The junction is usually near the top surface of the cell. Electron–hole pairs recombine quickly in heavily doped material. Therefore, top surface material (N-type) is heavily doped, the junction is placed near the top surface, and the lower P-type material is comparatively lightly doped. The optimum solar cell parameters are highly guarded by cell manufacturers for economic reasons. The most important condition of solar cell manufacturing is efficiency in light collecting. The ultimate design is toward efficiency.

The Photoelectric Effect

Albert Einstein, in 1905, received the Nobel prize for his wave theory. Einstein stated that light travels in waves and particles (photons). The photons knock electrons from atoms of light-sensitive material. This dual nature of light may be the most important theory in the solar cell field. Einstein explained that the energy of the fastest photons is directly proportional to the frequency of the light.

Solar Cell Equivalent Circuit

The solar cell can be equated to a constant current generator. A simple equivalent circuit for the ideal solar cell is illustrated in Figure 5–26. The solar cell (a light-activated, constant-current source) is placed in parallel with an ideal diode. The amount of current generated by the light source is proportional to the intensity of the light. This current is called constant current or I_L. The ideal diode is placed in parallel with

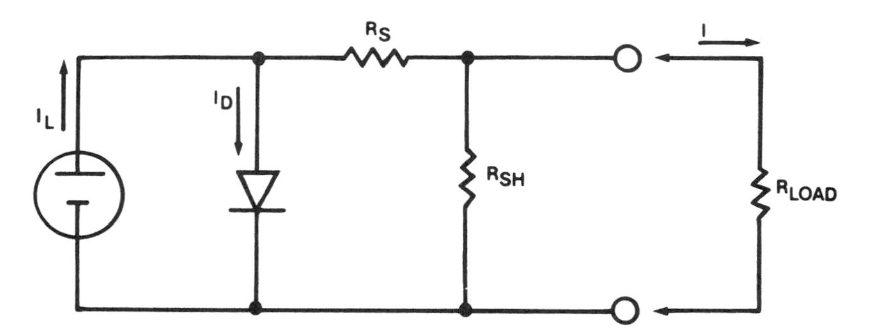

Figure 5–26 Solar cell equivalent circuit.

the constant-current generator. The resistance R_S is a series resistance. The resistance R_{sh} is a shunt resistance.

Figure 5–27 represents an ideal solar cell characteristic curve. The term V_{oc} at the base of the curve is open-circuit voltage. In a silicon solar cell the open-circuit voltage is nominally 570 millivolts. The load resistance R_{load} is extremely high (such as an open circuit) and current is minimum (zero). If the load resistance is lowered slowly, there will be a minimal decrease in voltage but a sharp increase in current. This situation will continue to point X on the curve. At point X the current is 750 milliamperes and 450 millivolts. The curve is at a knee. A further decrease in load resistance will not see any further current increase but the voltage will drop drastically toward zero. When the resistance is lowered to zero the circuit is considered to be open. This point is I_{sc} or current, short circuit. Voltage will be zero because you cannot have a voltage drop across a short circuit.

Power in an Ideal Solar Cell

To define the relationship of power, voltage, and current on an ideal solar cell we superimpose a power trace (in milliwatts) on the voltage–current characteristic curve shown in Figure 5–27. The power curve illustrates zero power at the two limits of open-circuit voltage and short-circuit current (see Figure 5–28). An increase in power is made as the current increases.

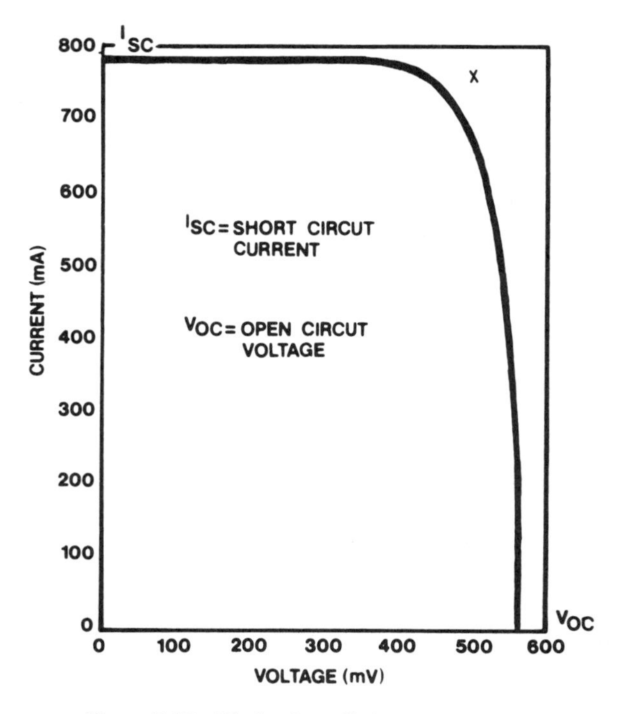

Figure 5–27 Ideal solar cell characteristic curve.

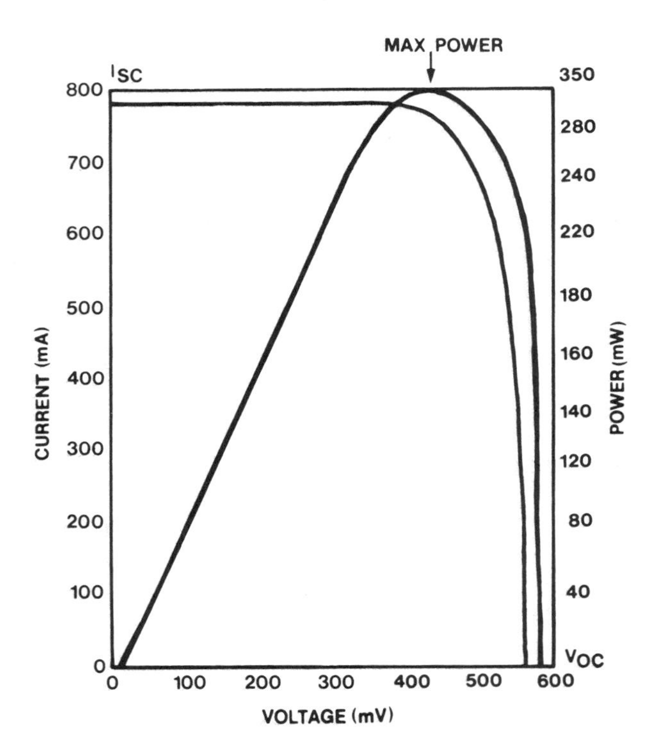

Figure 5–28 Power in an ideal solar cell.

The maximum power is derived at the knee of the voltage–current curve. This point usually occurs at 450 milliwatts. As the curve defines the maximum power, the resistance that provides this maximum power should be maintained at a point that provides the greatest efficiency.

Silicon Solar Cell Spectral Response

The spectral response curve is one that compares the relative response in percent of light to different wavelengths. The spectral response of a silicon solar cell is illustrated in Figure 5–29. The curve illustrates that silicon responds to a broad range of wavelengths. The range begins at the edge of ultraviolet, passes through the entire visual range, and into the infrared. The silicon cell will indeed function well under many conditions. Even on cloudy days, the silicon cell will respond to the light of the sun.

This curve is typical of the response curves that industry supplies. The shape of the curve is modified by engineering inputs. For instance, a shift in the curve may be made by varying the bulk resistivity and the junction depth. Larger bulk resistivity provides high short-circuit currents and low open-circuit voltages. A decrease in

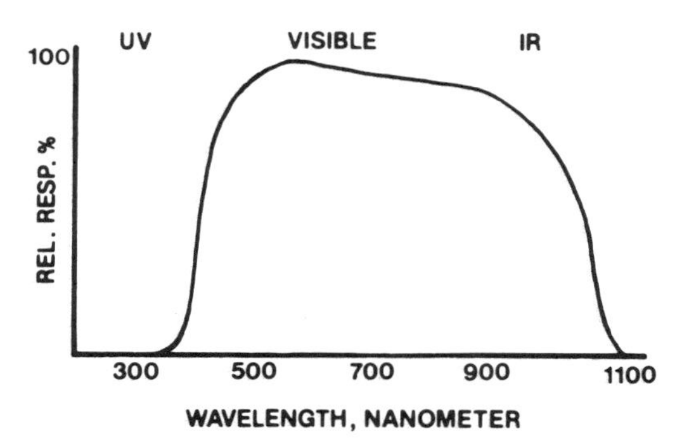

Figure 5-29 Silicon solar cell spectral response curve.

junction depth will increase short-circuit current at the expense of series resistance. Shallow junctions will in general have greater visual range response than deep junctions. Cells with a high resistivity will respond better to infrared. In essence, the cell may be made to have a wide variation in spectral response by varying the bulk characteristics and the junction depth. Response can be expanded by coating the cell with antireflection material; that is, up to a point. Thickness is carefully controlled because additional coating may not aid response but deter it.

Cell Efficiency

According to whose book you are reading, the maximum efficiency of solar cells may vary as much as 1 to 2%. The best guess to the maximum efficiency of a silicon solar cell the author can make is around 25%. Silicon solar cell conversion efficiencies are being met at 14% plus in regular production. It would be easy to get into comparisons of source data. It would be more meaningful to provide the reader with consideration on what affects efficiency.

The first possibility of loss occurs when photons are reflected from the surface of the cell. Reflection may occur directly from the cell surface. It may also occur due to reflection from bond material that carries current to the load. Antireflection coatings may aid this problem.

Construction of the cell has much to do with efficiency. Construction techniques are closely watched engineering secrets. Some of these are junction depth, N-layer thickness, minority carrier concentration lifetimes, diffusion mobilities, surface recombination velocities, and absorption coefficient.

As the collection efficiency is increased by shallow junction depth, the series resistivity is increased. This will show up as a power loss. As a general rule the impurity concentration of P layers is controlled to compromise between thickness and series resistance. Resistance may be decreased by placing metallic grids over the P layer. Resistivity of solar cells varies with temperature. Resistivity of silicon increases with temperature. Therefore, a change can be expected in power output.

Solar Cell Material

By far the most utilized material in the solar cell industry is silicon. The future of solar cells seems to be leaning toward the use of silicon for some time. Other materials called intermetallic compounds are also being studied and used. Some of these are as follows:

indium—phosphide	cadmium—sulfide
indium—antimonide	cadmium—selenide
gallium—antimonide	cadmium—telluride
gallium—arsenide	zinc—sulfide

Selection of material has to do with photon strength. Only photons with enough strength to produce electron–hole pairs are useful.

Terrestrial and celestial success in the solar energy field has been best served with silicon. Celestial success may be optimum using gallium arsenide or cadmium telluride.

Gallium arsenide provides an open-circuit voltage of 800 to 900 millivolts, whereas silicon has an open-circuit voltage of just under 600 millivolts.

It probably will be some time before compounds will come into general use, if ever. The high cost of the compound materials is further complicated by their fragile nature.

BASIC SOLAR CELL JUNCTIONS

There are three basic solar cell junctions: the homojunction, the heterojunction, and the Schottky barrier. All other cells utilize some modification of these basic junctions. The first three cells have been discussed previously but are repeated here for the use of those working or training in photovoltaics.

Homojunction Cells

The most used of all the solar cell types is the homojunction cell (see Figure 5–30a). The cell is all silicon, which accounts for the name homojunction. A large area called P-type material is doped with acceptor atoms such as boron. A very thin layer called N-type material is doped with donor atoms such as phosphorous. The barrier region, which is extremely thin, is created between the N- and the P-types of material. Electrons from N-type material move to fill the holes in the P-type material. The N-type material is left with a net positive charge. The extra electrons fill the holes in the P-type material with an excess of electrons. In this manner a thin static electric charge is developed across the junction between the N- and P-type materials. The barrier resists any movement of electrons through it. Therefore only electrons of high energy can make it through. The barrier is called by at least two other names, the depletion zone and the space charge. Every solar cell must have a barrier. The homojunction

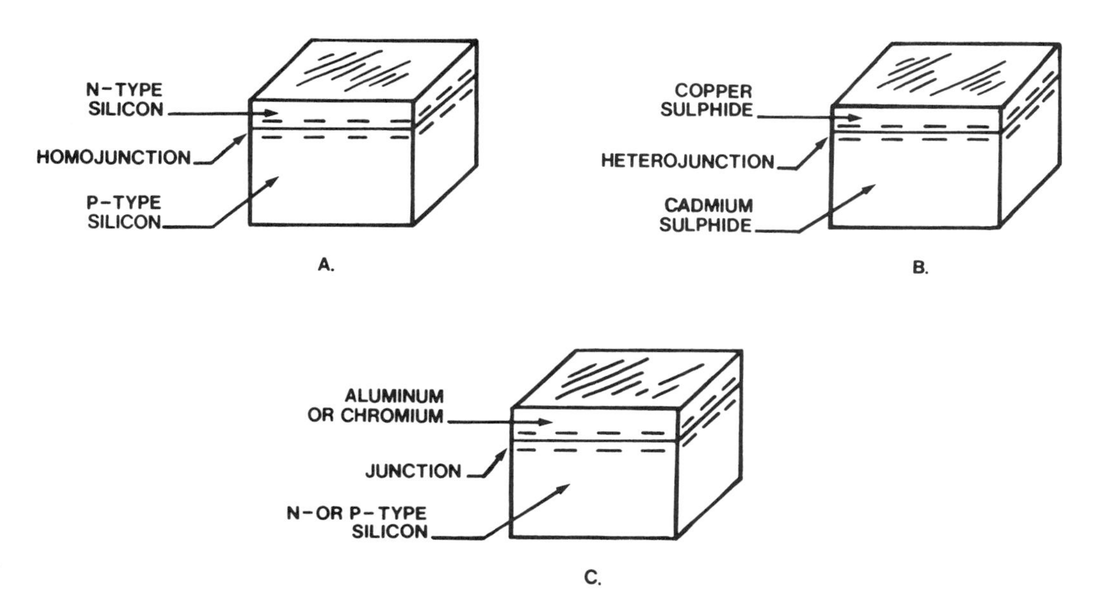

Figure 5-30 Basic solar cell junctions.

cell functions are indicative of today's solar cell. This being the case, one might feel that there are really no other types which are used. On the contrary, the entire discipline of photovoltaics is still in its infancy and cell types are numerous and are continuously being identified.

Heterojunction Cells

Since solar cells can be made of different materials than silicon, another cell type, the heterojunction (see Figure 5-30b), has found prominence. Solar cells of this type are made from dissimilar metals. One such type is the cadmium sulfide–copper sulfide cell. As with semiconductors, one metal, such as copper sulfide, is deposited on the second metal, such as cadmium sulfide. The two metals, on contact, form a junction called the *heterojunction.* The heterojunction is widely used in the photodetection field. In the terrestrial solar energy field, heterojunction devices are a struggling second cousin to the silicon homojunction cell. However, in the celestial field, the heterojunction is a strong contender.

The heterojunction cell falls into three categories: the single heterojunction (SH), the double heterojunction (DH), and the separate confinement heterojunction (SCH). The single heterojunction (SH) has only one heterojunction so that carriers are confined by the junction at only one boundary of the active area. The double heterojunctions (DH) have boundaries on both sides of the active region. The separate confinement heterostructure's (SCH) carriers are confined in a barrier within a wavelength. The latter two, DH and SCH, are used in the fiber optic field and with injection laser diodes. The SH is used with solar cells.

Schottky Barrier Cells

The Schottky barrier cell is a cell consisting of metal and semiconductor material. The Schottky barrier cell is manufactured by depositing a thin metallic layer on silicon to create a diode.

In Figure 5–30c a layer of aluminum or chromium is deposited on a substrate of N-type or P-type silicon. A junction is established. The Schottky barrier operates extremely well with short wavelengths. Furthermore, the Schottky photodiode is manufactured at low temperatures. Metal is deposited rather than diffused.

In further development, the Schottky barrier cell is placed on top of a heterojunction cell. This provides two junctions back to back. Photovoltaic conversion occurs at both junctions, top and bottom.

POLYCRYSTALLINE CELLS

Polycrystalline means many crystals. The polysilicon cell or polycrystalline silicon cell contains many crystals (see Figure 5–31). The material is granular. When melted, polysilicon is allowed to cool slowly, and crystallites form in large sizes. The large crystallites form grain boundaries as shown in the illustration. The grain boundaries are far enough away from each other to allow free electrons to find a path from N-type to P-type silicon. Grain boundaries are several millimeters in length. The function of the polycrystalline cell is the same as the homojunction cell. Polycrystalline cells, however, are much less expensive than single crystal cells. Furthermore, the process for growing polycrystals is faster. Polysilicon crystals can be grown in the form of ingots, cast in the form of cubes or rectangular blocks, and may be surface deposited in ribbon form. There are, as you may have determined, some problems with the polysilicon crystal. Electrons are inhibited by the grain boundaries. When moving through the material they often short out before reaching the external connections. Shorting may be reduced by adding hydrogen gas which is absorbed in the grain cracks. Hydrogen reduces shorts by insulating the grain edges. Polysilicon is

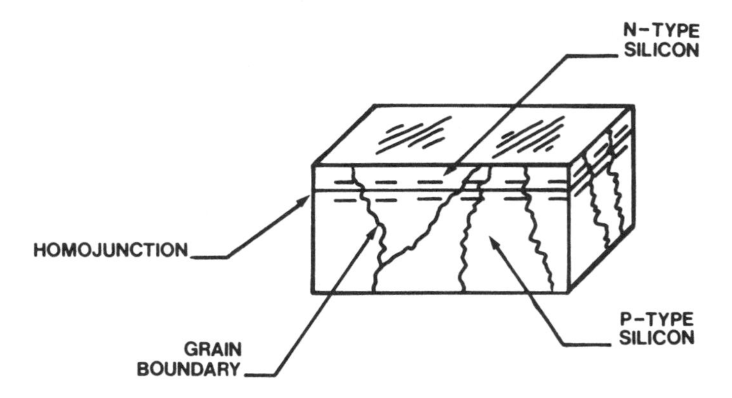

Figure 5–31 Polycrystalline cell.

not highly efficient but because of its speed and economy in manufacture may well prove to be a viable alternative to single crystal silicon for solar cell use.

INSULATOR CELLS

There are two types of insulator cells. These are the metal–insulator–semiconductor (MIS) and the semiconductor–insulator–semiconductor (SIS). The purpose of using the insulator cells is to reduce the short-circuit current slightly while yielding a higher open-circuit voltage.

The metal–insulator–semiconductor (MIS) cell (see Figure 5–32a) is constructed as a modification of the Schottky barrier. A thin layer of 10 to 20 angstroms of metal such as titanium oxide is placed between the metal and semiconductor material.

Another insulator cell is the semiconductor–insulator–semiconductor (SIS) cell (see Figure 5–32b). This cell is a modification of the MIS cell. A thin layer of insulating metal of 10 to 20 angstroms separates a pair of semiconductors. The efficiency of the metal insulator cells is somewhat higher than the Schottky or oxide semiconductor cells.

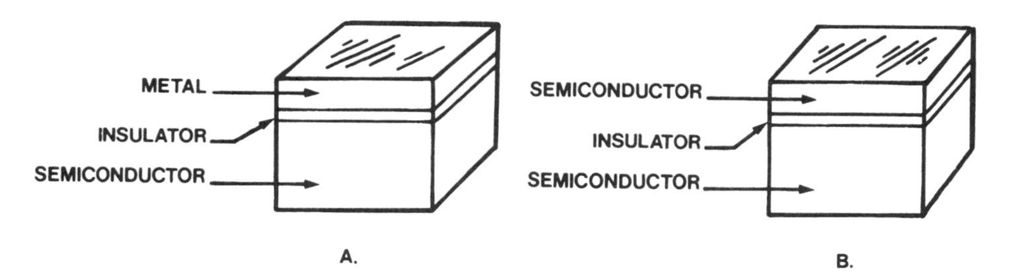

Figure 5–32 Metal insulator cells.

CADMIUM SULFIDE CELLS

Cadmium sulfide is a reasonable substitute for silicon in solar cells. These cells are reasonable in price and easy to manufacture but have a low comparative efficiency. Two types are in general use, the metal substrate and the glass substrate.

The metal substrate is usually copper (see Figure 5–33a). This type is also called the front wall cell because the sun's energy is directed into the front of the cell rather than the substrate. In Figure 5–33a, a thin layer of metal such as zinc is deposited on a substrate of copper. This is followed by a thicker layer of N-type cadmium sulfide and a thin layer of P-type copper sulfide. A separate, very thin layer of tin oxide separates the copper sulfide from the metallic contacts.

The glass substrate is also called the back wall cell because the sun's energy is directed through the glass substrate rather than the front of the cell. In Figure 5–33b, a thin layer of metal such as tin oxide is deposited on a substrate of glass. This is followed by a thicker layer of N-type cadmium sulfide and a thin layer of P-type

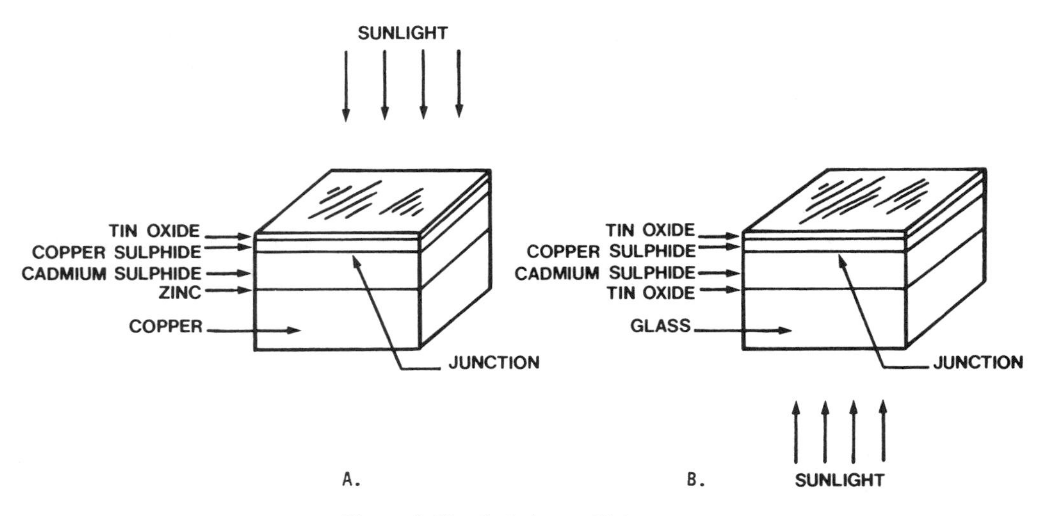

Figure 5-33 Cadmium sulfide cells.

copper sulfide. A separate, very thin layer of tin oxide separates the copper sulfide from the metallic contacts.

The cadmium sulfide cells are versatile in that they may be made in thicknesses between a few microns and up to 35 or 40 microns. Other combinations using cadmium sulfide have been made. These are: copper sulfide on cadmium zinc sulfide, indium phosphide on cadmium sulfide, and cadmium telluride on cadmium phosphide.

GALLIUM ARSENIDE CELLS

The gallium arsenide cells are used in concentrations. Gallium arsenide cells are the most efficient of all the cell types but also the most expensive. The use of gallium arsenide provides efficiencies of 25% and over for thick-film cells and up to 15% for thin-film cells. Efficiency is high for gallium arsenide because it absorbs a broad range of the solar spectrum. A second effect which stands out is that gallium arsenide can withstand high temperatures without losing efficiency. This attribute accounts for the use of this material with concentrators.

In Figure 5-34 P-type gallium arsenide is diffused on N-type gallium arsenide.

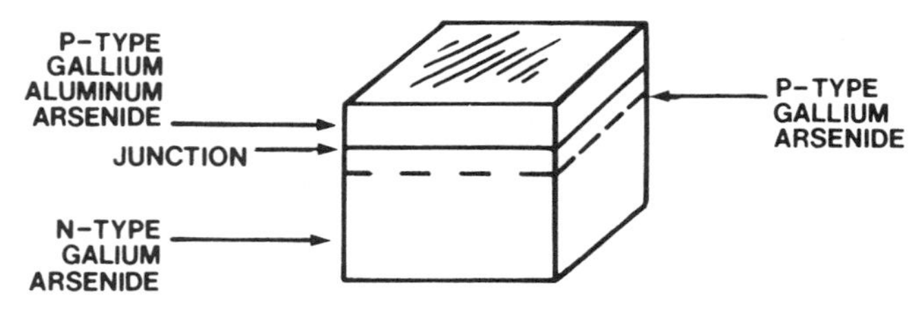

Figure 5-34 Gallium arsenide cell.

A second layer of P-type gallium aluminum arsenide is diffused over that layer. Gallium arsenide may be manufactured as single-crystal or polycrystalline material.

AMORPHOUS SILICON CELLS

The word *amorphous* means noncrystalline or no crystal structure. This material is used in thin-film processes. It is pure silicon but has no crystal alignment. It is inexpensive because no special growth processes are required. It is also very inefficient.

ELECTROCHEMICAL CELL

The electrochemical cell is unique in that one of the cell layers is a liquid electrolyte (see Figure 5–35). This cell is under research by several major corporations. The cell has two major attributes. First, the electrical energy generated may be extracted for external use. Second, the electrical energy generated may be used within the cell itself to create useful by-products such as hydrogen. Then later the hydrogen may be extracted and stored.

Electrolytes are aqueous, molten salt, and solids. Semiconductors being tested are single and polycrystalline, and amorphous types.

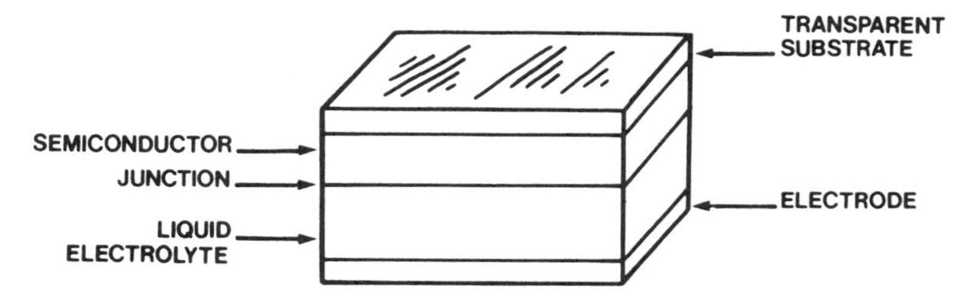

Figure 5–35 Electrochemical cell.

SELF-CHECK QUESTIONS

Self-check questions allow students to evaluate how well they have learned what they are studying.

1. Define the term *photodetection*.
2. Define the term *photoemission*.
3. Define the term *photoconduction*.
4. Define the term *photovoltaic*.
5. Name several types of photodetectors.
6. Name several material types that are utilized to construct photodetectors.
7. What are some of the curves used as specifications for detectors?
8. What are some of the curves used as specifications for phototransistors?

9. What are some of the many uses of photodetectors?
10. Define the function of the vidicon camera tube.
11. Define the function of the solid-state camera array.
12. What are several advantages of the solid-state camera array over the vidicon tube?
13. Why are opto-isolators used?
14. Describe the construction of an opto-isolator.
15. How can an opto-isolator be used in a linear circuit?
16. How are high voltages obtained from solar cell arrays?
17. How are high currents obtained from solar cells?

Six

Laser Optics

Since most laser technology deals in some way with optics, it seems appropriate to take at least a cursory look at some of these devices. Optics are used throughout the laser industry in various applications. Some of these are as mirrors, reflectors, filters, collimators, polarizers, beamsplitters, and magnifiers. Each time an optic is used, consideration must be made as to its size and application. Optics are coated or uncoated, ground or unground, high or low quality. They may be required to withstand high temperatures, and may be manufactured singularly or in mass production using a variety of materials from glass to germanium.

OBJECTIVES

After studying this chapter and completing the self-check questions at the end of the chapter, the reader will be able to:

1. Identify optics according to their geometry.
2. State the function of general-purpose windows.
3. State the function of the Brewster window.
4. Describe several uses of mirrors by their shape.
5. State the function of surface reflectors.
6. Compare the shapes and functions of various polarizers.
   a. Beamsplitter
   b. Linear polarizer

c. Beamsplitter cubes
d. Prisms
7. Present a general method for cleaning laser mirrors.
8. List the necessary inspections required for laser optics.
9. Define the term *flashlamp*.
10. Compare the three methods of triggering a flashlamp.
a. External triggering
b. Series triggering
c. Parallel triggering
11. Identify typical flashlamps.

OPTIC GEOMETRY

Figures 6-1 and 6-2 illustrate the geometry of some of the available optics on the markets. Dimensions usually include the length, height, thickness, and/or diameter. Other parameters are the angle at which they are mounted, the angle from the front

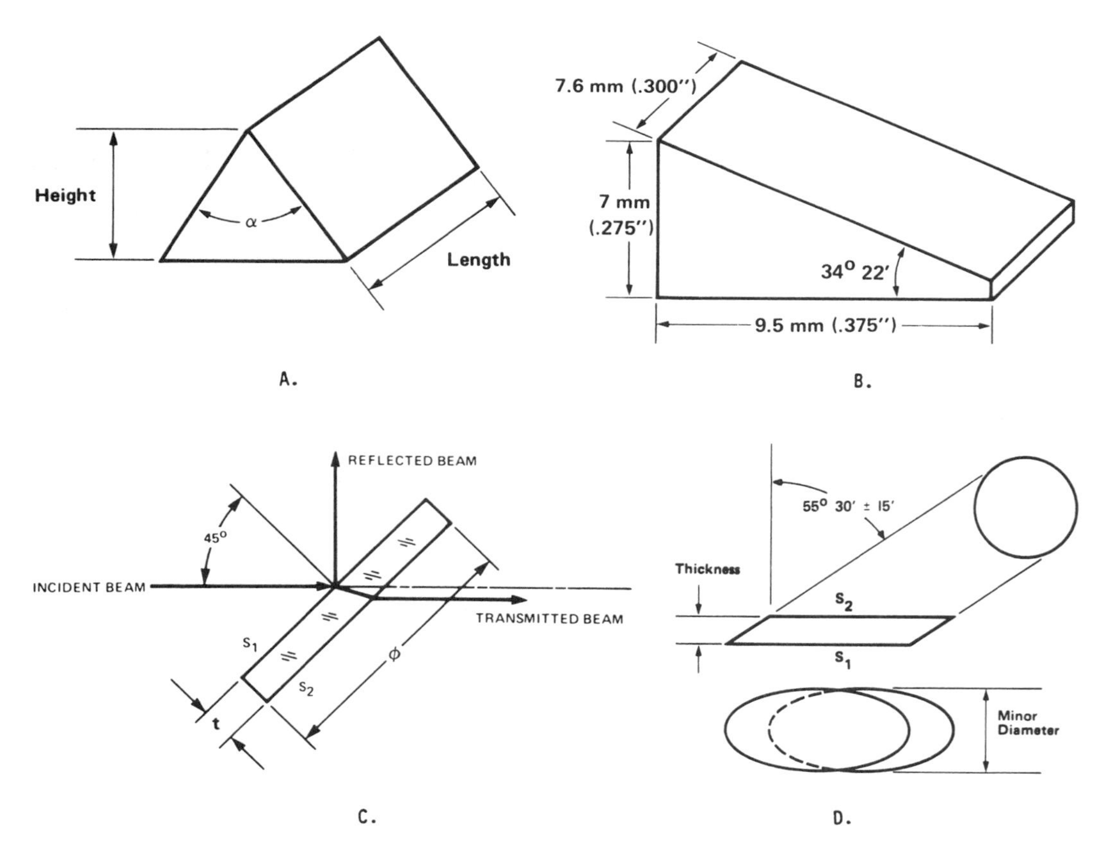

Figure 6–1 Unmounted optic geometrics (*courtesy of Coherent, Inc.*).

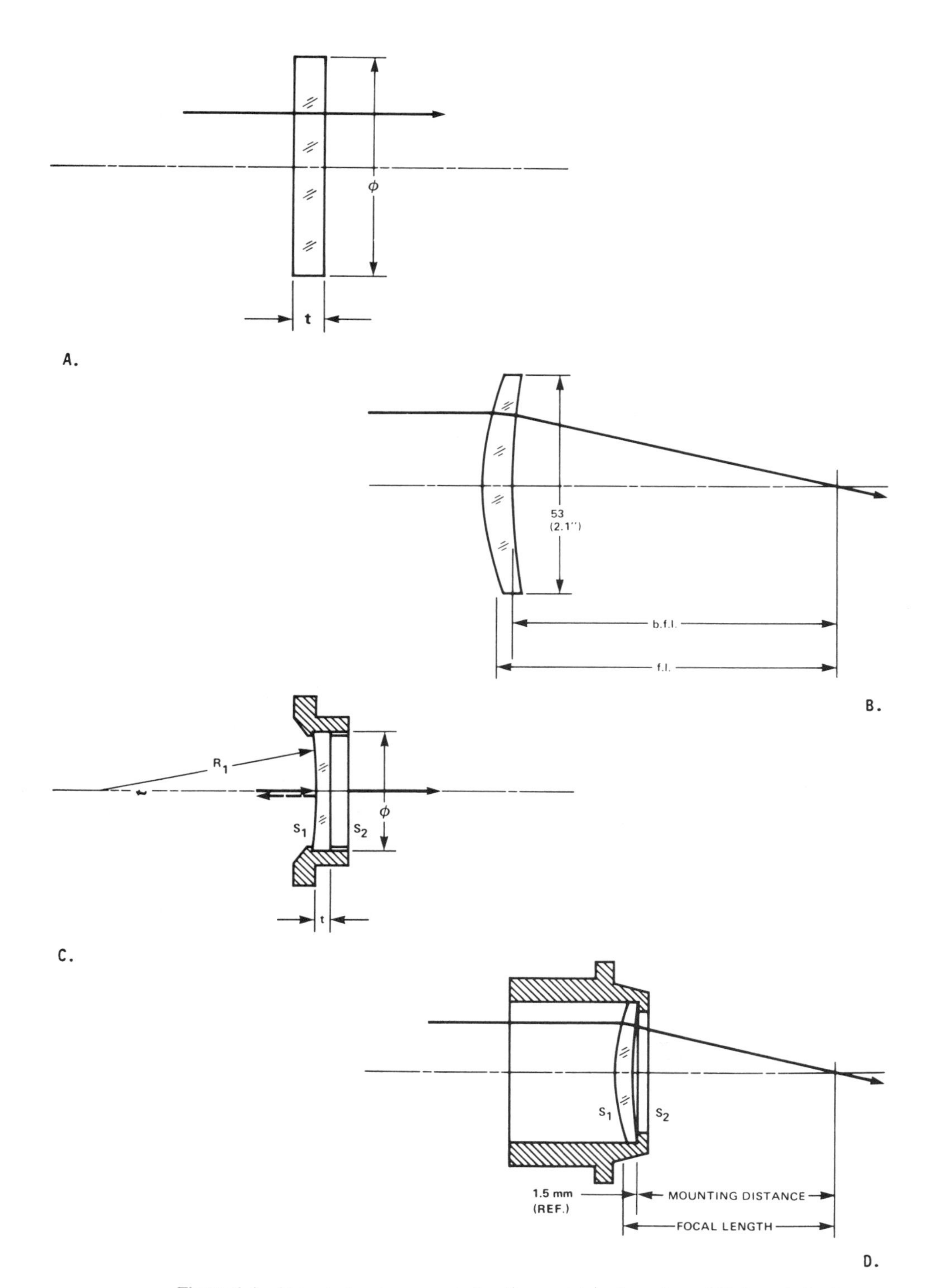

Figure 6–2 Mounted and unmounted optic geometries (*courtesy of Coherent, Inc.*).

to the back diameters, and the focal length. Specifications also include whether they come mounted or unmounted and if the mount is or is not part of its calibration or alignment. During design, the engineer may choose an off-the-shelf optic or demand a specially developed lens for a specific job. In any event, the choice of optics that is available, along with the designs that can be made, provide the user with an unlimited catalog for a variety of uses.

WINDOWS

The general-purpose window is chosen for its high light transmission and its surface qualities. For uses in ultraviolet to near infrared wavelength ranges, windows are made from fused silica. Infrared red windows are made from zinc selenide. The windows are flat with slight wedges of about 10 arc sec angle. A wedge is the small angle from the front to the back side of the mirror on its perimeter. Specifications for the window include the diameter of the front and back surfaces, the thickness of the window, and whether or not the surface is antireflection coated. Surfaces are polished to provide low scattering and to prevent etalon effects. Etalon effects indicate a change in optical path differences due to a mechanical displacement or a difference of refractive index (see Figure 6–3).

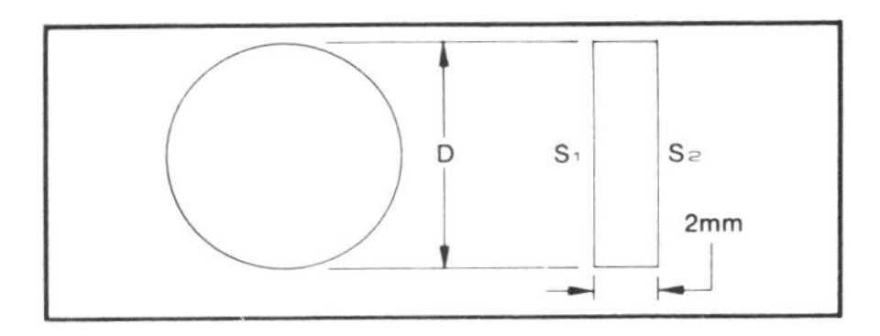

Substrate Specifications

Material: Schlieren Grade Fused Silica

Surfaces: S_1 Plano
$\lambda/10$ 10–5 Laser quality
S_2 Plano
$\lambda/10$ 10–5 Laser quality

Thickness: 2mm ± .1mm

Diameter Tolerance: +.00mm −.25mm

Wedge: <10 arc seconds

Figure 6–3 General-purpose windows (*courtesy of Spectra-Physics, Inc.*).

Brewster-Angle Windows

The Brewster window (see Figure 6–4) is a window placed on the ends of a plasma tube at an angle that allows maximum light power to pass through the active medium, which is usually gas. A plane parallel window would reflect much of the light. Brewster windows are made from top-grade fused silica. The windows are flat with a slight wedge from the front surface to back surface of less than 10 seconds wedge angle. Edges are machined to an elliptical circumference at Brewster's angle, which is around 55°32′. However, this is not a standard and can vary around this angle depending on the manufacturer. Specifications also include the front and rear diameters along with the thickness. As with general-purpose windows the Brewster windows are chosen for their high transmission and surface qualities. Grinding and shaping of the Brewster windows usally takes place prior to polishing. Windows are polished to provide low scattering.

MIRRORS

In general, laser mirrors are utilized either external to the laser cavity or within the cavity. The standard mirrors are designed for long life and durability. General-purpose mirrors are made from optical-grade fused silica. Dielectric metal oxides may be

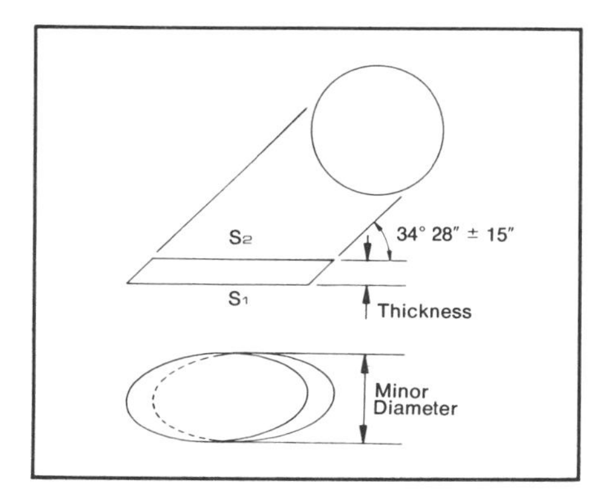

Substrate Specifications

Material: Schlieren Grade Fused Silica

Surfaces: S_1 Plano; $\lambda/20$; 10–5; Laser quality
S_2 Plano; $\lambda/20$; 10–5; Laser quality

Wedge: <10 seconds

Figure 6–4 Brewster angle windows (*courtesy of Spectra-Physics, Inc.*).

deposited on fused silica substrates to produce laser mirrors that can withstand plasma discharges in laser tubes. The dielectric material must also be able to withstand temperatures at levels between, say, −260 to +260 °C. The specific substrate/coating combination is a matter of wavelength demand. The selection is dependent on the wavelength of the application. Mirror testing includes humidity, water immersion, abrasion, and adhesion tests (see Figure 6–5).

Mirrors are shaped according to their use. Specifications include the radii of the first and second surfaces, along with the thickness and diameter of the substrate. A wedge between front and rear surfaces is usually less than a 30-minute angle. Mirrors are chosen to provide low scattering, low temperature distortion, and to maintain high transmission properties. Substrates that are used within a laser cavity must also be durable and are thick to ensure further stability. Mirrors usually come in sets that include the high-reflectance (100%) mirrors and the partial-reflectance transmission mirrors. In the case of the dye laser, the set would include a folding mirror and a pump mirror.

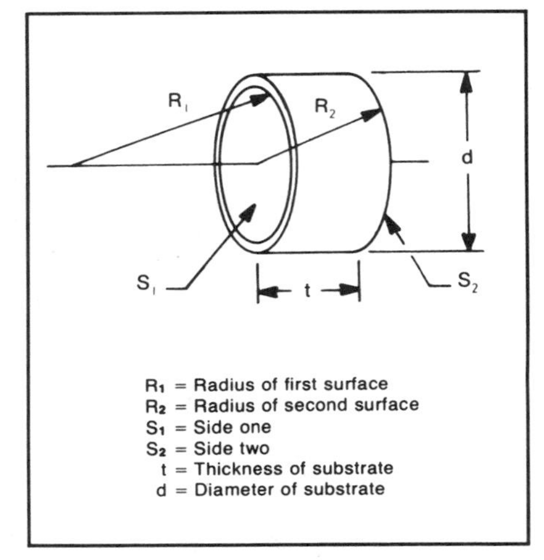

Substrate Specifications:

Material: Optical Grade Fused Silica

Surfaces: S_1 $\lambda/20$ 10–5 Laser quality
S_2 $\lambda/10$ 10–5 Laser quality

Wedge: less than 30 minutes

Diameter:	**7.75mm**	**15.0mm**	**25.4mm**
Diameter Tolerance:	+.00mm −.20mm	+.05mm −.15mm	+.05mm −.15mm
Edge Thickness Tolerance:	±.3mm	±.5mm	±.5mm
Radius Tolerance:	less than 1% from nominal		
Clear Aperture:	6mm	10mm	15mm

Figure 6–5 Laser mirror substrates (*courtesy of Spectra-Physics, Inc.*).

FRONT SURFACE REFLECTORS

Front surface reflectors (see Figure 6–6) are primarily designed for beam steering outside the laser. However, they may be used for general-purpose laboratory work not connected with lasers, which require light reflection. Front surface reflectors are made

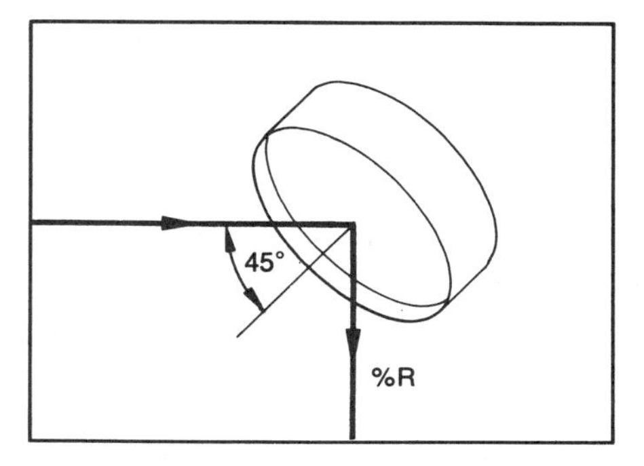

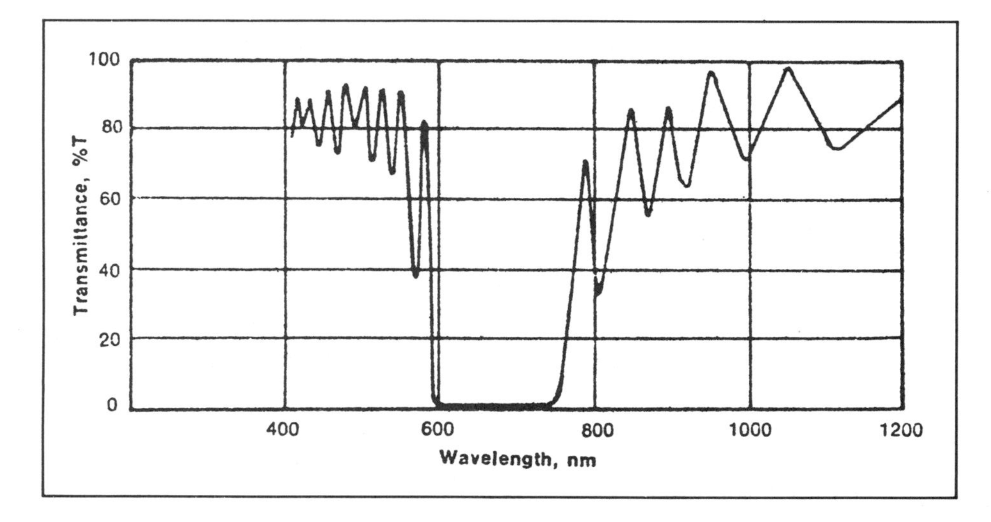

Coated Substrate Specifications

Material: Pyrex 7740; fine annealed
Surfaces: S_1 plano; λ/10; 60–40 per Mil-0-13830A; S_2 fine ground
Edges: As molded
Sizes: 25.4mm diameter × 12.7mm thick (G0050–000); 50.8mm diameter × 12.7mm thick (G0051–000)

Coating Specifications
at nominal wavelength:

Angle of incidence: 45°
Normal bandwidth reflectors: $R_{avg} > 99.0\%$ (Typical performance $R_{avg} > 99.5\%$)
Wide bandwidth reflectors: $R_{avg} > 98.0\%$
R_{avg} is the arithmetic average of reflectance in parallel and perpendicular planes of polarization.

Figure 6–6 Front surface reflector (*courtesy of Spectra-Physics, Inc.*).

from Pyrex with a dielectric coating. The front surface reflector is polished one side only. They have high reflectivity within the range of designed wavelengths. Operation of the reflector is around ±5% about the 45° design angle. Reflection is nominally around 95% at design wavelengths. Specifications for the reflector include the diameter of the reflector and its thickness. Other parameters include the angle of incidence in the application (usually 45°) and the device wavelength.

POLARIZERS

Incident Beam Polarization

When optics are used at some angle of incidence other than normal to the surface, the polarization of the incident beam must be taken into consideration (see Figure 6–7). A polarized coherent beam is characterized by an electric vector E in a plane perpendicular to the direction of the beam. When the beam strikes a surface a plane of incidence is formed by the beam and the normal to the surface. For the special case where the *E* vector is perpendicular to this plane the polarization state is called *S*. If the *E* vector lies in this plane the polarization is called *P*. Note that the polarization state depends on the orientation of the component relative to the *E* vector of the beam. The direction of the *E* vector of a lasar beam is usually determined by Brewster windows or other polarizing elements.

Beamsplitters

The beamsplitter, when placed at a 45° angle of incidence, will reflect a beam from the optic at 90° and transmit a beam through the optic. The reflection/transmission is near 50/50, ±5%. The transmitted beam is parallel to the incident beam. Beamsplitters for ultraviolet to near-infrared wavelengths are made from optic-grade fused silica. Infrared beamsplitters are made of polycrystalline germanium. Beamsplitters rear surface has an antireflection coating to minimize surface reflections. All coatings are low loss. The beamsplitters are designed for use at a 45° angle of incidence and for polarization normal (*S*) or parallel (*P*) to the plane of incidence. Specifications

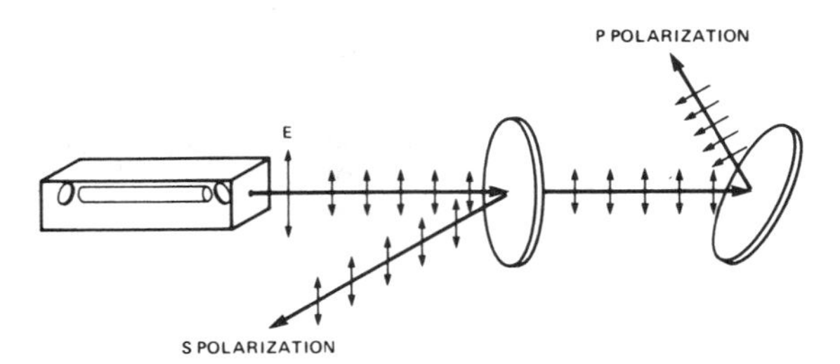

Figure 6–7 Incident Beam Polarization (*courtesy Coherent, Inc.*).

for the beamsplitter include its diameter and thickness. Other specifications are its wavelength and whether it is polarized normal (*S*) or parallel (*P*) to the plane of incidence.

In Figure 6–8 transmission curves which relate transmittance in percent to wavelength in nanometers are shown. The two curves represent polarized normal (*S*) or parallel (*P*). Transmission for unpolarized light is the average of the (*S*) and (*P*) values. The beamsplitter can be applied to beam manipulation, beam deflection, interferometers, imaging, and recording.

Coated Substrate Specifications

Material: Schlieren grade fused silica

Surfaces: S_1 plano; $\lambda/20$; 40-20 per Mil-0-13830A
S_2 plano; $\lambda/10$; 40-20 per Mil-0-13830A

Wedge: 30 min ± 5 min.

Sizes: 25.4mm diameter x 4.2mm thick **(G0020-000)**
50.8mm diameter x 8.3mm thick **(G0022-000)**

Specific Polarization Control

Coating specifications at nominal wavelength

Angle of incidence: 45°

S_1: Transmission tolerance for "S" polarization: ±3%
Transmission tolerance for "P" polarization: ±6%

S_2: Anti-reflection coating
$R_{avg} < 0.6\%$

Useable clear aperture at 45° incidence equals 53% of diameter

Scatter and absorption loss
$< .2\%$ 440nm and higher
$< .5\%$ below 440nm

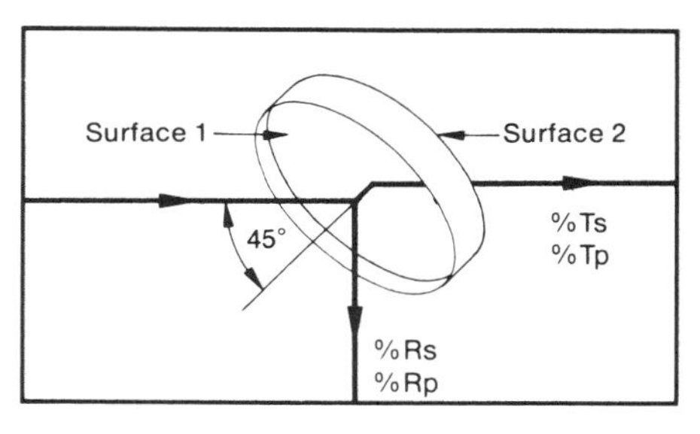

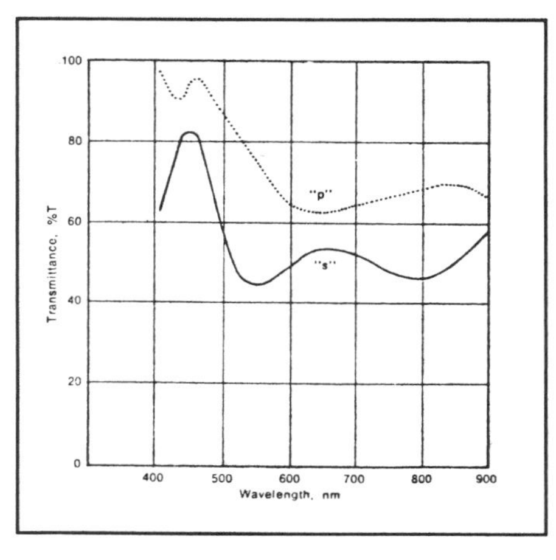

Figure 6–8 Beamsplitter (*courtesy of Spectra-Physics, Inc.*).

Linear Polarizers

Linear polarizers consist of multilayer dielectric coatings deposited on the interface between two cemented prisms. The polarizer coating interacts with randomly polarized light to produce two linearly polarized beams. In Figure 6–9 a linear polarizer is illustrated. In Figure 6–9b, *S* polarization is shown perpendicular to the plane of incidence. *P* polarization is shown parallel to the plane of incidence. In Figure 6–9c, the plane of incidence contains the light ray and the normal to the surface the light ray is falling on. Linear polarizers selectively transmit a *P* polarized beam, while reflecting an *S* polarized beam at approximately 40° to the exit beam. The transmitted beam has low wavefront distortion relative to the input beam. The rejected component does not have a controlled wavefront specification. The polarizer coatings are designed for broad spectral ranges, and can be used in both laboratory and production applications. Specifications for the linear polarizer include the transmittance of the beam, the wavefront distortion, and angular beam deviation.

Specifications

Transmitted beam only

Transmittance: $T > 95\%$ over the wavelength range

Wavefront Distortion: $< \lambda/2$

Angular Beam Deviation: $<$ one minute of arc

Clear Aperture: 5mm diameter, centered

Acceptance Angle: $\pm$ 1° over full wavelength range

Anti-reflection coatings: Entrance and exit face $R_{av} < 0.5\%$

Damage Resistance: to 100W/cm² CW; 1 megawatt/cm² peak

Extinction Ration (T_p/T_s): 1000:1

A.

s

p

B.

INPUT

OUTPUT

C.

Figure 6–9 Linear polarizer (*courtesy of Spectra-Physics, Inc.*).

Polarization Beamsplitter Cubes

Polarization cubes (see Figure 6–10) divide an input beam into two linearly polarized beams at right angles to each other. The transmitted beam has the *P* polarized component and the reflected beam has the *S* polarized component. Polarized cubes are manufactured from high-quality optical glass to exacting tolerances that maintain low wavefront distortion of the entrance beam. Other specifications include transmittance of the beam, wavefront distortion of the exit beam, and transmitted beam deviation.

Applications for the beamsplitter cube are illustrated in Figure 6–11. The applications include (a) isolation of the laser cavity energy from light reflected from an optical system, (b) combination of two beams of the same wavelength but opposite polarizations, (c) read/write equipment, and (d) coupled-cavity experiments in nonlinear optic applications.

Prisms

Prisms generally are manufactured in three basic forms: the right angle, the Littrow, and the Brewster angle. These three forms are illustrated in Figure 6–12 and a, b, and c, respectively. The right-angle lens permits general purpose beam bending with good optical figures. The Littrow prism is designed for use as a dispersion element in laser cavities. They are used in applications that require low scatter. The Brewster-

Specifications

Transmittance: $T \geq 98\%$ of total energy
Wavefront Distortion of Exit Beams: $< \lambda/2$
Reflected Beam: $90° \pm 30$ min from incident beam
Transmitted Beam: Deviation < 5 milliradians
Clear Aperture: 8 mm centered for 12.7mm cube; 16mm centered for 25.4mm cube
Anti-reflection Coating: All entrance and exit faces $R \leq 0.25\%$
Damage Resistance: To 100W/cm^2 CW; 1 megawatt/cm^2 peak
Extinction Ratio: Specified as minimum in both exit beams; T_p/T_s transmitted beam; T_s/T_p reflected beam
Angle at which maximum extinction occurs: $0 \pm 1°$ incidence

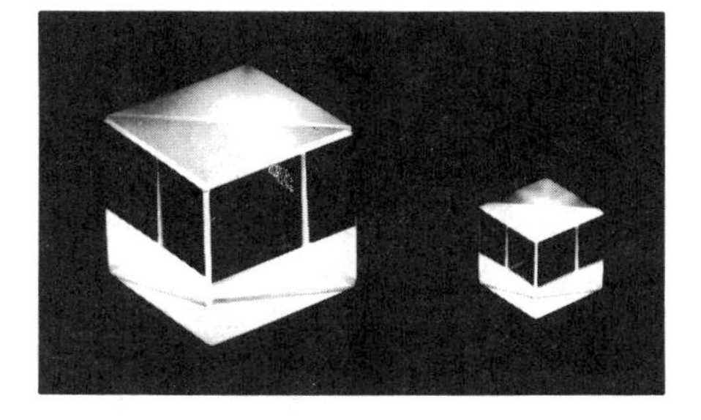

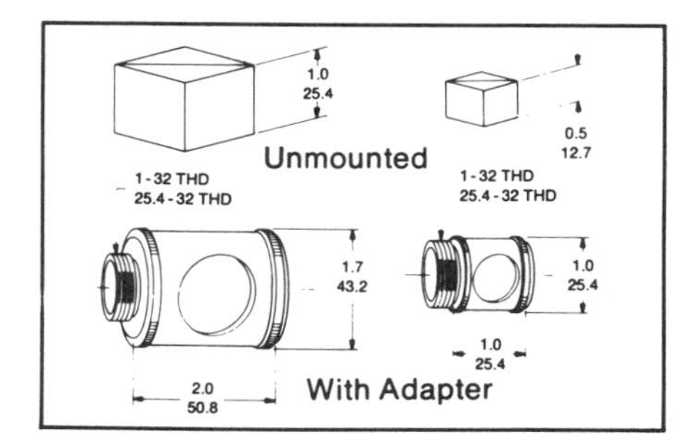

Figure 6–10 Polarization beamsplitter cubes (*courtesy of Spectra-Physics, Inc.*).

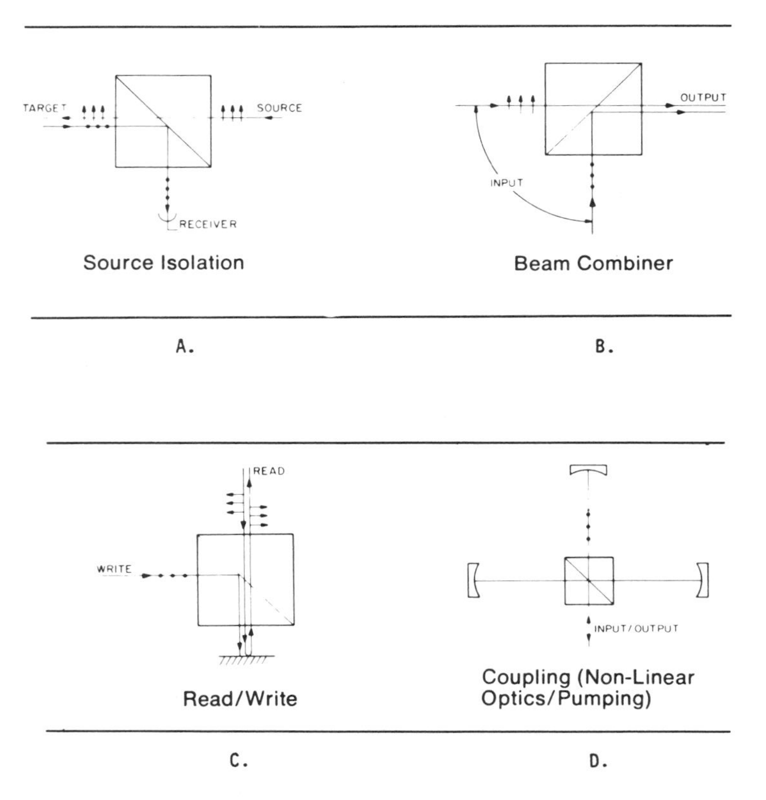

Figure 6–11 Polarization beamsplitter cube applications (*courtesy of Spectra-Physics, Inc.*).

angle prism is also used as a dispersion element in laser cavities. Light enters the prism and is dispersed as it passes through the element and exits at Brewster's angle.

Specifications for prisms are surface quality and angular tolerances. Prisms are made from high-quality optical fused silica. Coating may or may not be desirable as the application demands.

CLEANING LASER MIRRORS

The cleaning of laser mirrors is, as you may have determined, a simple but extremely important task. Mirrors must be free of contamination and dust. The difference between a clean and a dirty optic may be a loss in power output or a change in the beam profile. The procedures listed here are for general maintenance. There may be a case where optics have collected contamination beyond normal cleaning maintenance. These may entail the use of more stringent methods such as ultrasonic cleaners and strong detergents. Procedures vary somewhat for different types and manufacturers, but generally utilize a procedure such as the following:

Substrate Specifications

Material: BK-7 or equivalent
Surfaces: S_1 Plano $\lambda/10$
40–20 per Mil-O-13830A
S_2 & S_3 Plano $\lambda/8$
40–20 per Mil-O-13830A
Angular Tolerance: ± 10 minutes
Sides Ground and all edges beveled
Tolerances on dimensions ±.25mm

A.

Substrate Specifications

Material: Schlieren Grade
Fused Silica
Surfaces: S_1 & S_2 $\lambda/20$
10–5 Laser quality
All others ground
Physical Tolerances: ±.4mm
Pyramid Tolerance: ±5 min.
Size: 11.81mm High
12.70mm Wide
8.07mm Deep
Apex Angle: 34° 28 min ± 15 min.

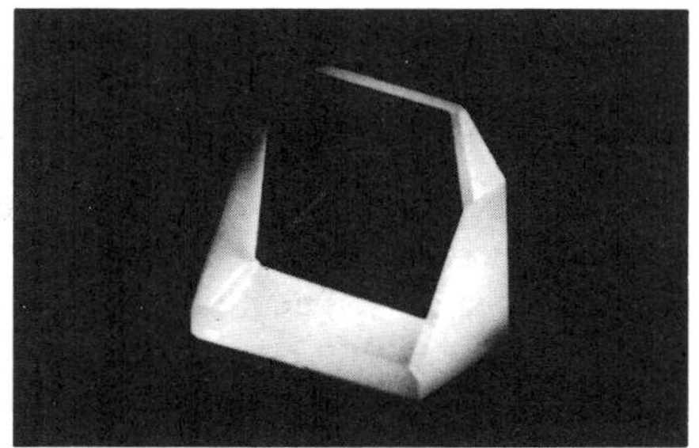

B.

Substrate Specifications

Material: Schlieren Grade
Fused Silica
Surfaces: S_1 & S_2 $\lambda/20$
10–5 Laser quality
All others ground
Base Angles: 55° 30 min ± 15 min
Physical Tolerances: ± .12mm
Size: 11.46mm High
14.38mm Wide
7.92mm Deep

C.

Figure 6–12 Prisms (*courtesy of Spectra-Physics, Inc.*).

1. Select a clean, dust-free atmosphere and a flat working surface.
2. Place a light pencil mark on the top center of the laser mirror before removing from laser head.
3. Remove the laser mirror from the laser head.
4. Place the laser mirror on the flat working surface with the coated side of the optical working surface in the up position.

5. Using a hemostat or other plier-type tool, pick up a lens tissue (folded several times). Do not touch the lens tissue with your fingers.
6. Soak the lens tissue in methanol or acetone. Shake off excess fluid.
7. Place the moist tissue on the coated surface of the lens and make one sweep. Throw away the tissue.
8. Repeat as necessary using a new tissue for each sweep.
9. Place the laser mirror in the laser head aligning with pencil mark made in step 2.

When cleaning filters, etalons, and special lenses, use precautions that are peculiar to the laser. It is advised that the manufacturer's precautions be read carefully, for the successful operation power output and beam profile of the laser are dependent on clean optics.

INSPECTION OF OPTICAL SUBSTRATES

In any manufacturing field, inspection is an integral part of the ongoing process. Inspection of optics fall into several major areas. Surface quality is the first of these inspections. The numerical designation of the maximum number and size of surface defects within the clear aperture is defined by MIL–0–13830A. These are referred to as scratch-dig defects and are represented by numbers such as 80–50 and 20–10, etc. Standard quality refers to defects that are visually apparent when illuminated with a standard 40-watt incandescent bulb. Laser quality means those defects that are virtually apparent under high-intensity illumination of 10,000 to 15,000 footcandles as obtained from a focused American Optical microscope illuminator or its equivalent. Figure 6–13 illustrates an optic under inspection. A high-intensity illumination ensures conformance to laser quality standards.

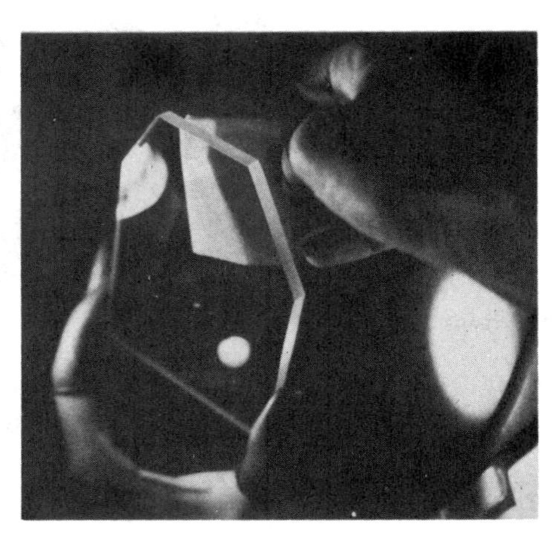

Figure 6–13 Inspection of an optical substrate under intense magnification (*courtesy of Spectra-Physics, Inc.*).

Component clear aperture is the next inspection. The area of the polished surface or surfaces of an optical component over which the technical specifications apply is usually defined in linear terms. These terms may be diameter, length, width, thickness, and so on. Surface figure tolerance is surface flatness or surface accuracy. The maximum departure of a polished surface from the ideal specified surface is monitored. This tolerance is expressed in wavelengths of light. A typical specification would be 632.8 nanometers wavelength.

The allowable deviation of the radius of curvatures from the specified value is expressed in units of length. A radius tolerance would be ±1%. Edge chips are allowable as long as they do not encroach the clear aperture.

Specifications of polarization are assumed to refer to linearly polarized light in which the electrical (E) vector is parallel, (P) polarization to the plane of incidence, perpendicular, (S) polarization to the plane of incidence, or at an angle of 45° (average polarization) to the plane of incidence. The plane of incidence is defined by the incident beam and a normal to the surface.

FLASHLAMPS

Flashlamps are gas discharge devices, usually filled with xenon or krypton, designed to produce pulsed radiation (see Figure 6–14). Although there are higher brightness sources available (vortex lamps, plasma pinch devices, etc.), tubular flashlamps offer a combination of high brightness, high efficiency, long life, and convenient operation that makes them universally used for pulsed solid-state lasers.

Flashlamps are available in many sizes and shapes, including linear, helical, and U-shaped. Modern pulsed solid-state lasers rely largely on linear flashlamps. These typically range between 2.5 and 20 cm in arc length and between 1 and 15 mm in bore diameter. A wide variety of designs exist, each stressing some particular aspect of lamp performance. The range of performance is extremely broad—electrical input energy may be varied from 10^{-2} to 10^5 joules per pulse, pulse length from 10^{-3} to 10^5 microseconds, and pulse repetition rate from single shot to 5×10^4 Hz.

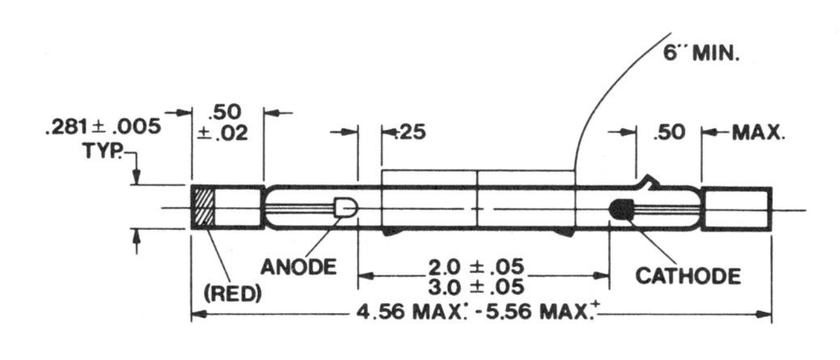

Figure 6–14 Flashlamp in cross section (*courtesy of ILC Technology*).

FLASHLAMP OPERATION

In characterizing the electrical behavior of tubular rare gas flashlamps it is convenient to deal with several distinct operating regimes separately. The first regime is triggering and initial arc formation. This is followed by a second, the regime of unconfined discharge. The final regime is that of the wall-stabilized plasma at high current.

Triggering and Initial Arc Formations

Triggering is defined as the initiation of a complete discharge through a flashlamp (see Figure 6–15). An important requirement of any system using an arc discharge tube is that lamp triggering be controlled in a manner such that it occurs repeatably and reliably.

Overvoltage. If the initial bias voltage V_0 across the lamp is itself sufficiently high to break down the gas in the tube and begin the discharge, the lamp is normally isolated from it by means of a switch. Otherwise, uncontrolled self-triggering occurs at random values of voltage as V_0 is approached. The switch used is usually a mercury ignitron, a hydrogen thyratron, or a triggered spark gap. Each of these devices is itself able to be triggered. This closes the switch and results in the triggering of the lamp due to "overvoltage."

Series and external triggering. In solid-state lasers, the initial bias voltage V_0 across the tube is usually much lower than the voltage required to break down the gas. The majority of lamps incorporated into these systems use either series or external triggering. Each of these methods of triggering has its advantages. External triggering provides greater design flexibility since the secondary winding of the transformer is not in the main discharge circuit. Transformers used for external triggering are less expensive and smaller than those used for series triggering. However, series triggering offers better reliability than external triggering. Series triggering is almost always used in CW solid-state lasers using high-pressure krypton arc lamps because reliable triggering takes place at lower initial bias voltage, high voltage is not exposed on the outside of the lamp, and triggering is less likely to cause EMI problems. For series-triggered flashlamps used in pulsed lasers, the required inductance in the pulse-forming network is often designed into the trigger transformer, eliminating one circuit component.

Parallel triggering. The characteristics of parallel triggering and series triggering are similar to one another because both involve the imposition of a high-voltage spike directly to the lamp electrode. Parallel triggering is attractive because it is possible to use a small "external trigger-style" transformer while obtaining most of the advantages of series triggering. One disadvantage is that a component must be selected and incorporated into the trigger circuit that both passes the high-voltage trigger pulse and blocks the resulting current discharge from flowing through the transformer secondary. Another disadvantage exists because the main discharge circuit impedance

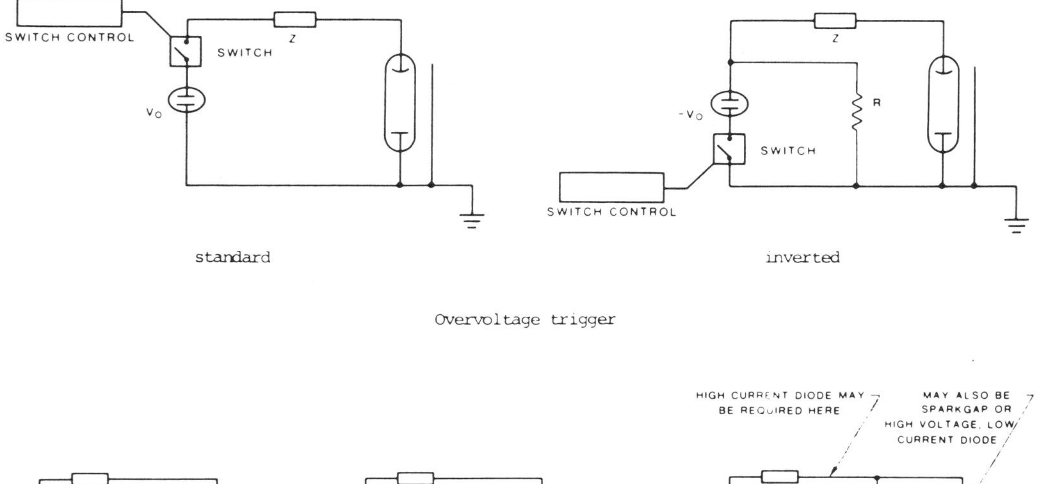

HIGH CURRENT DIODE MAY BE REQUIRED HERE
MAY ALSO BE SPARKGAP OR HIGH VOLTAGE, LOW CURRENT DIODE

Series trigger External trigger Parallel trigger

Figure 6–15 Flashlamp triggering circuits (*courtesy of ILC Technology*).

is frequently too low. When this is the case, a diode or diode train must be placed in the main current discharge path to block the trigger pulse from being absorbed by this circuit. It is not unusual to discover that the requirements for the diodes are near or beyond the limits of diode technology. The diodes are either prohibitively expensive or cannot be obtained at all.

Ground plane. For triggering to take place most repeatably at fixed values of bias voltage and trigger voltage an electrically conductive surface that is a stable voltage reference, usually referred to as a "ground plane," is required near the lamp. This is frequently a "trigger wire"; a wire wrap scanning the lamp arc length consisting of several turns of wire wound on the outside of the lamp envelope. The ground plane is important regardless of the type of triggering used. Triggering occurs at the lowest values of voltage when the ground plane makes intimate contact with the surface of the lamp envelope throughout the region of the arc length.

Regime of Unconfined Discharge

After undergoing triggering, a lamp is operating in the regime of unconfined discharge when most of the characteristics of the arc are not influenced by the existence of the wall of the lamp envelope. Either of the following conditions exists:

1. The lamp arc is undergoing rapid expansion while approaching equilibrium during startup.
2. The lamp is in a steady-state condition at currents in the region of negative slope on its voltage–current characteristic curve.

Once triggering has occurred, the length of time and nature of a lamp's excursion through the regime of unconfined discharge is dependent on the amount of charge available from the power source that initially stands at the bias voltage V_0 and the value of impedance in the circuit through which this charge is driven.

Simmer mode operation. The majority of pulsed solid-state lasers incorporate flashlamp driving circuits designed to maintain a steady state of partial ionization in the lamp during the time between flashes. This important technique is called "simmer mode" operation (see Figure 6–16a). It usually involves establishing and maintaining a low-current dc arc between the lamp electrodes. It is not unusual to observe lamp voltage fluctuations as great as 50 volts at fixed simmer current over a period of tens of seconds. Arc wander in the lamp bore and rapid changes in the position of arc attachment on the electrode tips are the cause of this behavior.

It is important to understand the transient behavior of the lamp voltage–current relationship during the period immediately following the main high-power discharge, while the plasma is recovering its steady-state characteristics. During this recovery period, voltages significantly higher than the time-averaged steady-state simmer voltage are required to sustain the lamp arc. If the simmer power supply and simmer circuit are unable to supply the lamp's voltage–current requirements during this transient period, the arc self extinguishes during the interpulse interval. The simmer circuit and dc simmer power supply must be designed to drive the lamp into the regime of unconfined discharge during lamp startup and to stably maintain the simmer streamer until the high power flash discharge is initiated. They must also prevent self-extinguishing of the arc immediately after the high power flash discharge during the time the plasma is recovering its steady-state simmer characteristics. Figure 6–16 shows the typical simmer circuit for a flashlamp.

The value of steady-state simmer current in a flashlamp should be chosen between 25 milliamperes and several amperes. Simmer currents at the high end of this range require more massive simmer power supplies and dissipate more power but lead to higher electrical stability and less difficulty with regard to problems during recovery. Simmer currents at the low end of this range require relatively high simmer power supply open-circuit voltages and are frequently troubled by instabilities due to "hopping" of the simmer arc at the point of attachment on the tip of the lamp cathode.

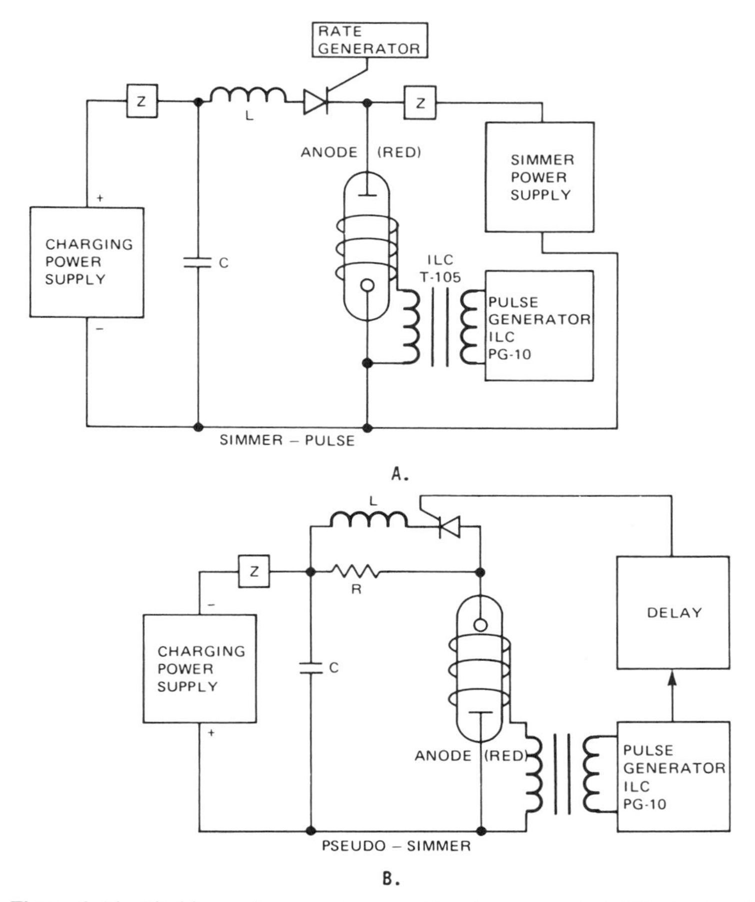

Figure 6–16 Flashlamp simmer pulse operation (*courtesy of ILC Technology*).

Pseudo-simmer. At low pulse repetition rates, when maintaining a simmer discharge during the entire interpulse interval is prohibitively wasteful of power, many of the advantages of simmer operation can be obtained by incorporating a pseudo-simmer mode of operation. The typical circuit is shown in Figure 6–16b. The lamp is ignited by a conventional trigger signal, but current flow is limited by R to approximately 0.050 ampere for 100 to 200 microseconds. R is then shorted out by the SCR and the main power discharge occurs.

Pulsed discharge in the regime of unconfined discharge. The plasma in a flashlamp enters unconfined discharge regime almost instantaneously after being triggered or, under conditions of simmer operation, after the main high power discharge switch is closed. This period of time is characterized by rapidly increasing current and rapidly falling voltage, usually occupying the first 15 to 20% of the rise time of the current pulse through the lamp. This is an interval of rapid arc expansion during which the energy required to bring the plasma into wall stabilization is being dissipated. When a sufficient amount of energy has been delivered to bring the arc into a realm of much slower, controlled growth, the discharge enters the regime of the wall-stabilized plasma at high current.

The advantages of simmer-pulse operation are:

1. Higher Nd:YAG pumping efficiency
2. Increased lamp life
3. Improved pulse to pulse reproducibility of flashlamp output
4. Reduced time jitter
5. Higher repetition rate capability

Many of the advantages of simmer-pulse operation can be achieved in small, light weight systems by using the pseudo-simmer technique. The lamp is ignited by a conventional external trigger signal, but initial current flow in the discharge circuit is limited to approximately 50 mA by a series resistance. After a delay of approximately 100 to 200 microseconds, the limiting resistor is shorted out by an SCR switch and the main discharge occurs.

The Wall Stabilized Plasma at High Current

In practice this regime is only encountered in pulsed discharges because of the high current and power densities it encompasses. This regime is arbitrarily defined as a range of operation in a pulsed light source where the arc is stabilized by the inside wall of the lamp envelope. The arc is either expanding at a rate that is determined by its proximity to the envelope wall or expansion has ceased.

TYPICAL FLASHLAMPS

Typical flashlamps are those illustrated in Figures 6–17, 6–18, and 6–19. The first, in Figure 6–17 are a group of lamps used in industrial applications. The thin lamp on the top is a krypton flashlamp used in about 75% of all laser resistor trimmers and semiconductor wafer scribers. The center lamp in Figure 6–17 is a krypton flashlamp used in high average power industrial lasers that perform a wide variety of welding and drilling operations. The lamp with the cable ends is a xenon flashlamp used in flashlamp pumped dye lasers and in lasers utilizing unstable resonators.

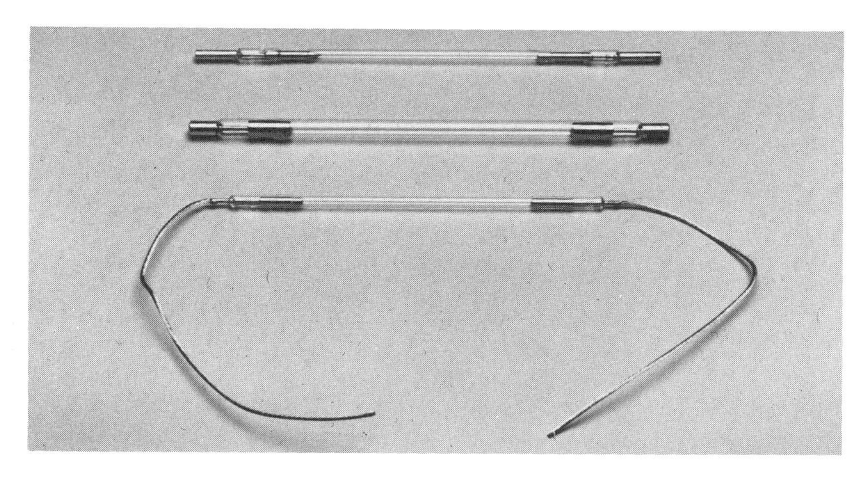

Figure 6–17 Industrial laser lamps (*courtesy of ILC Technology*).

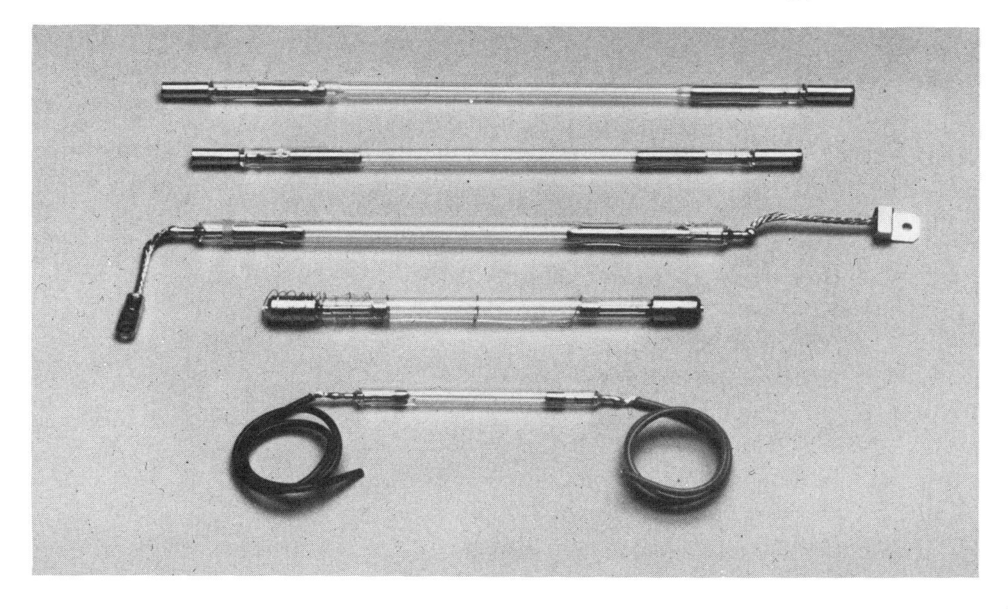

Figure 6–18 Military laser lamps (*courtesy of ILC Technology*).

In Figure 6–18 are lamps for military lasers. All the lamps here are used in military rangefinders and target designators. In Figure 6–19 are lamps used for fusion lasers. These lamps have high voltage (60 kV) insulation systems for use in air and nitrogen filled cavities. Lamp reliability is assured by microscopic inspection during and after manufacture and by severe functional tests.

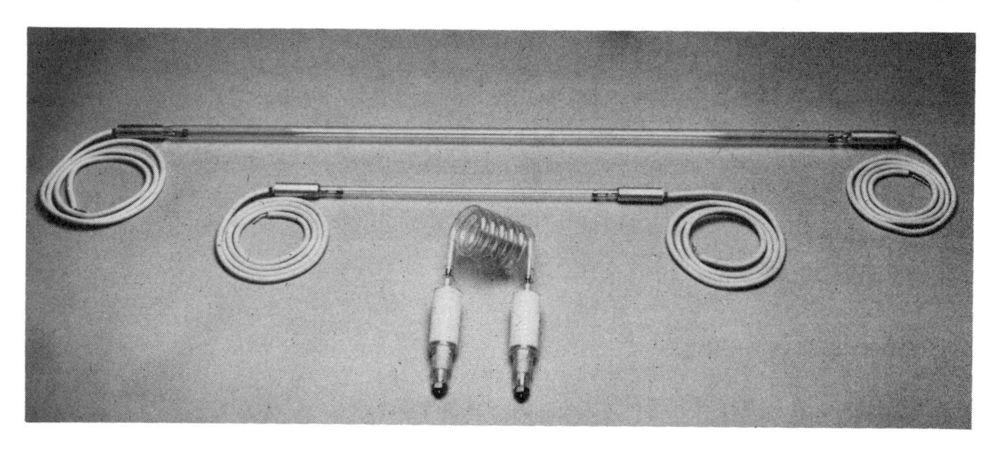

Figure 6–19 Lamps used for fusion lasers at high voltages (*courtesy of ILC Technology*).

SELF-CHECK QUESTIONS

Self-check questions allow students to evaluate how well they have learned what they are studying.

1. Name some of the applications of laser optics.
2. How are optical windows chosen?
3. How are optical mirrors chosen?
4. What is the purpose of a reflector?
5. What is a beamsplitter?
6. Polarizers are used for what purpose?
7. Name the three prism types.
8. Why must optics be kept clean?
9. What inspections are made on optical devices?
10. Define the term *flashlamp*.
11. How may a flashlamp be triggered?

Seven

Optical Fibers, Cables, and Couplings

Optical fibers are transparent, dielectric cylinders surrounded by a second transparent dielectric cylinder. The fibers are light waveguides used to transmit energy at optical wavelengths. The light is transported by a series of reflections from wall to wall of an interface between a core (inner cylinder) and its cladding (outer cylinder). The reflections are made possible by a high refractive index of the core material and a lower refractive index of the cladding material. Refractive index is a measure of the fiber's optical density. The abrupt differences in the refractive indexes causes the lightwave to bounce from the core–cladding interface back through the core to its opposite wall (interface). Thus the light is transported from a light source to a light detector on the opposite end of the fiber.

In this chapter we look at fibers in detail. We discuss types, classes, and manufacturing procedures. We make an in-depth study of problems of misalignment and the concepts of alignment. We study the fiber specifications and consider the advantages and disadvantages of properties such as dispersion, attenuation, bandwidth, and tensile strength. An informative look at cables will then be made. The final area of coverage in this chapter is the interface of fibers and cables with couplers and splicing techniques.

OBJECTIVES

After studying this chapter and completing the self-check questions at the end of the chapter, the reader will be able to:

1. Explain the basic operation of light transmission into a fiber.

2. Define the six major properties of fibers.
   a. Numerical aperture
   b. Dispersion
   c. Attenuation
   d. Bandwidth parameters
   e. Rise time
   f. Fiber strength
3. Describe the double crucible and chemical-vapor deposition processes of fiber manufacture.
4. Compare the several fiber types.
   a. Single-mode step index
   b. Multimode step index
   c. Multimode graded index
   d. Multimode, wideband graded index
   e. Large core, plastic clad, multimode
5. Compare the several fiber cable types.
   a. Light duty
   b. Single-fiber strengthened
   c. Two-fiber strengthened
   d. External strength member, heavy duty
6. Analyze and explain the transmission angles necessary to design a fiber optic system.
   a. Cone of acceptance
   b. Angle of acceptance
   c. Angle of refraction
   d. Angle of reflection
   e. Graded-index fiber angles
7. Identify and explain the causes of coupling loss in the use of fibers.
   a. Core irregularity
   b. Lateral misalignment
   c. Gap loss
   d. Angular loss
   e. Loss in transmission
   f. Coupling attenuation
8. Describe the several philosophies that have to do with fiber alignment.
   a. Point of contact
   b. Overlapping surface
   c. Resilient mechanisms
9. Explain the concepts used in fiber alignment mechanics.
   a. Opposed lens alignment
   b. Transfer molded alignment
   c. Multirod alignment
   d. Resilient self-centering alignment

e. Double eccentric alignment
f. Watch jewel alignment
g. Three-sphere alignment

10. Examine a typical fiber-end preparation technique.
11. Compare several fiber-splicing techniques.
    a. Fused splice
    b. Precision tube splice
    c. Loose tube splice
    d. Double metal plate splice
    e. V-groove splice
12. Identify several fiber connector types.
13. Examine a typical fiber connector installation procedure.
14. Describe several fiber testing methods and/or requirements.
    a. Fiber tensile strength test
    b. Proof test
    c. Dynamic short-length strength test
    d. Static short-length strength test
    e. Attenuation test
    f. Dispersion test
    g. Numerical aperture (NA) measurements

FIBER GENERALITIES

The discussion of how light is launched into a fiber and arrives on the other end of the fiber is easy to understand. The details involved in transmission and the mechanics of coupling, however, are manifold. Let us first discuss in lay terms how transmission takes place.

Light is propagated through optical fibers in a series of reflections from one side wall to opposite wall. The acceptance angle i must be small enough so that all of the signal is reflected. In Figure 7–1, a fiber core is covered by a layer called the *cladding*. The core has a refractive index higher than that of the cladding. Index of refraction is determined by the manufacturer when the fiber is produced. When the light enters the optic fiber, it reflects from wall to wall of the cladding back into the core. The fiber illustrated is a step-index fiber. That is, there is an abrupt refractive index change between the core and the cladding. In order that the light beam may bounce freely from wall to wall without loss, the cladding was added.The cladding acts as a mirror to reflect without power loss. A second use of the cladding is to prevent light from escaping to nearby fibers in the event of bundling several together. The cladding also contributes to the strength of the fiber.

The acceptance angle i is one-half the total cone of acceptance. The cone of acceptance is the light-gathering area at the fiber input. The mechanics of transmission angles are handled in detail later in this chapter.

∠i = INPUT HALF ANGLE
N1 = CORE INDEX OF REFRACTION
N2 = CLADDING INDEX OF REFRACTION

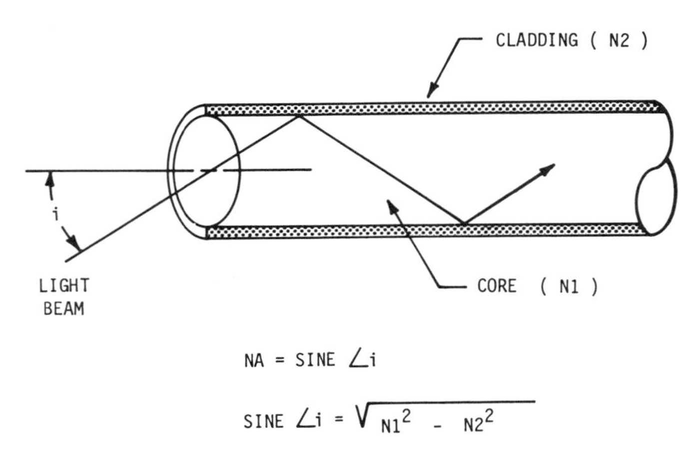

Figure 7–1 Basic optical fiber (*Mike Anaya*).

Fiber Properties

There are six major properties involved when choosing the correct fiber for an optical system. These are:

1. *Numerical aperture (NA).* Numerical aperture is the mathematical sine of the angle of acceptance. Half the cone of acceptance is the angle *i* in Figure 7–1.

   $$\text{NA} = \sin \angle i$$

   $$\text{NA} = \sqrt{n_1^2 - n^2_2} = \sin \angle i$$

   where
   n_1 = refractive index of the core
   n_2 = refractive index of the cladding
   $\angle i$ = angle of acceptance

   NA is a trigonometric function which represents the sine of the angle *i*. If the NA increases, the angle *i* must have also increased and the fiber sees more light. This may seem like a good thing. A problem enters when there is a greater spread in propagated light rays. The increase in ray velocity limits the bandwidth. The effect is called intermodal dispersion. The NA of a fiber can never be greater than 1.0. This is mathematical maximum sine value. Most NA values are low, between 0.20 and 0.60, and seldom near the maximum sine value. This is a range and must not be judged. For specific values, the manufacturers' data sheets must be consulted.

2. *Dispersion.* The second property to contend with is *dispersion.* Dispersion is the spreading or widening of light rays. In fiber optic systems, dispersion is

either intermodal or intramodal. The intermodal dispersion (also called *multimode*) is the propagation of rays of the same wavelength along different paths through a fiber. This results in the wavelengths arriving at the opposite end of the fiber at different time periods. The second dispersion, intramodal is determined by three separate factors: material dispersion, waveguide dispersion, and cross-product dispersion. Material dispersion is due to variations of the index of refraction of the core and the cladding materials. Waveguide dispersion is a phenomena caused by the bandwidth of the signal and the waveguide configuration. Cross-product dispersion is leakage of optical energy from one material, fiber, cable, or connector to other materials. Wavelength and cross-product dispersions are usually small and can be ignored. Dispersion is a function of the refractive index of the material the fiber is made of and the wavelength or mode of the light traveling through the fiber. It is measured in nanoseconds per kilometer (ns/km).

3. *Attenuation.* Attenuation is the loss or reduction in amplitude of transmitted energy. Intrinsic losses are due to light scattering of transmitted energy. Losses of light due to the differences of refractive indexes of two mediums at their interface are called *Fresnel losses.* Intrinsic losses are due to light scattering within the fiber. The most prevalent type of light scattering is Rayleigh. Rayleigh scattering is power loss due to the molecular structure of the fiber material. Other power loss may be due to imperfections or bubbles in the fiber material. Man-made power losses from scratches or dirt are common.
4. *Bandwidth parameters.* Attenuation curves are always provided by manufacturers of fibers to allow the designer to choose the correct fiber for a particular use. Curves or equivalent tabulated data are expressed in decibels per kilometer (dB/km). The curve in Figure 7–2 plots wavelength against decibels. This provides the designer with the fiber's transmission qualities. These parameters are developed during manufacture.
5. *Rise time.* The rise time is a parameter that tells the designer that selected parts will operate at speed required. Rise time identifies the fiber's dispersion properties. Rise time is determined by intermodal (multimode) dispersion and/or intramodal (material) dispersion. To find the total rise time of a system you simply add the rise time required for each time-critical component, then add some tolerance. In their technical Note R–5, ITT Electro-Optical Products Division uses this simple calculation. The overall system rise time is 1.1 times the root sum square of all system components.

$$T_{\text{system}} = 1.1\ \sqrt{T_I^2 + T_M^2 + T_{\text{etc}}^2}$$

where T_I = intermodal rise time
T_M = material intramodal rise time

6. *Fiber strength.* Another fiber property is tensile strength, which is governed by manufacturing intricacies. These techniques eliminate flaws and microcracks

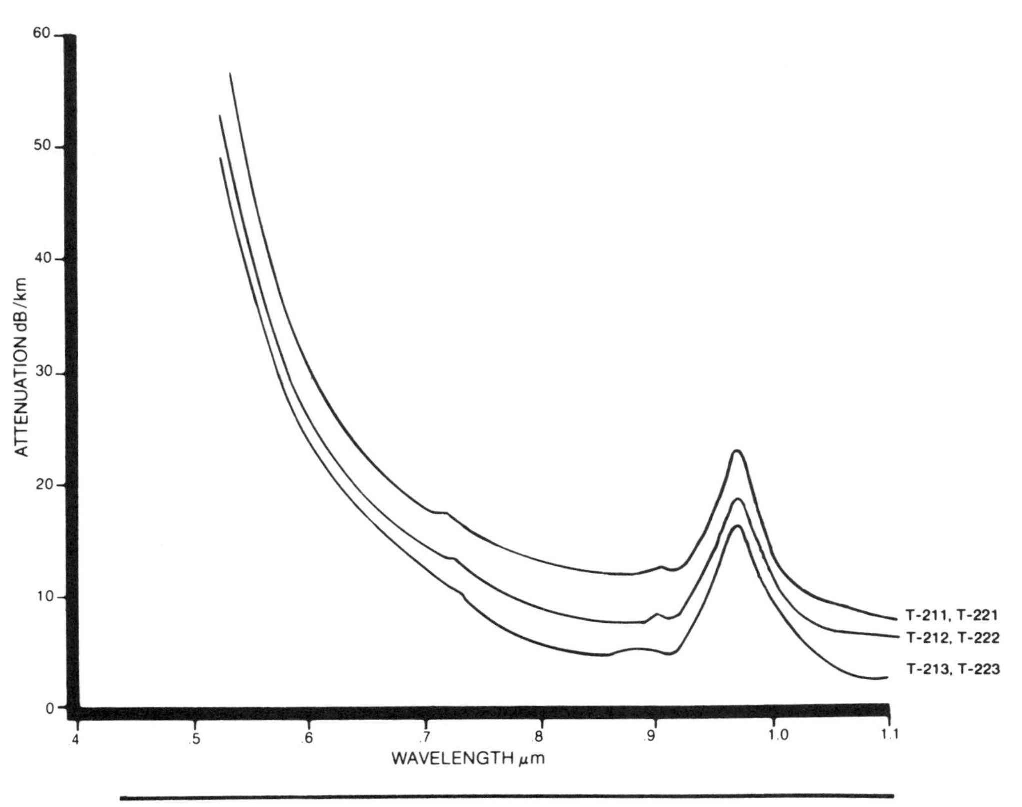

TYPICAL SPECTRAL ATTENUATION

Fiber Type	Attenuation @ (85 μm)	@(1.06 μm)
T-211, T-221	12 dB/km	8 dB/km
T-212, T-222	8 dB/km	5 dB/km
T-213, T-223	5 dB/km	3 dB/km
Numerical Aperture	.25	
Fiber Type	**Dispersion** (−3 dB Intermodal)	
T-211, T-212, T-213	2 ns/km	
T-221, T-222, T-223	1 ns/km	

Figure 7–2 Wavelength versus attenuation curve (*courtesy of Alcatel Cable Systems Group*).

in the fiber. Flaw-free fiber cores, claddings, and surfaces are the ultimate attempt in the fabrication of excellent fibers.

It must be noted that these six parameters (numerical aperture, dispersion, optical attenuation, bandwidth and frequency curves, rise time, and fiber strength) are necessary requirements. Additional considerations during design may be size, shape, and bend radius, among others.

FIBER MANUFACTURE

Material Involved in Fiber Manufacture

Most fibers are made from fused mixtures of metal oxides, sulfides, or selinides. The optically transparent mixtures used in fibers are called *oxide glasses.* The oxide glasses are common silica (SiO_2); lead silicate, such as crystal; and sodium calcium silicate, such as plate glass. Materials involved in the process would be sodium carbonate, calcium carbonate, boric oxide silica (sand), and lead oxide. The chemical mixtures are extremely complex and must satisfy all the requirements of attenuation, strength, and so on, that were just discussed.

General Techniques

The manufacturing techniques involved in the production of optic fibers are highly sophisticated processes. Most of these processes are secret or at the least proprietary. Considering the competitive nature of the product this is understandable. There are three general techniques used in the fabrication process. The first to consider is to melt glass in one or two containers and pour the glass together to form the fiber. The second is create a glass core or cladding, then melt a second glass to liquid or glass and deposit the second as a core or cladding on the original. The third is to begin with an exceptional rod of glass and coat it with a plastic cladding. The two processes that we shall discuss are the double-crucible and the chemical-vapor-deposition (CVD) processes.

Double-crucible method of fiber fabrication. In early processes, the rod-in-tube method was used (see Figure 7–3). This method exudes cladding glass over core

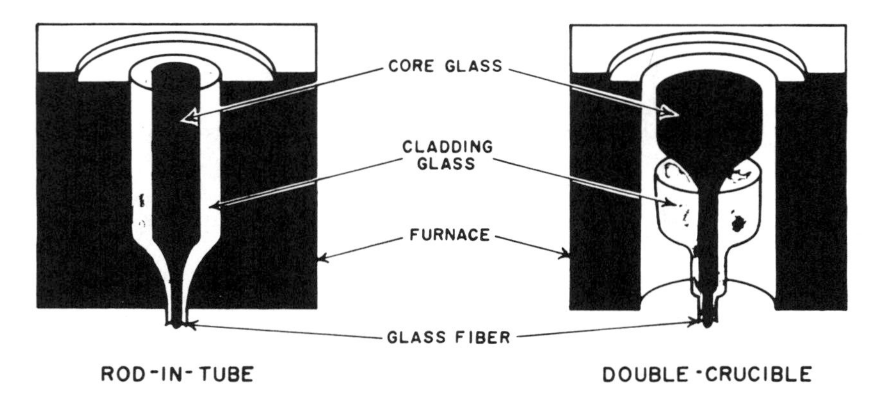

Figure 7–3 Double-crucible method of fiber fabrication (*courtesy of Corning Glass Works*).

glass. The double-crucible furnace is built so that two crucibles holding cladding glass and core glass are vertically aligned. The crucibles are made from platinum or other heat-resistant material. Simple silica crucibles may be used, but they have a tendency to shatter when cooling to room temperature. Each crucible has a narrow aperture at its base. As the cladding glass and core glass are heated, they flow symmetrically through the crucible apertures so that a composite glass flow is produced. The composite glass may be pulled to form a perfect or near-perfect fiber. The dimensions of the two crucibles are very exact, to ensure that core-to-cladding ratios are met. The viscosities and densities of the materials must be considered in flow restrictions, but usually are very similar in both cladding and core. The atmosphere around the crucibles is carefully controlled to prevent contamination and introduction of oxygen, which will cause bubbles in the glass and consequent high scatter losses.

The chemical-vapor-deposition (CVD) process. The CVD method was first used by Corning Glass Works of Corning, New York, when they developed the first 20-decibel-per-kilometer fiber. The technique was called the *soot process.* Figure 7–4 is a line drawing of two doped deposited silica (DDS) fabrication processes. On the left is an inside vapor-phase oxidation (IVPO) process. This process is used to create single-mode optical waveguides. The outer layer of glass is a fused silica tube used as the cladding. Gaseous components react in a flame to produce a fine deposit of doped or undoped silica. The soot deposition is internal in this case. It is deposited as a doped layer to form the core. The redraw process involves heating the glass by

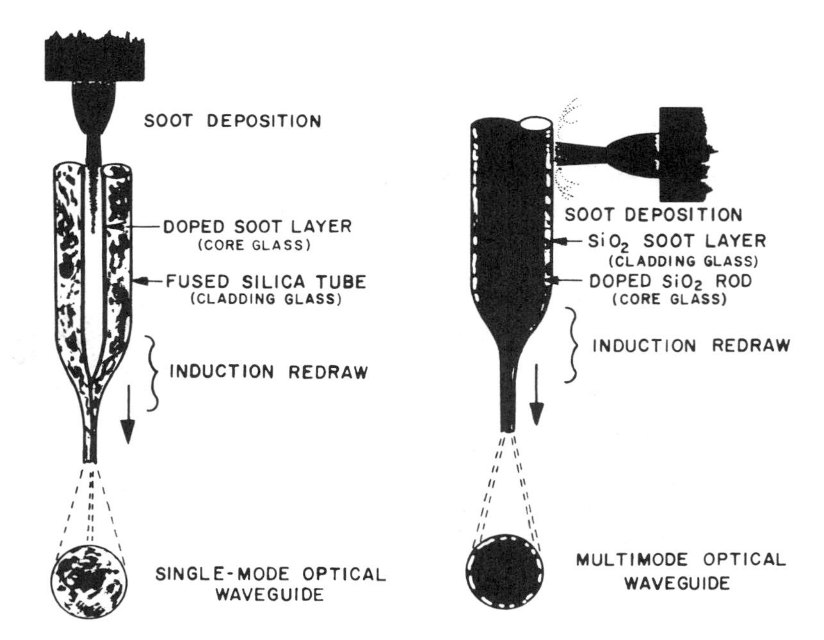

Figure 7–4 Inside (IVPO) and outside (OVPO) vapor-phase oxidation processes (*courtesy of Corning Glass Works*).

induction and drawing it to form the single mode optical waveguide. On the right in Figure 7–4, the outside vapor-phase oxidation (OVPO) process is illustrated. The inner layer provides the core glass. The outer soot layer (that was deposited), establishes the cladding. The redraw process involves heating the glass by induction and drawing it to form the multimode optical waveguide.

The doped-deposited-silica (DDS) process shown in Figure 7–5 allows the introduction of carefully controlled amounts of dopants to the silica tube. The dopants may be increased or decreased to raise or lower the refractive indexes of the host glass, which carefully defines the waveguide (fiber). This makes the DDS process uniquely suited for graded-index waveguide fabrication. Graded-index waveguides require a continuously varying, precisely controlled profile of the refractive index within the waveguide core.

A second class of methods leading from soot to chemical vapor deposition (CVD) produced a glassy layer rather than sootlike layer on the surface of a silica tube or the inside of the tube.

The CVD process is by far the most widely utilized process in developing and manufacturing of waveguides. The process begins with an extremely high-quality tube of quartz, such as that shown in Figure 7–6. The glass tube's length is near a meter and its diameter is about 20 millimeters, depending on the fiber being processed and the manufacturing technique. The tube is placed in a special lathe (see Figure 7–7) to create a preform which has the properties of the fiber. Heat is applied. A constant gas flow through the tube deposits boron-doped silica on the inner wall of the tube. The boron-doped silica has a lower refractive index than pure silica. This process

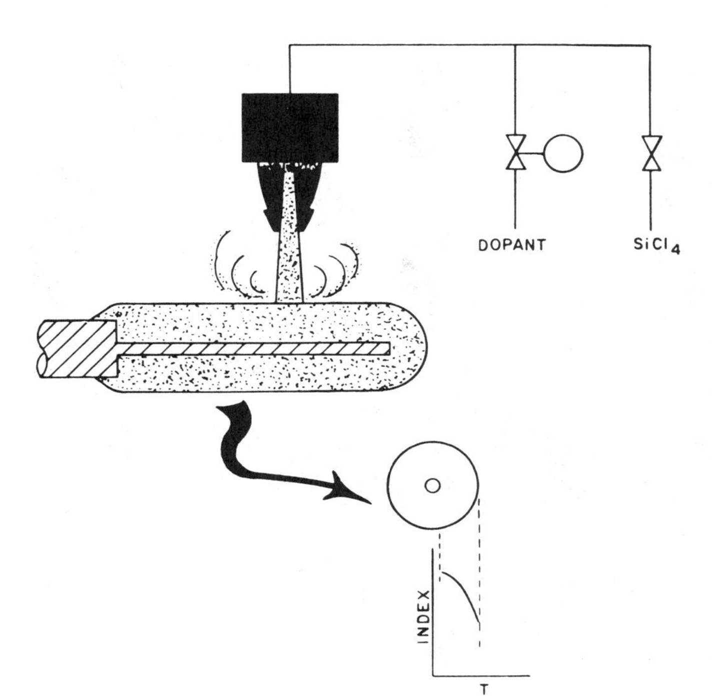

Figure 7–5 Dope-deposited silica (DDS) process (*courtesy of Corning Glass Works*).

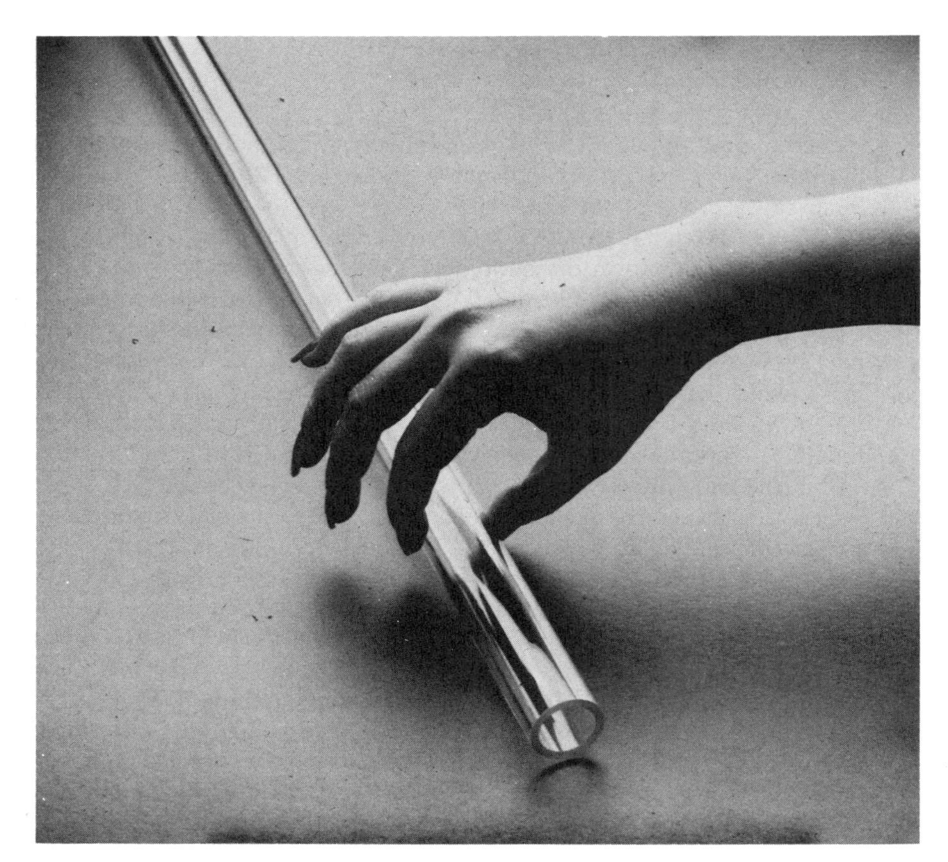

Figure 7–6 High-quality quartz tube (*courtesy of AT&T*).

Figure 7–7 Preform lathe (*courtesy of AT&T*).

provides a core material for the fiber. The borosilicate soot is uniformly deposited by the maintenance of a constant level of gas flow and a regulated flame along the tube. By repeating the process with varying amounts of dopant, graded or step index core regions may be built up. Germanium and phosphorus dopants are used to increase refractive indexes. Boron is used to lower refractive indexes.

Figure 7–8 is a quartz glass tube that has been through the deposition process. After a suitable number of layers of borosilicate has been deposited, the tube is collapsed to form the fiber preform (see Figure 7–9). The preform is then placed into a specially designed lightguide drawing machine where the quartz preform is heated

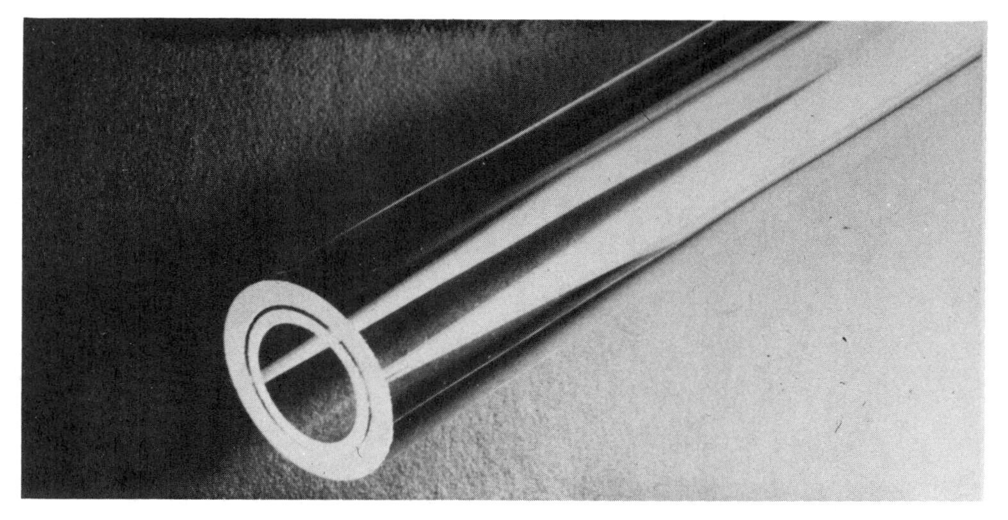

Figure 7–8 Quartz tube after deposition process (*courtesy of AT&T*).

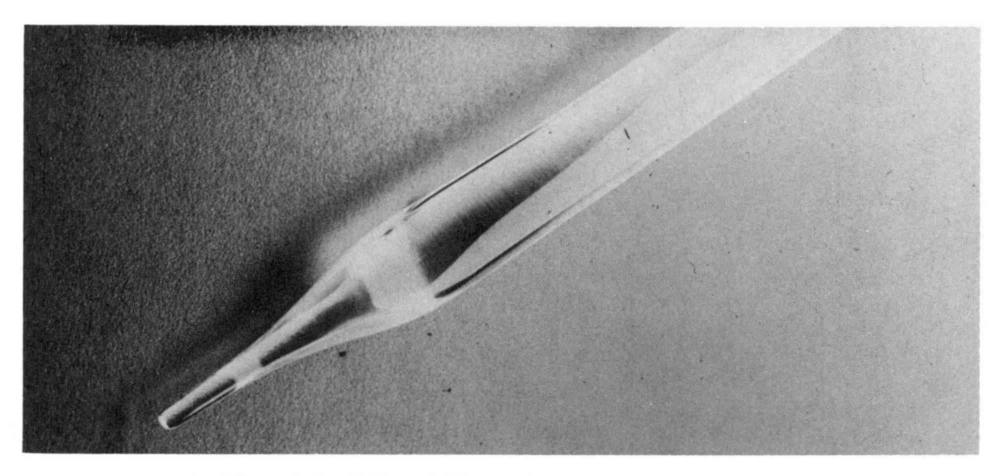

Figure 7–9 Collapsed fiber preform (*courtesy of AT&T*).

in an induction furnace or by CO_2 laser. In the machine, the preform is pulled into several kilometers of hair-thin lightguides. The lightguide drawing machine is a precision instrument. Figure 7–10, is a photograph of one of the drawing machines developed by Bell Laboratories and AT&T.

It must be noted that the fabrication processes described here are simplified, because the scope of this book is introductory. The mechanics of fiber fabrication are critical and complex. Glass melting, flow rate, diffusion methods, geometry of containers, cooling, crucible loading and feeding procedures, atmosphere control, pulling/drawing, and winding are just a few of the exacting details involved in the fabrication process.

Figure 7–10 Lightguide (fiber) drawing machine (*courtesy of AT&T*).

FIBER TYPES AND SPECIFICATIONS

Fibers are constructed to fill the need of a particular requirement. The major need is in the area of bandwidth, that is, low, medium, high, or ultrahigh. In addition, a fiber may be constructed for its extremely low attenuation per kilometer. In some cases tensile strength is the most desired characteristic. Two operating classes of fibers are utilized, single mode and multimode. Single-mode fibers are fibers that carry one mode. Single-mode fibers must be designed to accommodate a specific wavelength, otherwise large attenuation will result. An attractive feature of the single-mode fiber is that it is not sensitive to microbending. These are losses induced by local lateral displacements of the fiber. Multimode fibers allow intermodally dispersed wavelengths. That is, the propagation of rays of the same wavelength follow different paths through the fiber, causing different arrival times.

There are two types of optical fibers, step index and graded index. Cross-sectional views along with the typical light-ray path of these fiber types are illustrated in Figure 7–11. The step-index fiber has an abrupt refractive index between the core and the cladding. The graded-index fiber has a variation of refractive index between the core and the cladding. This minimizes dispersion and allows a greater bandwidth of information to be transmitted.

Fibers are constructed from plastic or glass or a combination of the two: glass

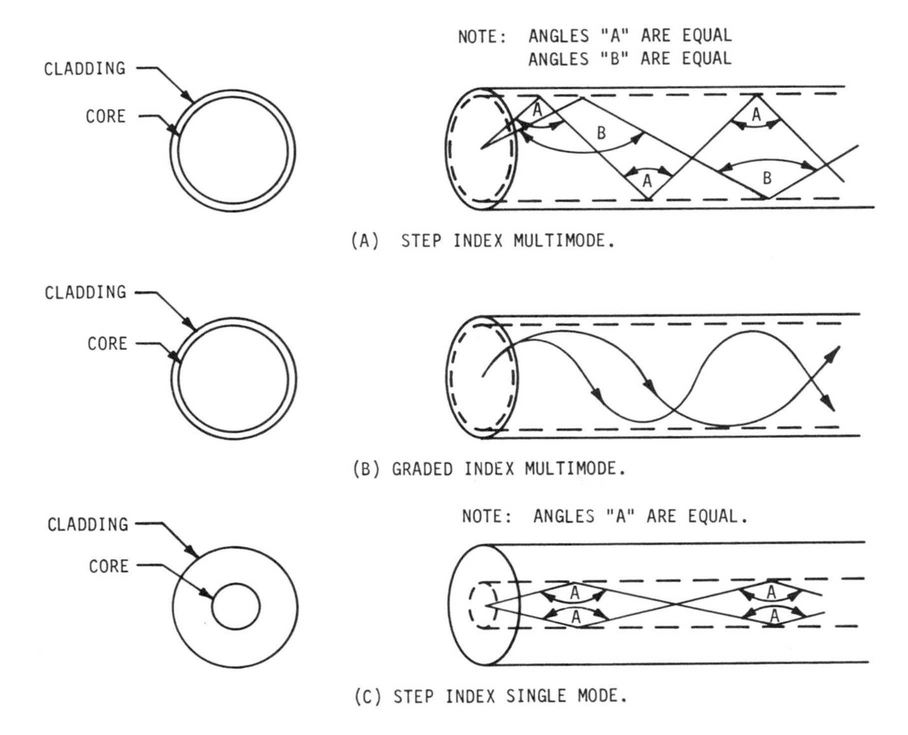

Figure 7–11 Fiber types (*Mike Anaya*).

core–glass cladding, plastic core–plastic cladding, or glass core–plastic cladding. The choice of a fiber type has to do with the quality, the parameters required, and economics. Can the system live with dispersion or must we use a better fiber at a higher price? In real life there are trade-offs in design. You cannot always receive lower cost without lower performance. A lower attenuation per kilometer may also mean a very high price tag.

In the next several paragraphs we shall take a look at several fiber types and their specifications. As you are comparing fiber types, consider what makes each more or less desirable than others.

Single-Mode Step-Index Optical Fiber

The single-mode step-index designated is designed for use in very high bandwidth (500 MHz or more), single-fiber data systems, or in other applications that require single-mode propagation of light. The fiber consists of a doped silica core and a borosilicate cladding with a step-index profile. The cladding is coated with silica. A pair of jackets (one plastic) is extruded onto the fiber to provide mechanical and environmental protection (see Figure 7–12).

SINGLE-MODE FIBER, STEP-INDEX SPECIFICATIONS

Specifications	Nominal
Attenuation @ 0.63 μm and 0.85 μm	4–20 dB/km
Numerical aperture (NA)	0.1
Fiber core diameter	4.5 μm
Fiber outer diameter	80 μm
Jacket outer diameter	500 μm
Minimum bend radius	0.5 cm

Courtesy of Alcatel Cable Systems Group.

Step-Index Multimode Fiber

This glass step-index fiber (see Figure 7–13) is designed for use in medium- to high-bandwidth single-fiber data transmission systems. The fiber consists of a doped silica core and a borosilicate cladding with a step-index profile. The cladding is coated with silica. A pair of jackets (one plastic) is extruded onto the fiber to provide mechanical and environmental protection.

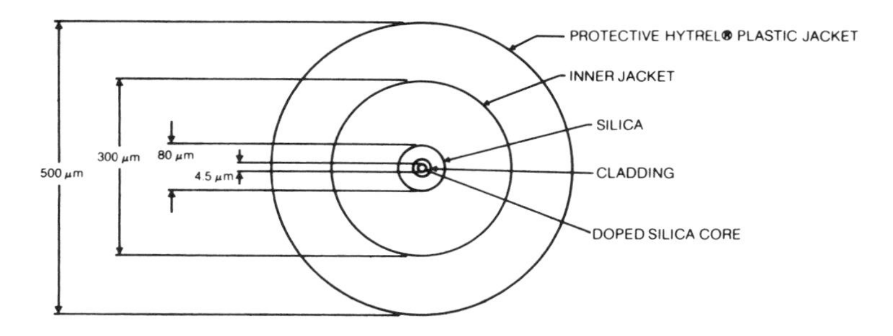

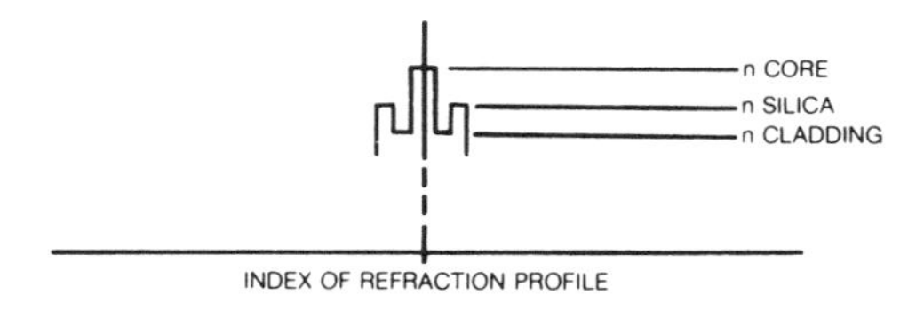

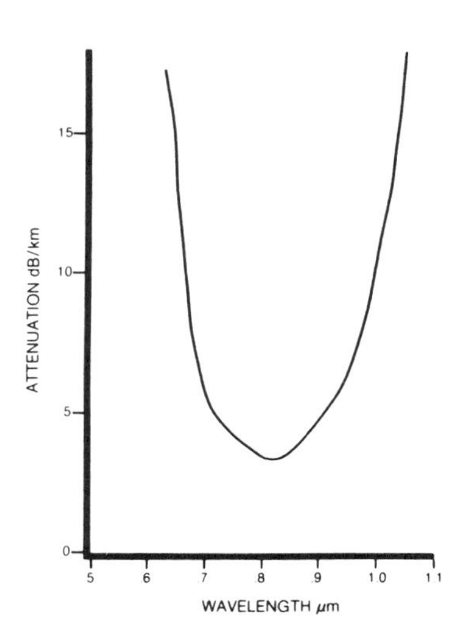

SAMPLE SPECTRAL ATTENUATION-SINGLE MODE STEP-INDEX OPTICAL FIBER

Figure 7–12 Single-mode, step-index optical fiber (*courtesy of Alcatel Cable Systems Group*).

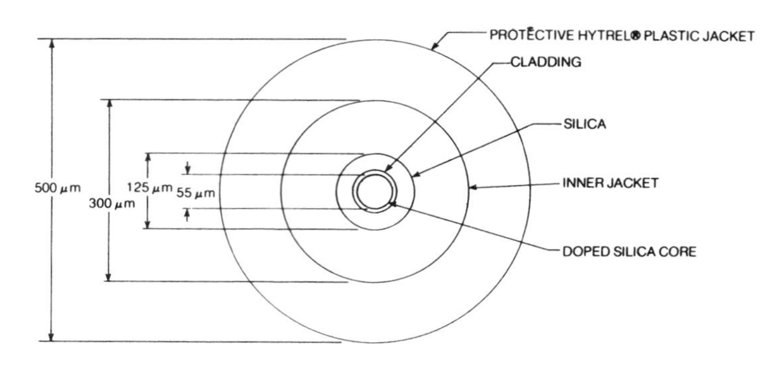

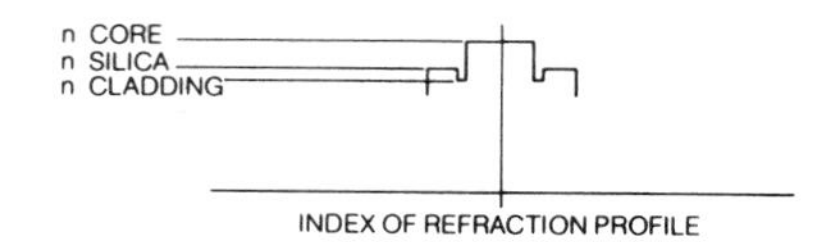

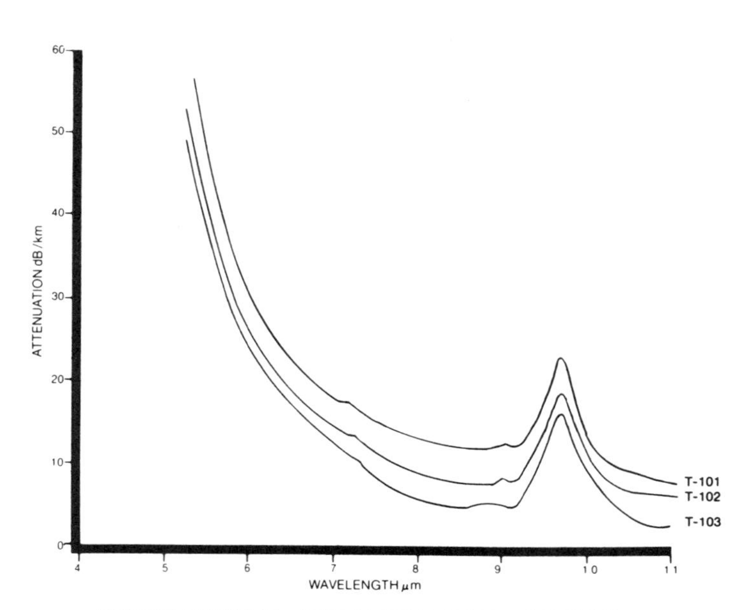

Figure 7–13 Step-index multimode fiber (*courtesy of Alcatel Cable Systems Group*).

MULTIMODE FIBER, STEP-INDEX SPECIFICATIONS

Specifications	Nominal
Attenuation	
@ 0.85 μm	12 dB/km — T101 As selected
	8 dB/km — T102 As selected
	5 dB/km — T103 As selected
@1.06 μm	8 dB/km — T101 As selected
	5 dB/km — T102 As selected
	3 dB/km — T103 As selected
Numerical aperture (NA)	0.25
— 3dB intermodal dispersion	15 ns/km
Core index of refraction	1.48
Fiber core diameter	55 μm
Fiber outer diameter	125 μm
Jacket outer diameter	500 μm
Tensile strength	5×10^5 psi
Minimum bend radius	0.5 cm

Courtesy of Alcatel Cable Systems Group.

Graded-Index Multimode Optical Fiber

This glass graded-index fiber is designed for use in high-bandwidth single-fiber data transmission systems. The fiber consists of a doped silica core with a graded index-of-refraction profile and a borosilicate cladding. The graded-index profile significantly reduces dispersion, thus permitting larger bandwidth signals to be transmitted. The cladding is coated with silica. A pair of jackets (one plastic) is extruded onto the fiber to provide mechanical and environmental protection. The fiber may be used singly or may be incorporated into multichannel fiber optic cables (see Figure 7–14).

GRADED-INDEX MULTIMODE FIBER SPECIFICATIONS

Specifications	Nominal
Attenuation	
@ 0.85 μm	12 dB/km — T201 As selected
	8 dB/km — T202 As selected
	5 dB/km — T203 As selected
@ 1.06 μm	8 dB/km — T201 As selected
	5 dB/km — T202 As selected
	3 dB/km — T203 As selected
Numerical aperture (NA)	0.25
—3 dB intermodal dispersion	3.5 ns/km
Core index of refraction	1.48
Fiber core diameter	55 μm
Fiber outer diameter	125 μm
Jacket outer diameter	500 μm
Tensile strength	5×10^5 psi
Minimum bend radius	0.5 cm

Courtesy of Alcatel Cable Systems Group.

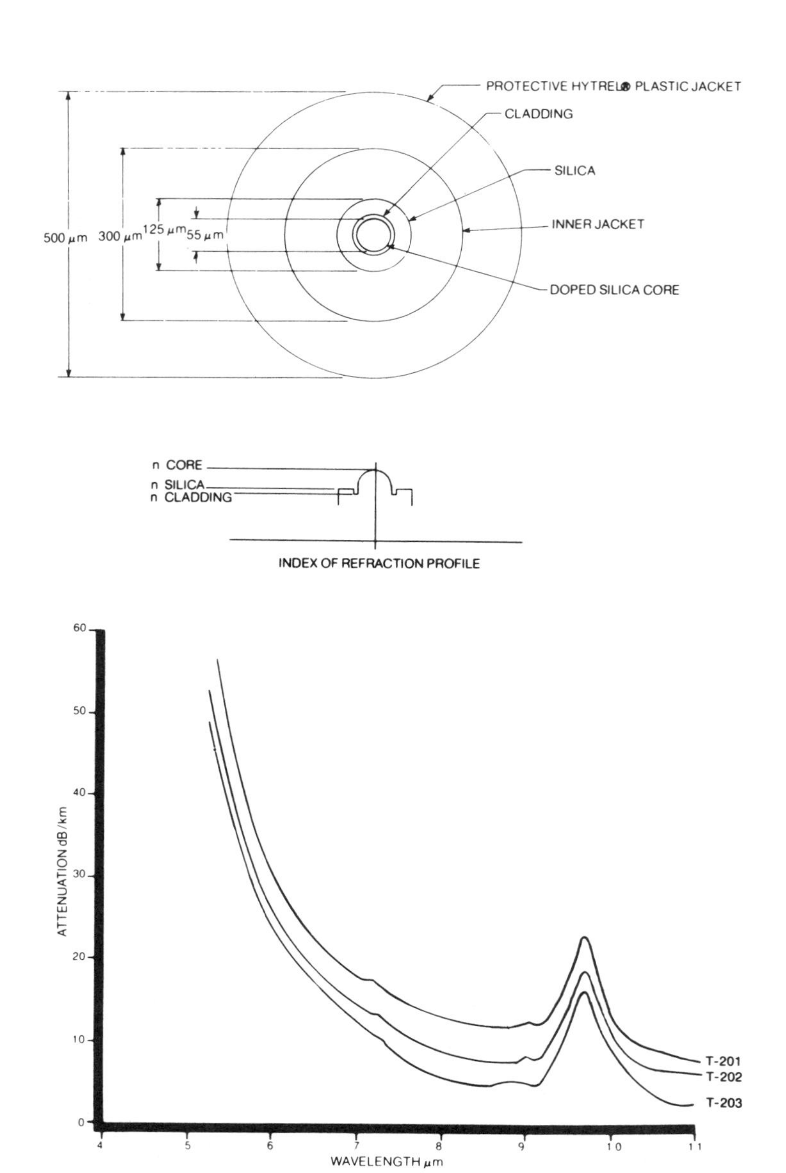

Figure 7–14 Graded-index, multimode optical fiber (*courtesy of Alcatel Cable Systems Group*).

Wideband Graded-Index Multimode Optical Fiber

Figure 7–15 shows a wideband glass graded-index fiber that is designed for use in applications demanding the ultimate in channel capacity that present fiber optic technology can provide. This ultrahigh-bandwidth fiber consists of a silica core doped

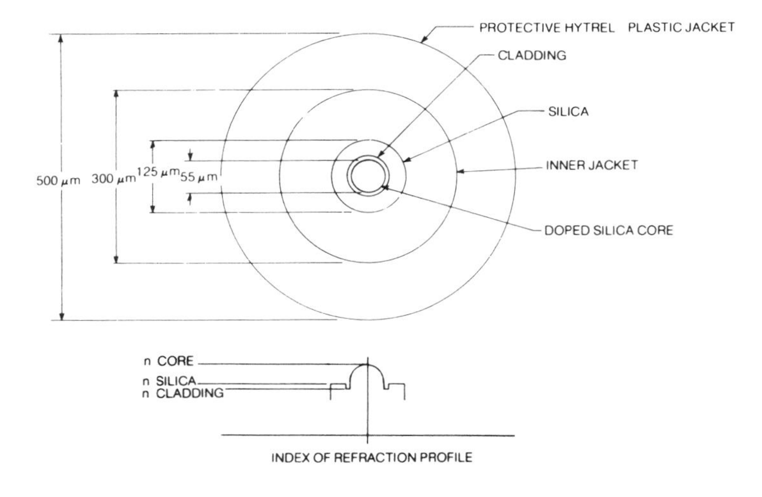

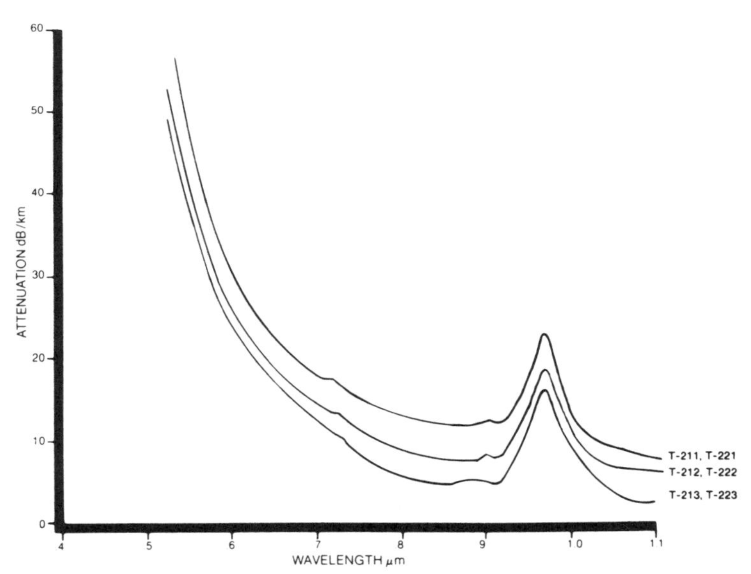

TYPICAL SPECTRAL ATTENUATION — WIDEBAND GRADED INDEX OPTICAL FIBER

Figure 7–15 Wideband graded-index, multimode optical fiber (*courtesy of Alcatel Cable Systems Group*).

to produce a graded refractive index profile, and a borosilicate cladding. The cladding is coated with silica. A pair of jackets (one plastic) provides mechanical and environmental protection. Bandwidths of 400 MHz can be achieved over 1-kilometer lengths of this fiber with narrow spectral-width sources. The fiber may be used singly or incorporated into multichannel fiber optic cables.

WIDEBAND, GRADED-INDEX MULTIMODE FIBER SPECIFICATIONS

Specifications	Nominal
Attenuation	
@ 0.85 μm	12 dB/km — T211, T221 As selected
	8 dB/km — T212, T222 As selected
	5 dB/km — T213, T223 As selected
@ 1.06 μm	8 dB/km — T211, T221 As selected
	5 dB/km — T212, T222 As selected
	3 dB/km — T213, T223 As selected
Numerical aperture (NA)	0.25
—3 dB intermodal dispersion	2 ns/km — T211, T212, T213
	1 ns/km — T221, T222, T223
Core index of refraction	1.48
Fiber core diameter	55 μm
Fiber outer diameter	125 μm
Jacket outer diameter	500 μm
Tensile strength	5×10^5 psi
Minimum bend radius	0.5 cm

Courtesy of Alcatel Cable Systems Group.

Large-Core Plastic-Clad Silica Multimode Optical Fiber

Large-core plastic-clad silica fiber is designed for use in economical, moderate distance, medium-bandwidth, single-fiber data transmission systems. The 200-micrometer core diameter offers greater source coupling efficiencies, which are useful when higher power levels are required at the receiver than could be provided by 125-micrometer-core plastic-clad silica fibers. The fiber consists of a homogeneous pure silica core surrounded by a plastic optical cladding and a plastic protective jacket. The core and plastic cladding form the optical waveguide, while the outer coating provides mechanical and environmental protection. The fiber may be used singly or incorporated into multichannel fiber optic cables (see Figure 7–16).

LARGE-CORE, PLASTIC-CLAD SILICA OPTICAL FIBER SPECIFICATIONS

Specifications	Nominal
Attenuation @ 0.79 μm	35 db/km — T321
	20 dB/km — T322
	10 dB/km — T323
Numerical aperture (NA)	0.3
—3 dB intermodal dispersion	30 ns/km
Core index of refraction	1.46
Fiber core diameter	200 μm
Jacket outer diameter	500 μm
Tensile strength	5×10^5 psi
Minimum bend radius	0.8 cm

Courtesy of Alcatel Cable Systems Group.

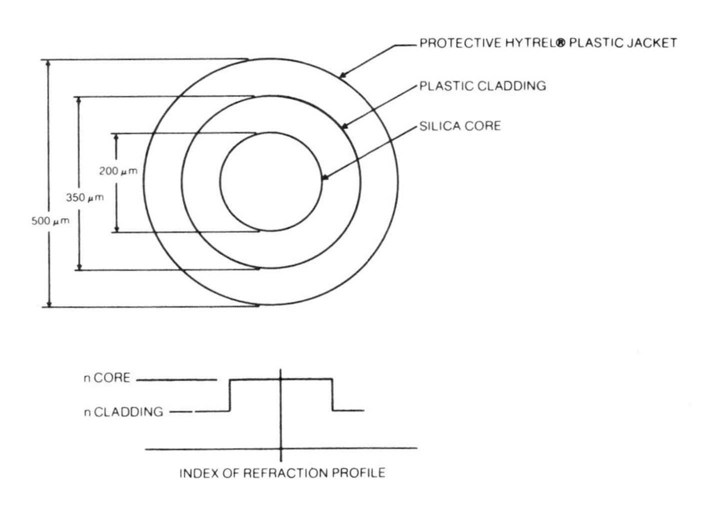

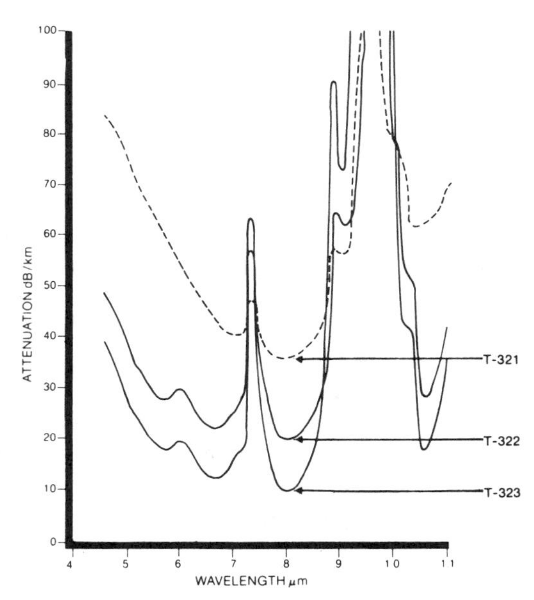

Figure 7-16 Large-core, plastic-clad silica optical fiber (*courtesy of Alcatel Cable Systems Group*).

Plastic-Clad Silica Multimode Optical Fiber

Plastic-clad silica fiber is designed for use in economical, medium-bandwidth, medium-distance, single-fiber data transmission systems. The fiber consists of a homogeneous pure silica core surrounded by a plastic optical cladding and a plastic protective jacket. The core and the plastic cladding form the optical waveguide, while the outer coating provides mechanical and environmental protection. The fiber may be used singly or incorporated into multichannel fiber optic cables (see Figure 7–17).

PLASTIC-CLAD SILICA MULTIMODE OPTICAL FIBER SPECIFICATIONS

Specifications	Nominal
Attenuation @ 0.79 μm	35 db/km — T301 20 dB/km — T302 10 dB/km — T303
Numerical aperture (NA)	0.3
—3 dB intermodal dispersion	30 ns/km
Core index of refraction	1.46
Fiber core diameter	125 μm
Jacket outer diameter	500 μm
Tensile strength	5×10^5 psi
Minimum bend radius	0.5 cm

Courtesy of Alcatel Cable Systems Group.

CABLE TYPES AND SPECIFICATIONS

Cables of single or multiple fibers are usually manufactured for protection of the fiber. Fibers are encased within the cable using protective jackets (usually polyurethane). Strength members surround the fibers. Cables are constructed to feature light weight, high flexibility, resistance to kinks, strength, and resistance to crushing. Some tests imposed on cables in the development stage are impact, bend, twist, fatigue under load, high and low temperature, and even storage losses.

As in the choice of fibers, certain trade-offs are often necessary. However, the reasons for cable development are centered around strength and environmental conditions. These should not be traded off in the search for economics. The next several paragraphs contain information on cable types. The reader should compare each type and determine what the benefits are for using a particular cable.

Light-Duty Optical-Fiber Cable

The light-duty optical-fiber cable (see Figure 7–18) is designed for use in single-fiber-per-channel transmission systems where a high-strength, rugged cable is not required. The standard cable contains up to 19 optical fibers. The optical fibers are encased in an extruded polyurethane jacket resulting in a small, flexible, lightweight cable. The cable is available with either plastic-clad silica, glass step-index, or glass graded-index fibers.

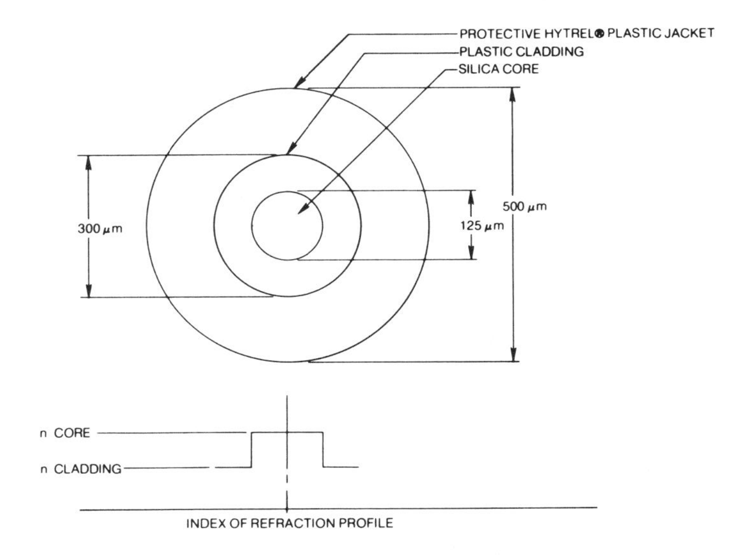

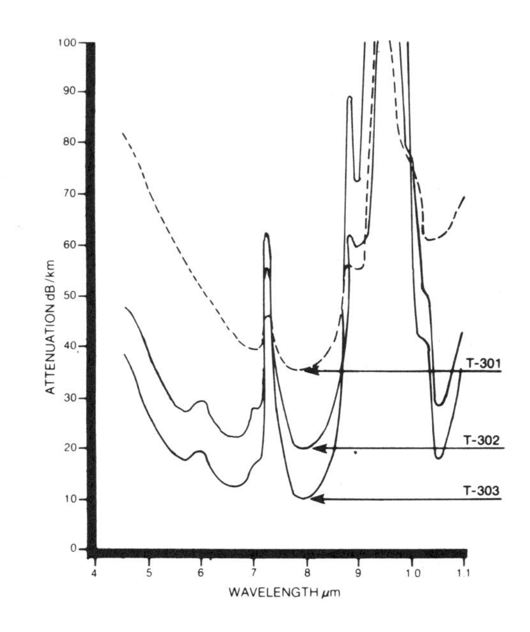

TYPICAL SPECTRAL ATTENUATION - PLASTIC CLAD SILICA FIBER

Figure 7–17 Plastic-clad silica optical fiber (*courtesy of Alcatel Cable Systems Group*).

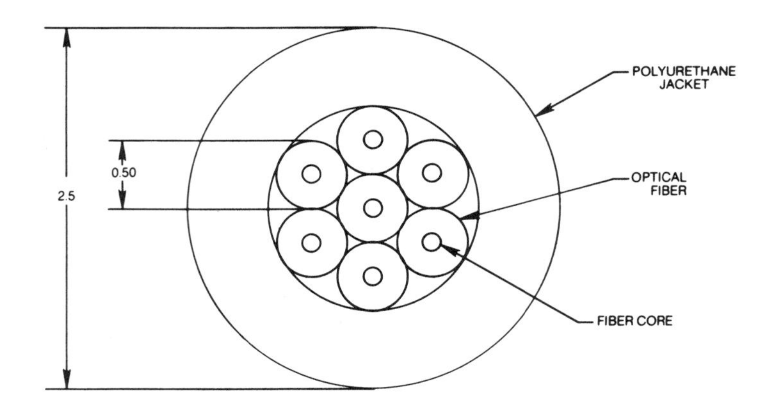

All dimensions in millimeters

*Other Configurations Available Upon Request.

OPTICAL SPECIFICATIONS (NOMINAL)*

CABLE TYPE+ NUMBER**	OPTICAL FIBER DESCRIPTION	ATTENUATION (dB/km)	INTERMODAL DISPERSION (—3dB) (ns/km)	FIBER TYPE
		(.85 μm)		
T-461-XX	Glass Step	15		T-101
T-462-XX	Index Fiber	10	15	T-102
T-463-XX	(55 μm Core)	6		T-103
		(.85 μm)		
T-471-XX	Glass Graded	15		T-201
T-472-XX	Index Fiber	10	3.5	T-202
T-473-XX	(55 μm Core)	6		T-203
		(.79 μm)		
T-561-XX	Plastic Clad	35		T-301
T-562-XX	Silica Fiber	20	30	T-302
T-563-XX	(125 μm Core)	10		T-303
		(.79 μm)		
T-571-XX	Plastic Clad	35		T-321
T-572-XX	Silica Fiber	20	30	T-322
T-573-XX	(200 μm Core)	10		T-323

*Better performance available on a custom order basis.
**The last two type number digits indicate the number of active fibers in the cable.
+Formerly designated Type LD.

Figure 7–18 Light-duty optical-fiber cable (*courtesy of Alcatel Cable Systems Group*).

LIGHT-DUTY OPTICAL-FIBER CABLE SPECIFICATIONS

Specifications	Nominal
Number of fibers	Up to 19
Jacketed fiber diameter	500 μm
Cable diameter	2.5 mm
Weight	6.4 kg/km
Minimum bend radius	2.5 cm

Courtesy of Alcatel Cable Systems Group.

Single-Fiber Strengthened Optical-Fiber Cable

The single-fiber strengthened optical-fiber cable (see Figure 7–19) is a small, strong, flexible, lightweight cable design, suitable for many general-purpose single-channel fiber optic links, such as between buildings, vehicles, or computers. The cable is suitable for installation in conduit, cable ducts, cabling trays, and for limited field use. The cable consists of a single optical fiber surrounded by six strength members and is encapsulated in an extruded polyurethane jacket. The cable is available with either plastic-clad silica, glass step-index, or glass graded-index fibers.

SINGLE-FIBER STRENGTHENED OPTICAL-FIBER CABLE SPECIFICATIONS

Specifications	Nominal
Jacketed fiber diameter	500 μm
Cable diameter	2.5 mm
Weight	6 kg/km
Tensile strength	45 kgf
Minimum bend radius	2.5 cm

Courtesy of Alcatel Cable Systems Group.

Two-Fiber Strengthened Optical Cable

The two-fiber strengthened optical cable (see Figure 7–20) is designed for dual or duplex fiber-optic links within or between buildings, vehicles, or computers. It features small size, light weight, high strength, and flexibility. The cable is suitable for installation in conduit, cabling ducts, or cabling trays, and for limited outdoor use. The cable consists of two single-fiber cables, each composed of a central optical fiber. Each fiber is surrounded by six strength members and an extruded jacket. The cables are encapsulated in a black extruded polyurethane jacket. The cable is available with either plastic-clad silica, glass step-index, or glass graded-index fibers.

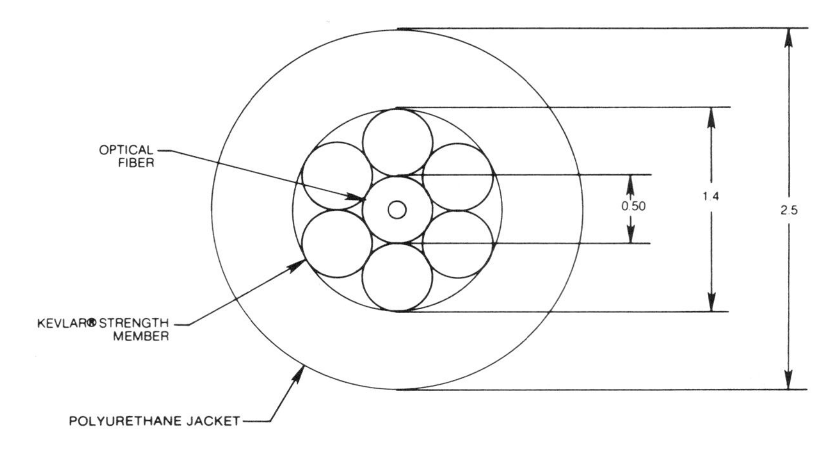

All dimensions in millimeters

OPTICAL SPECIFICATIONS (NOMINAL)*

CABLE TYPE NUMBER+	OPTICAL FIBER DESCRIPTION	ATTENUATION (dB/km)	INTERMODAL DISPERSION (−3dB) (ns/km)	FIBER TYPE
		(.85 μm)		
T-421	Glass Step	15		T-101
T-422	Index Fiber	10	15	T-102
T-423	(55 μm Core)	6		T-103
		(.85 μm)		
T-431	Glass Graded	15		T-201
T-432	Index Fiber	10	3.5	T-202
T-433	(55 μm Core)	6		T-203
		(.79 μm)		
T-521	Plastic Clad	35		T-301
T-522	Silica Fiber	20	30	T-302
T-523	(125 μm Core)	10		T-303
		(.79 μm)		
T-531	Plastic Clad	35		T-321
T-532	Silica Fiber	20	30	T-322
T-533	(200 μm Core)	10		T-323

*Better performance available on a custom order basis.
+Formerly designated S1.

Figure 7–19 Single-fiber strengthened optical-fiber cable (*courtesy of Alcatel Cable Systems Group*).

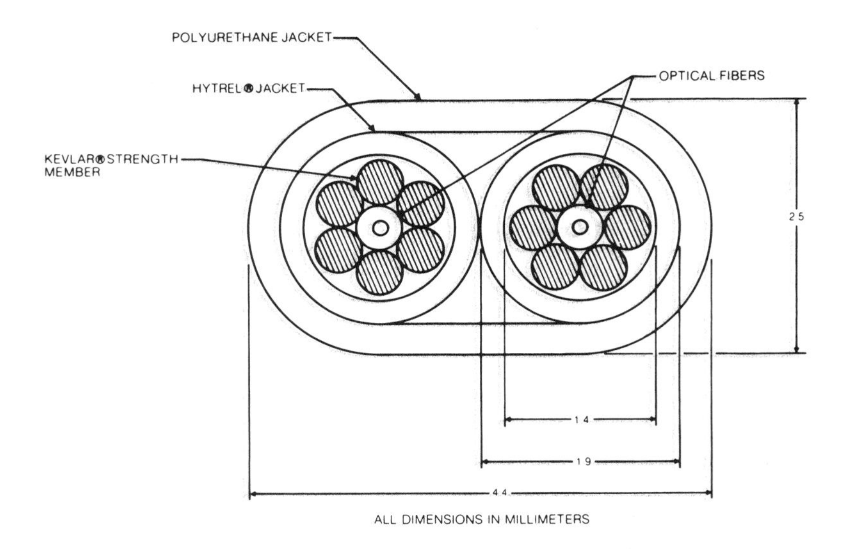

OPTICAL SPECIFICATIONS (NOMINAL)*

CABLE TYPE NUMBER	OPTICAL FIBER DESCRIPTION	ATTENUATION (dB/km)	INTERMODAL DISPERSION(—3dB) (ns/km)	FIBER TYPE
		(.85 μm)		
T-441	Glass Step	15		T-101
T-442	Index Fiber	10	15	T-102
T-443	(55 μm Core)	6		T-103
		(.85 μm)		
T-451	Glass Graded	15		T-201
T-452	Index Fiber	10	3.5	T-202
T-453	(55 μm Core)	6		T-203
		(.79 μm)		
T-541	Plastic Clad	35		T-301
T-542	Silica Fiber	20	30	T-302
T-543	(125 μm Core)	10		T-303
		(.79 μm)		
T-551	Plastic Clad	35	30	T-321
T-552	Silica Fiber	20		T-322
T-553	(200 μm Core)	10		T-323

*Better performance available on a custom order basis

Figure 7–20 Two-fiber strengthened optical cable (*courtesy of Alcatel Cable Systems Group*).

TWO-FIBER STRENGTHENED OPTICAL-FIBER CABLE SPECIFICATIONS

Specifications	Nominal
Jacketed fiber diameter	500 μm
Cable dimensions	4.4 mm × 2.5 mm
Weight	13 kg/km
Tensile strength	90 kgf
Minimum bend radius	3.5 cm

Courtesy of Alcatel Cable Systems Group.

External-Strength-Member Heavy-Duty Optical-Fiber Cable

The external-strength-member heavy-duty optical-fiber cable (see Figure 7–21) is designed for use in single-fiber-per-channel transmission systems that require high cable strength and crush resistance. In addition, its high flexibility and kink resistance make it ideally suited for use in conduit, cable trays, and a variety of intravehicle applications. The cable contains up to 19 optical fibers in an extruded polyurethane jacket surrounded by helically laid strength members and an outer extruded polyurethane jacket. The design provides ease of strength member and optical fiber termination, high strength, and abrasion resistance. The cable is available with either plastic-clad silica, glass step-index, or glass graded-index fibers.

EXTERNAL STRENGTH MEMBER, HEAVY-DUTY OPTICAL FIBER CABLE SPECIFICATIONS

Specifications	Nominal
Jacketed fiber diameter	500 μm
Cable diameter	6.0 mm
Weight	30 kg/km
Tensile strength	100 kgf
Minimum bend radius	5.0 cm

Courtesy of Alcatel Cable Systems Group.

TRANSMISSION ANGLES

Transmission and power are dependent on many specifications and parameters. Perhaps the most important of these are the transmission angles. There are three angles involved in a step-index fiber. These are the angle of acceptance, the angle of refraction, and the angle of reflection.

If you recall from Chapter 2, an incoming incidence ray, when reflected from

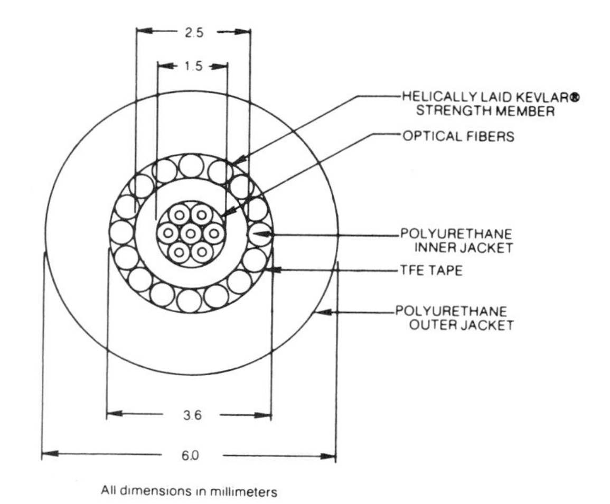

*Other Configurations Available Upon Request

OPTICAL SPECIFICATIONS (NOMINAL)*

CABLE TYPE+ NUMBER**	OPTICAL FIBER DESCRIPTION	ATTENUATION (dB/km)	INTERMODAL DISPERSION (—3dB) (ns/km)	FIBER TYPE
		(.85 μm)		
T-401-XX	Glass Step	15		T-101
T-402-XX	Index Fiber	10	15	T-102
T-403-XX	(55 μm Core)	6		T-103
		(.85 μm)		
T-411-XX	Glass Graded	15		T-201
T-412-XX	Index Fiber	10	3.5	T-202
T-413-XX	(55 μm Core)	6		T-203
		(.79 μm)		
T-501-XX	Plastic Clad	35		T-301
T-502-XX	Silica Fiber	20	30	T-302
T-503-XX	(125 μm Core)	10		T-303
		(.79 μm)		
T-511-XX	Plastic Clad	35		T-321
T-512-XX	Silica Fiber	20	30	T-322
T-513-XX	(200 μm Core)	10		T-323

*Better performance available on a custom order basis.
**The last two type number digits indicate the number of active fibers in the cable.
+Formerly designated Type ESM.

Figure 7–21 External strength member, heavy-duty optical fiber cable (*courtesy of Alcatel Cable Systems Group*).

a surface, reflects at the same angle as it strikes the surface. The angle of incidence equals the angle of reflection. The equal angles are also dependent upon the surface being smooth. The type of reflection is called *specular*. Light refraction occurs at the surface that separates two different media such as air and water or air and optical fiber. The index of refraction is a ratio of the speed of light in a vacuum to the speed of light in a specific material. Each material has an index of refraction. Indexes for common materials are listed in tables of refraction in physics books. In the case of fibers, the manufacturer can provide this information. The refraction index is important when considering the angle of the light source at the fiber end.

Snell's Law

$$n_a \sin \angle a = n_r \sin \angle r$$

where
n_a = index of refraction of air
n_r = index of refraction of fiber
$\angle a$ = angle of incidence
$\angle r$ = angle of refraction

The physics of refraction is explained in terms of Snell's law (see Figure 7–22). Rays intersecting a plane surface (fiber end) are refracted in the following relationship. The index of refraction of the air times the sine of the angle of incidence is equal to the index of refraction of the fiber times the sine of the angle of refraction. Now let's consider the angles related to optical fibers.

Cone of Acceptance

The cone of acceptance is the area in front of a fiber that determines the angle of light waves that will be accepted into the fiber (see Figure 7–23).

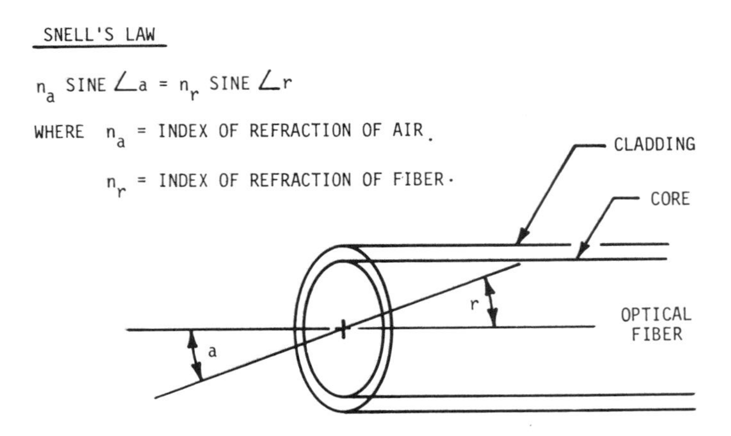

Figure 7–22 Snell's law (*Mike Anaya*).

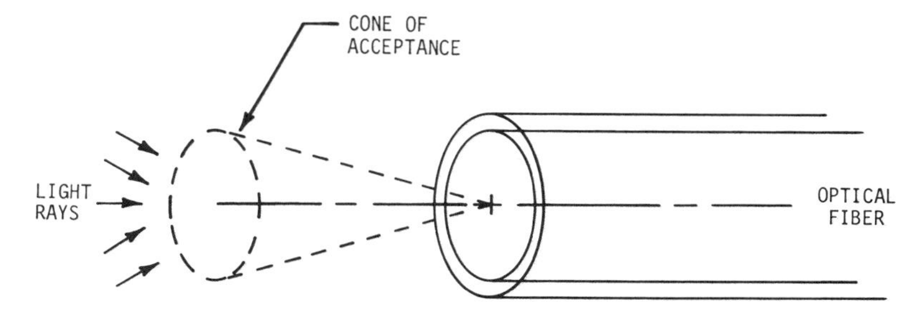

Figure 7–23 Cone of acceptance (*Mike Anaya*).

Angle of Acceptance

The angle of acceptance is an angle that represents the half-angle of the cone of acceptance (see Figure 7–24). The mathematical sine of the angle of acceptance is called the *numerical aperture* (NA).

$$\text{NA} = \sin \angle a$$

$\angle$a = ANGLE OF ACCEPTANCE.

SINE $\angle$a = NUMERICAL APERTURE (NA).

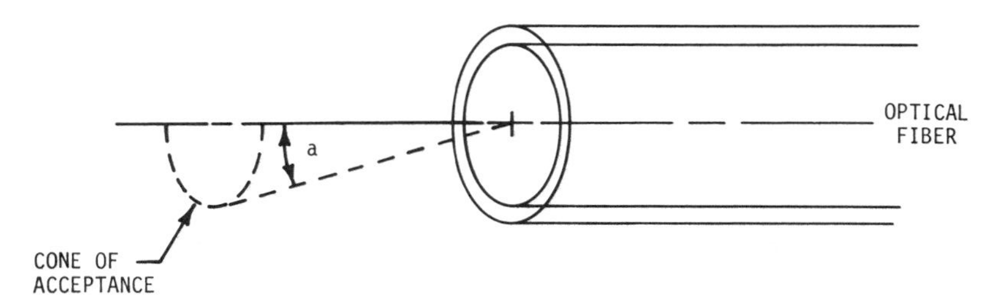

Figure 7–24 Angle of acceptance (*Mike Anaya*).

The reader will note that the angle a ($\angle a$) is usually designated $\angle \theta$ in engineering circles. It was felt, though, for explanation purposes, that we would use the first letter of the angle title. The numerical aperture (NA) can also be construed as the maximum acceptance angle which, in turn, defines the light-gathering power of the fiber. That is, the measure of the fibers ability to accept light waves and reflect them through the fiber.

Numerical aperture can be calculated by several means:

$$\text{NA} = \sin \angle \text{a}$$

or

$$\text{NA} = \sqrt{n_1^2 - n_2^2}$$

or

$$\text{NA} = (n_1^2 - n_2^2)^{\frac{1}{2}}$$

where

n_1 = index of refraction of the core

n_2 = index of refraction of the cladding

$\angle a$ = acceptance cone half angles

Angle of Refraction

The angle of refraction is the angle at which the rays are refracted from the cone of acceptance into the core (see Figure 7–25). Light waves are propagated from air at $\angle a$. These waves are refracted at the surface of the optic fiber into the core at $\angle r$. Therefore, the angle of incidence and the angle of refraction are related, using Snell's law. Calculations using Snell's law are as follows:

$$n_a \sin \angle a = n_r \sin \angle r$$

Therefore,

$$\sin \angle a = \frac{n_r}{n_a} \sin \angle r$$

Remember the numerical aperture NA = sin $\angle a$, which represents the maximum acceptance angle. Rays entering the fiber at a greater angle than $\angle a$ are lost and will not reflect. Rays entering the fiber at a smaller angle than $\angle a$ are totally reflected.

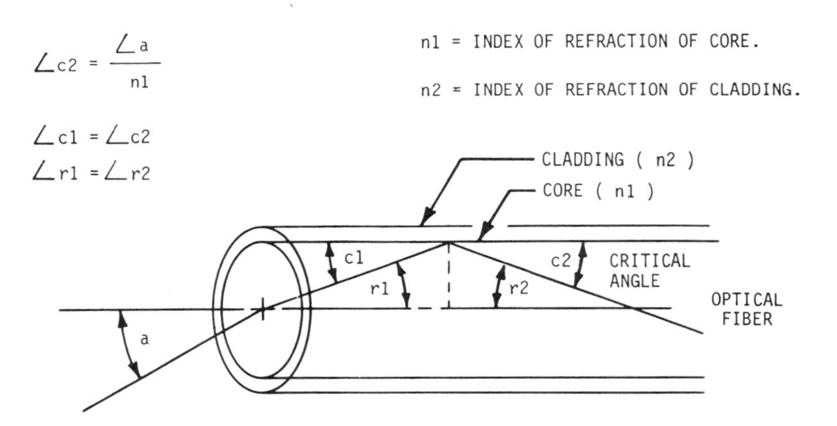

Figure 7–25 Transmission angles (*Mike Anaya*).

Angle of Reflection

Propagated waves at the angle of refraction ($\angle r$) strike the interface of the core and cladding. The angle of refraction at this point is actually an angle of incidence and the waves are reflected from the core–cladding interface at the same angle at which they strike the interface. In Figure 7–25, the angles r_1 and r_2 are equal and the angles c_1 and c_2 are also equal. The angles c_1 and c_2 are called internal circuit angles. Any mode that strikes the core–cladding interface at an angle greater than c will not be reflected but lost in the cladding. Modes that strike the interface at a smaller angle than c will be reflected through the fiber. The critical angles are also functions of the core and cladding indexes of refraction. The internal circuit angle c_2 is related to the fiber indexes of refraction in the following manner.

$$\angle c_2 = \frac{\arcsin \sqrt{n_1^2 - n_2^2}}{n_1}$$

Graded-Index Fiber Angles

A graded-index fiber is used to reduce dispersion within the fibers (see Figure 7–26). The numerical aperture (NA) is calculated in the same manner as with the step-index fiber. The graded-index core has an index of refraction that decreases with radial distance. The light waves travel faster near the cladding and slower near the core center. The phenomena allow the speed of all waves at all angles to be approximately the same. In actual practice there are three ray paths. Rays in the center of the fiber travel in high index of refraction (slow) material but travel the shortest path. Rays taking a sinusoidal path through the fiber cover a greater distance than the center rays, but travel in low-index-of-refraction (fast) material, part of the time. Finally, rays traveling in a helical path cover a greater distance, but travel in fast material (low index of refraction). This is an improvement over the step-index fiber in that larger angles in the step index take longer to reach the receiving end of the fiber.

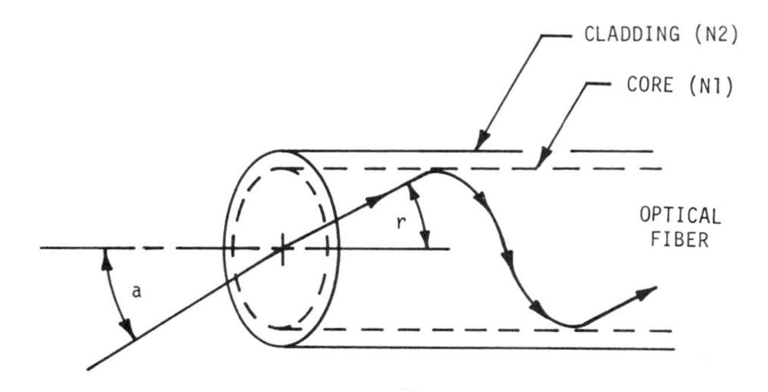

Figure 7–26 Graded-index fiber ray paths (*Mike Anaya*).

CAUSES OF COUPLING LOSS

Before we develop procedures defining alignment of optical fibers, it seems appropriate to discuss some of the causes of power loss in coupling. Power losses are rated in decibels (dB) and are additive regardless of their reason for existence. Some of these losses are as follows:

1. Heat is always a problem in dealing with any engineering effort. Optical fibers must be isolated with heat sinks tied to receptacles, mountings, connectors, and certain splices.
2. Fiber isolation, when in a cable or bundle, may cause some power loss. Sealing

off adjacent light sources is necessary. Opaque fiber coverings and sheathing are often added to protect against light sources and the environment.

3. If cable ends are allowed to vibrate, the vibration may actually cause modulation of the light rays.
4. Sand or dust on the critical mirror-finished fiber ends may act as an abrasive. Pitted finish may reflect light rays in a diffuse rather than a specular manner.
5. Water or humidity at the connection may actually improve the coupling. Water has an index of refraction much nearer the fiber core than air.

These are just some of the reasons for power loss. The most prevalent of coupling power loss is due to misalignment of the fiber. These are four major forms of fiber misalignment: core irregularity, lateral misalignment, gap loss, and angular loss. A description of each is provided in the next several paragraphs.

Core Irregularity

During manufacture, the core of the fiber may be placed off center within the fiber. In extremely long fiber lengths there may be short stretches of core that are off-center. In Figure 7–27, cores are shown off center. The amount is exaggerated for understanding.

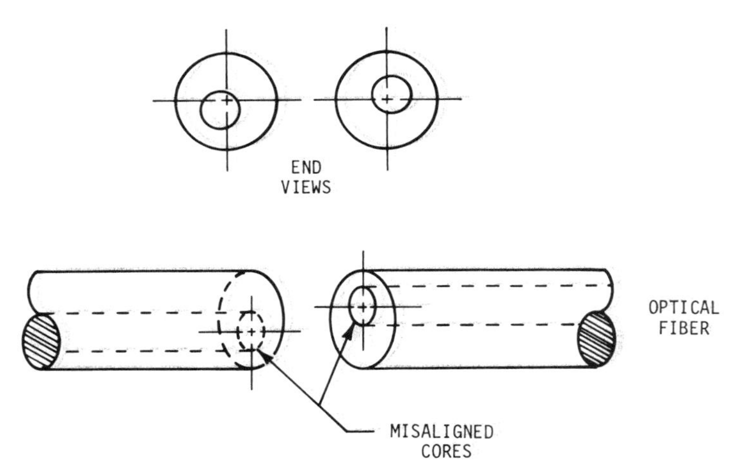

NOTE: EXAGGERATED CORE IRREGULARITY.

Figure 7–27 Core irregularity (*Mike Anaya*).

Lateral Misalignment

The centerlines of cores may be misaligned. Axial misalignment, as this is often called, is the lateral displacement between centerlines of the cores in the two mating fibers (see Figure 7–28). Power loss in decibels is plotted as an *L/D* ratio, where *L* is the lateral displacement and *D* is the core diameter. Centerlines should be between 0.0001 and 0.0002 inch as specified, depending on fiber diameter.

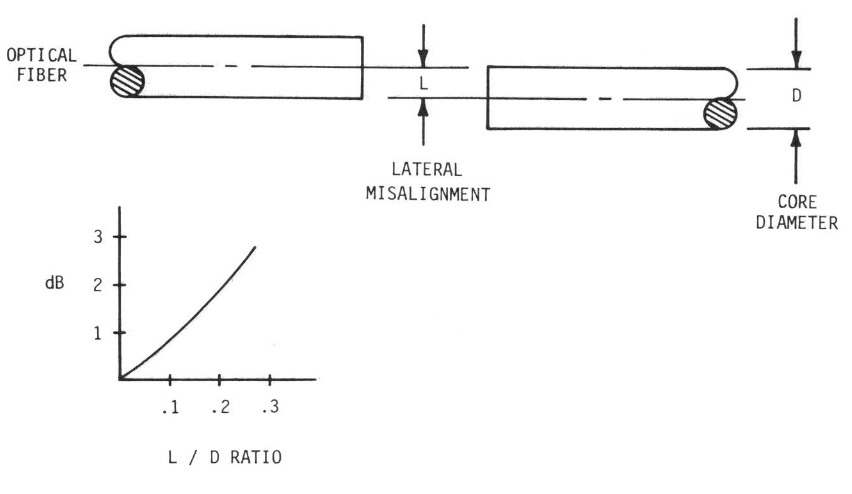

Figure 7–28 Lateral misalignment (*Mike Anaya*).

Gap Loss

Gap loss is the separation of the two mirror-finished fiber ends (see Figure 7–29). Gap loss is expressed in decibels and is determined by the NA of the fiber. Lower value NAs have less gap loss than higher NAs. Gap loss is plotted as a *G/D* ratio

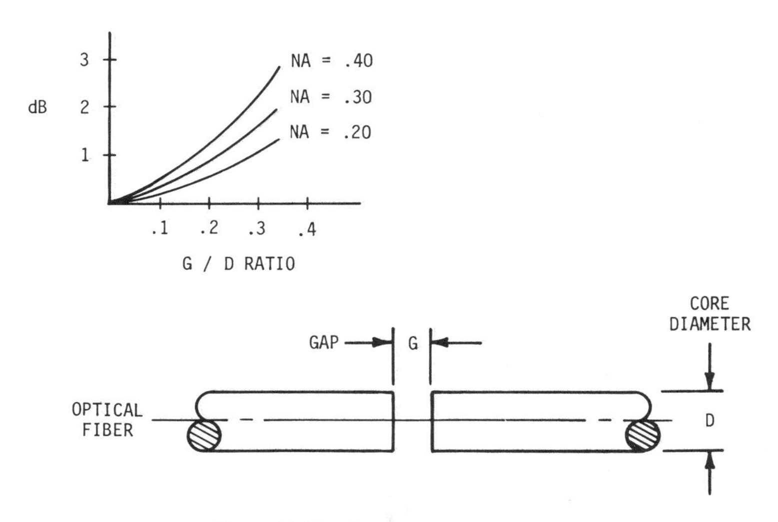

Figure 7–29 Gap loss (*Mike Anaya*).

against decibels where G is the gap and D is the core diameter. Typical allowable gaps are 0.0002 or 0.0003 or as specified.

Angular Loss

Angular loss is caused by cutting the fiber ends at an angle or by preparing the faces of the fibers so that they are not symmetric (see Figure 7–30). Again, the power loss is expressed in decibels. Angular losses are plotted as angles in degrees against decibels, where the angle is the shape between the two mirror fiber end faces. Typical allowable angle is less than 1/4°.

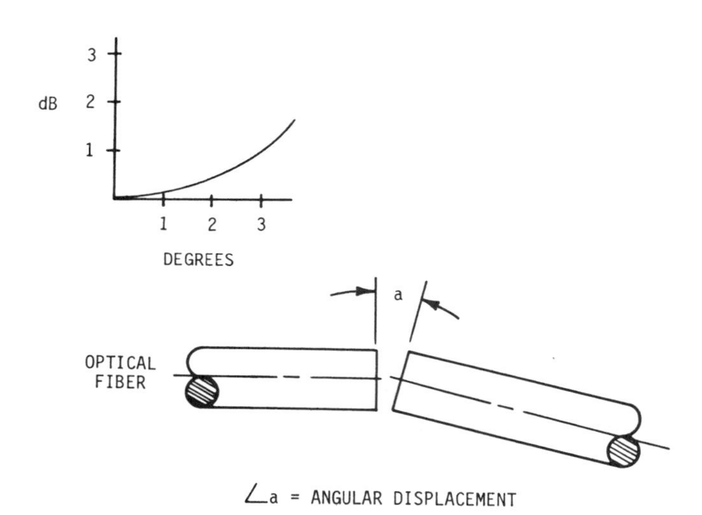

Figure 7–30 Angular loss (*Mike Anaya*).

Loss in Transmitting from a Large Fiber to a Smaller One

When two fibers of different diameters are mated together, a loss of optical energy will result. Calculation of this loss is made in the following manner:

$$\text{dB} = -10 \log \frac{D_r}{D_t}$$

where D_r = diameter of the receiving fiber
D_t = diameter of the transmitting fiber

Losses in Transmission Due to Changes in Numerical Aperture

When two fibers at different numerical aperture (NA) are joined together, the loss of optical energy may be calculated in the following manner:

$$dB = -10 \log \frac{NA_r}{NA_t}$$

where NA_r = numerical aperture of receiving fiber
NA_t = numerical aperture of transmitting fiber

Coupling Attenuation

Attenuation in coupling is established by two general criteria. These are intrinsic losses and extrinsic losses. Intrinsic losses are due to factors involved in the fiber and its manufacture. Intrinsic losses have nothing to do with the connector or its installation. Several producers of intrinsic loss are thus:

1. *Core diameter variations.* The core diameter variations are generally minute but can cause loss.
2. *Fresnel reflection.* Fresnel loss is due to the difference in the index of refraction of the air and the fiber at the fiber end.
3. *Transmission loss.* Transmission losses are due to the light waves lost in the cladding and not reflected from its wall.
4. *Absorption loss.* Absorption loss is due to nontransparent core material.

Extrinsic losses are usually attributed to the connector hardware and man-made misalignment. Connectors and the details of misalignment and/or alignment are covered under separate headings.

ALIGNMENT PHILOSOPHY

Optimum alignment must be made each time a connection is made or broken and remade. This alignment must be easily obtainable without the need of special equipment, special handling, or tuning of transmitter and receiver. An extensive fiber alignment philosophy was presented at an Electronic Components Conference, Anaheim, California, on April 24–26, 1978, by Terry Bowen of AMP, Incorporated, Harrisburg, Pennsylvania. This philosophy is generally recognized by the fiber optic industry. This philosophy is provided in the next several paragraphs. Copyright © 1978 by AMP, Incorporated, Harrisburg, Pennsylvania. All international rights reserved. Reproduced with the permission of the copyright holder.

Point of Contact Analysis

A primary alignment mechanism is formed when the common overlapping surface is in direct contact with the outer diameters of the fibers to be aligned (see Figure 7-31). As seen in Figure 7-32, optical fibers are generally round, formed of rigid material, and have outside diameters that vary in size as a result of manufacturing conditions. When brought into contact with a flat plane of rigid material, the axis of the fibers of different diameters will locate at different heights above the plane

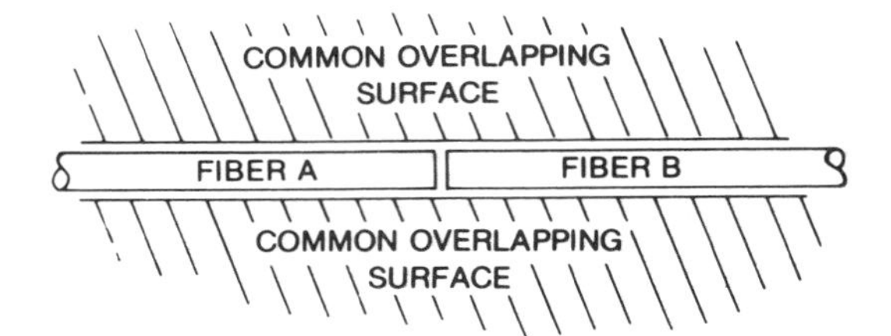

Figure 7–31 Primary alignment system (*courtesy of AMP, Incorporated*).

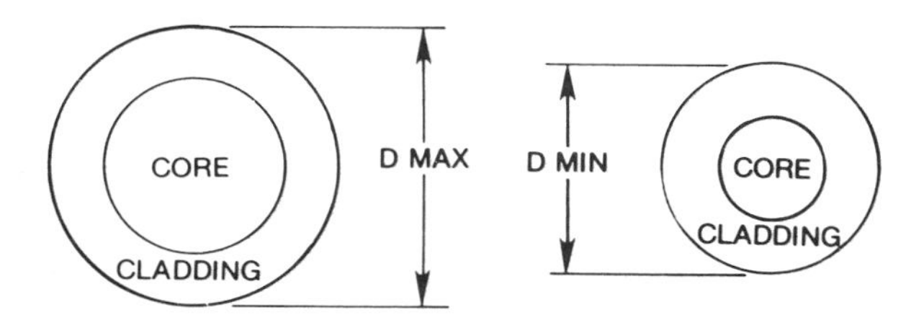

Figure 7–32 General shape of optical fibers (*courtesy of AMP, Incorporated*).

(see Figure 7–33). There is no confinement in the horizontal direction, so the fibers are free to roll about on the flat plane. Therefore, a flat plane does not provide a good alignment mechanism for optical fibers.

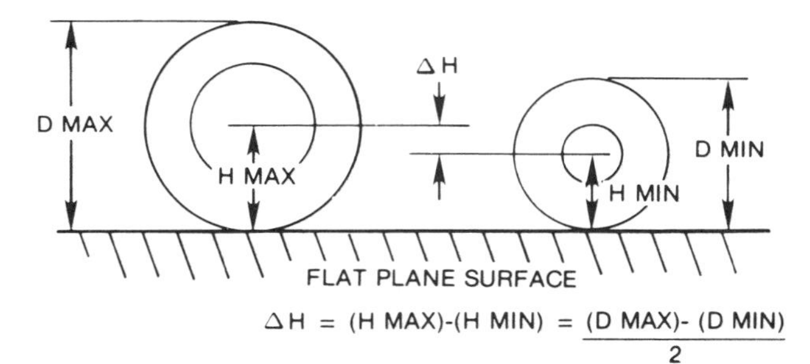

$$\Delta H = (H\ MAX)-(H\ MIN) = \frac{(D\ MAX)-(D\ MIN)}{2}$$

Figure 7–33 Core height misalignment (*courtesy of AMP, Incorporated*).

If two flat planes are made to intersect to form an included angle of 2θ, the resultant V-shape will prevent the round fibers from rolling (see Figure 7–34). By making line contact at two points in the fiber periphery, the V-shape places the axis of any size round fiber in a common vertical plane which bisects the V. One drawback continues: Different-diameter fibers will have their axes placed at different heights above the bottom of the V-shape.

Formulas for height (ΔH) variations and maximum/minimum diameters (D) are listed below.

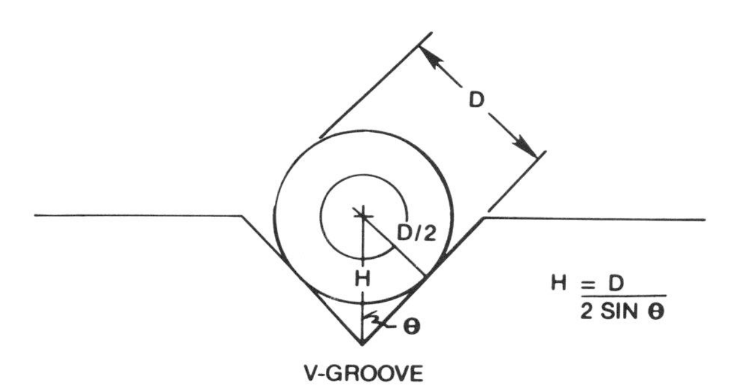

Figure 7–34 V-groove philosophy (*courtesy of AMP, Incorporated*).

$$H(\max) = \frac{D(\max)}{2 \sin \theta}$$

$$H(\min) = \frac{D(\min)}{2 \sin \theta}$$

$$\Delta H = H(\max) - H(\min) = \frac{D(\max) - D(\min)}{2 \sin \theta}$$

It can be seen that the sharper the V-shape (the smaller θ becomes), the greater the difference in the height of the fiber axis. When optical fibers have identical diameters, a V-shape is an excellent alignment mechanism.

Three lines of contact are established on the round fiber if a third plane is placed over a shallow V-groove as shown in Figure 7–35. In still another variation (see Figure 7–36), three round rods are placed in contact and the round fiber is fitted into the interstitial space formed.

Four lines of contact can be made with opposing shallow V-shapes (see Figure 7–37), or with four round alignment rods placed about the fiber. Attempts to form more than four lines of contact with circular-geometry tubes or sleeves produce, at best, a single-line contact effect with the fiber (see Figure 7–38). Here, the fiber outer diameter must be smaller than the inner diameter of the tube, so it will tend to contact the tube at one point only.

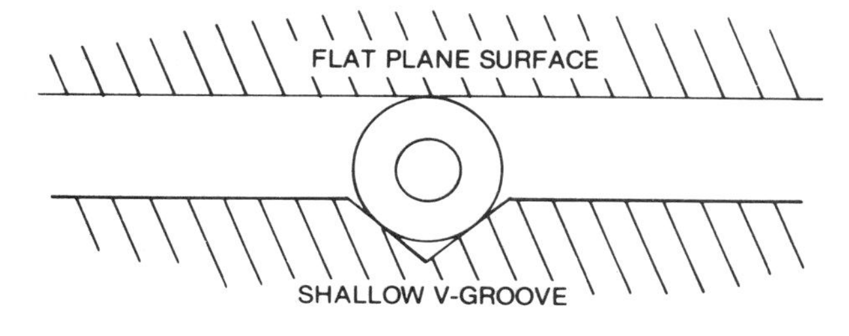

Figure 7–35 Three lines of contact (*courtesy of AMP, Incorporated*).

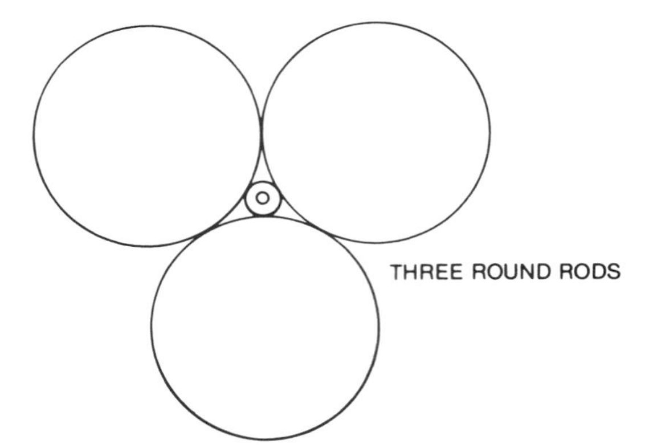

Figure 7–36 Round rod contact (*courtesy of AMP, Incorporated*).

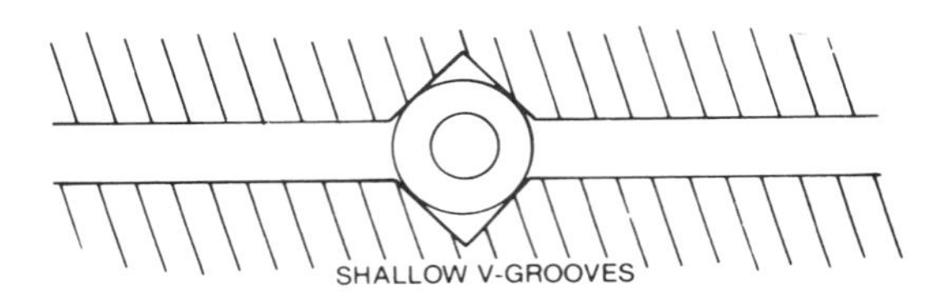

Figure 7–37 Four lines of contact (*courtesy of AMP, Incorporated*).

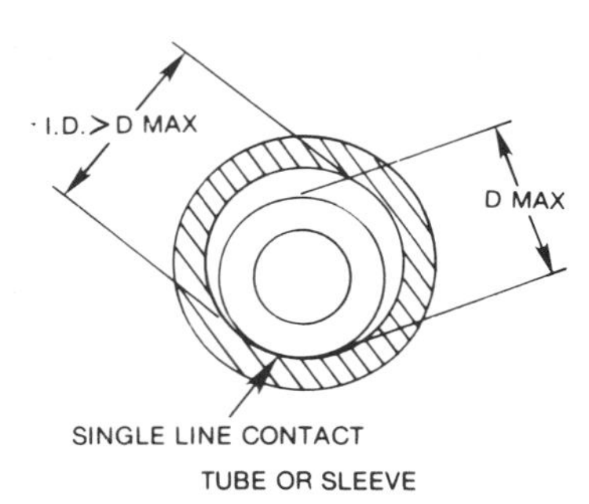

Figure 7–38 Circular tube single-line live contact (*courtesy of AMP, Incorporated*).

The common overlapping surfaces described are usually made of rigid material. A modified primary alignment mechanism can be made using resilient materials backed up with rigid materials (see Figure 7–39). The inclusion of nonrigid or resilient materials provides a higher order of connector-alignment mechanism that successfully joins different-sized fibers.

Ordinal Approach to Alignment Mechanism

Figure 7–40 permits us to discuss the advantages and disadvantages of primary, modified primary, secondary, and tertiary alignment mechanisms.

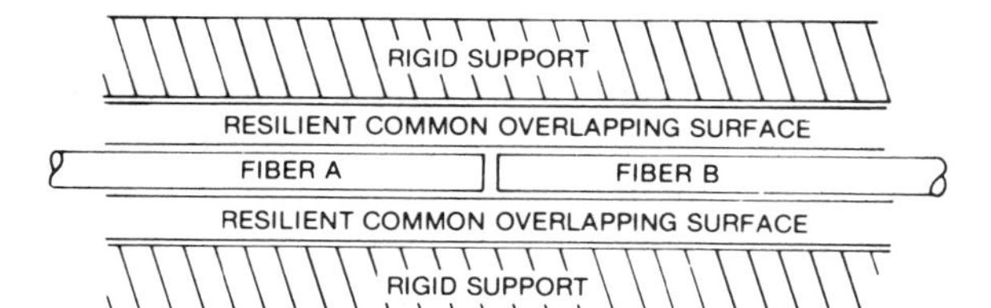

Figure 7–39 Modified primary alignment mechanism (*courtesy of AMP, Incorporated*).

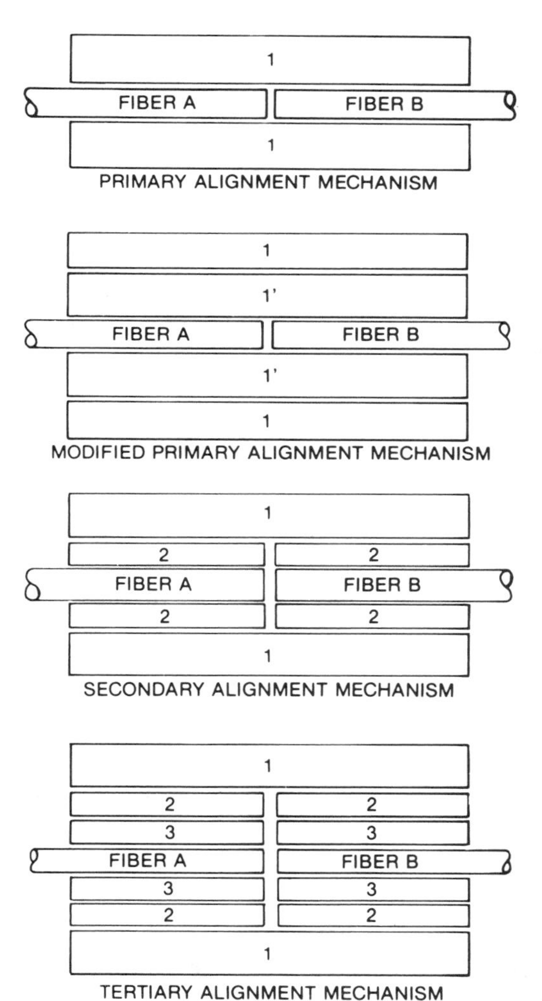

Figure 7–40 Ordinal approach to alignment mechanisms (*courtesy of AMP, Incorporated*).

Primary. A rigid common overlap surface (1) straddles the fibers to be joined. On the plus side is simplicity; only one interface is made and that is directly with the fiber. The only tolerance involved is that associated with the fiber diameter and the surface(s) of (1). End preparation is simple using the score-and-break technique. The disadvantages come into play if this mechanism is used for connectors that must undergo repeated mating. There is a high risk of damaging or degrading the exposed fiber that is required to be interacting with the hard overlapping surface(s) of (1). For this reason the primary mechanism is best suited to permanent splicing, where the fibers are inserted only once.

Modified primary. The optical fibers are first covered by a resilient common overlay surface (1′), then a rigid common overlay surface (1). This concept is still a simple one: Granted there are two interfaces; however, the 1-to-1′ interface is not critical. The key here is that the 1′-to-fiber A and B interface can absorb fiber diameter tolerances, and 1′ is less likely to damage the fibers than is 1. Also, the resilient member provides a sealing capability. There is one drawback: The fiber surfaces are still exposed and may be subject to degradation.

Secondary. The common rigid overlay surface (1) is still used; however, separate alignment components (2) are attached to fiber A and fiber B. These may be rigid components or resilient components. In either case, the connector can be more rugged and can provide better protection for the fiber.

With all rigid components, two rigid-to-rigid interfaces must be made, each having tolerances to be concerned with. Attempts are made to keep these tolerances to a minimum, which naturally increases manufacturing costs. A resilient component compensates for tolerance buildup just as in the modified primary alignment mechanisms, and is the subject of this section.

Tertiary. This class adds another component layer (3), which brings with it a third interface with tolerances to consider. The advantages and disadvantages of tertiary alignment mechanisms are simply extensions of those of the secondary approach.

Resilient Alignment Mechanism

The resilient alignment mechanism for a single optical fiber is a secondary alignment mechanism with the rigid fiber being protected by a resilient elastometric ferrule (see Figure 7–41).

Precise Hole Placement

The key to the success of the resilient alignment mechanism technique is the ultra-precise placement of this fiber centering hole—not in relationship to traditional alignment components, but actually in relationship to the true center that exists when the

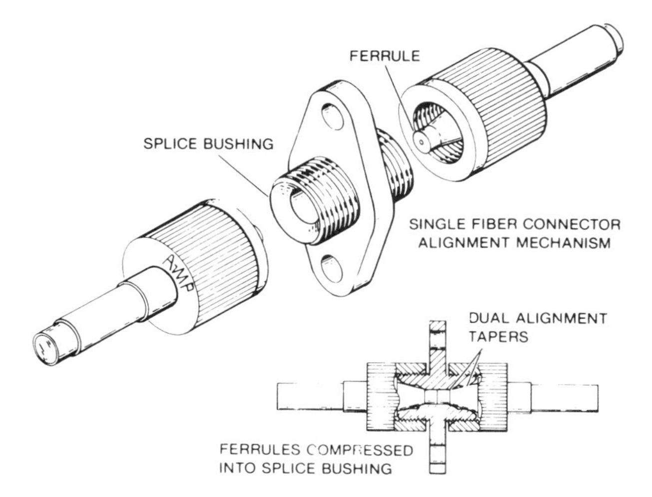

Figure 7–41 Single-fiber resilient alignment mechanism (*courtesy of AMP, Incorporated*).

connector is in service. A proprietary manufacturing technique assures that the hole is on true center regardless of the size of the fiber or the initial shape of the ferrule. Ferrules are molded with the front end initially blank, and the small fiber-receiving hole precisely drilled during a secondary operation (see Figure 7–42).

Basically, the blank ferrules are compressed into a circular drill fixture that forces the ferrules to assume their "in-service" shape. All irregularities in size and shape of the ferrule blank are transferred to variations in static loads at the periphery of the drill fixture.

The fiber-receiving hole is then drilled at the dead center of the blank. Ultraprecision required in the positioning of the drill fixture during the manufacturing setup is thereby introduced into the molded parts.

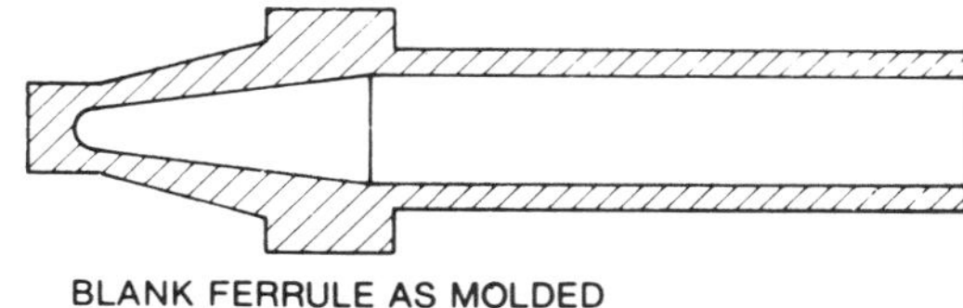

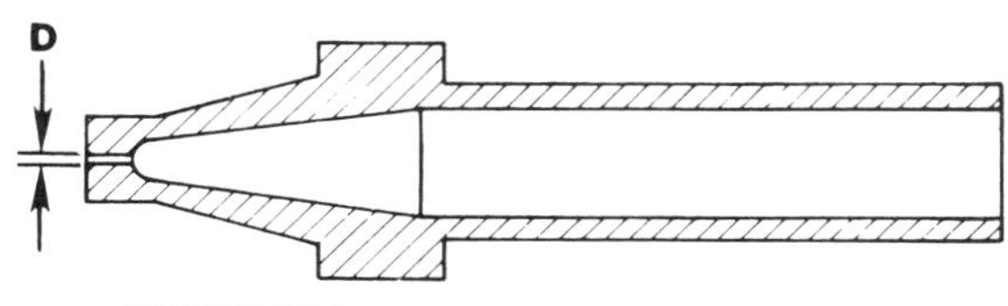

Figure 7–42 Ultraprecise ferrule drilling (*courtesy of AMP, Incorporated*).

Precise Hole Manufacturing Process

Figures 7–43 and 7–44 follow the manufacturing process. The process for precise hole drilling in Figure 7–43 is as follows:

A. Inexpensive ferrule blanks may be slightly out of round.
B. Ferrule blanks are compressed into drilling fixture which simulates "in-service" condition of ferrule.
C. Small hole is drilled, centered relative to drilling fixture; ultraprecise centering parts are now essentially identical.
D. Free state after drilling; ferrule is usable now that ultraprecise hole positioning has been accomplished.

The process for insertion of the optical fiber is illustrated in Figure 7–44.

A. Fiber within acceptable tolerances is inserted with adhesive added.
B. Ferrule is compressed to fit fiber size while adhesive sets.
C. Faces of fiber and ferrule are polished to optical finish or the fiber is "broken" to form a mirror surface.
D. The ferrule with fiber is placed into final alignment details.

By producing the fiber-receiving hole of the ferrule to be slightly larger than the maximum fiber diameter, only one hole size is required. Compressive forces on the outer surface of the resilient ferrule are used to reduce the inner diameter producing a tight grip on even the smallest fiber size. A tight fit is achieved at the fiber-

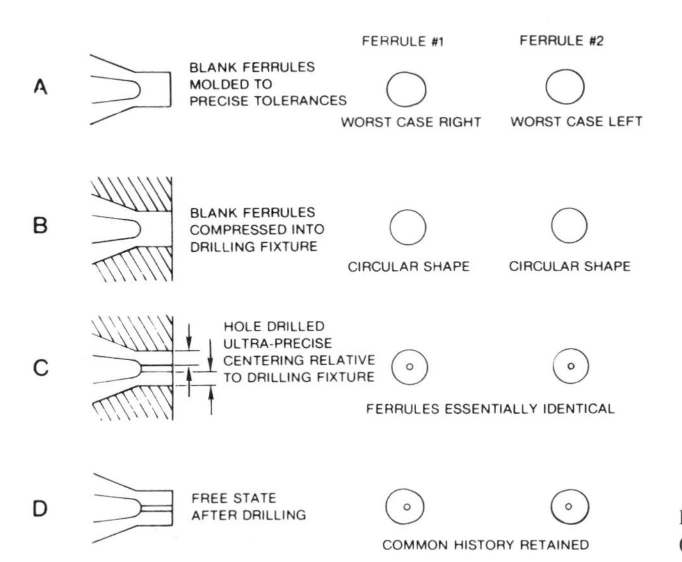

Figure 7–43 Precise hole shape (*courtesy of AMP, Incorporated*).

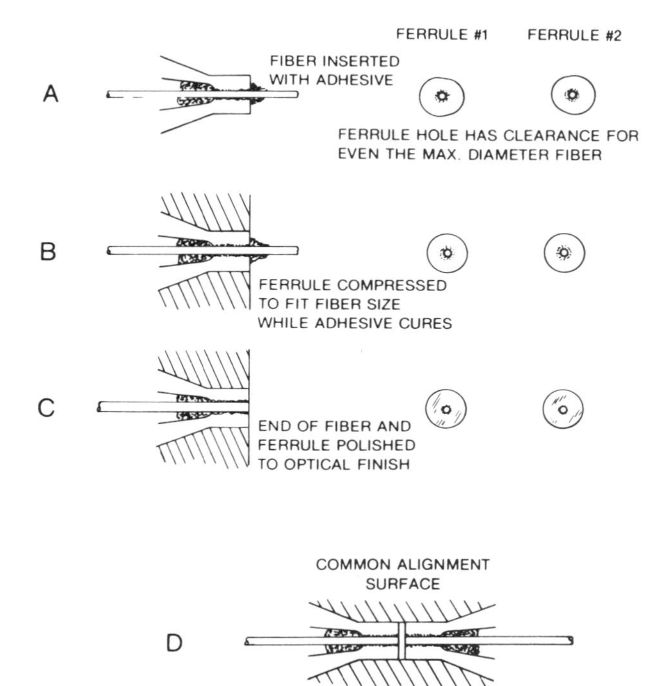

Figure 7–44 Insertion of optical fiber (*courtesy of AMP, Incorporated*).

to-resilient ferrule interface and at the resilient ferrule-to-rigid overlap interface. The amount of compression achieved is determined by the particular size of the fiber involved. In all cases, the compression is sufficient to seal off the optical interface.

Benefits provided by the resilient alignment mechanism include:

1. The effect of inherent initial tolerance differences in fiber size are eliminated.
2. Ultraprecision "centered" connectors—by drilling the hole while the elastometric component is in compression—can be economically produced in volume.
3. A sealed interface protects the optical junction from hostile environments.
4. The rugged connector will survive many matings.
5. Resilient member cushions the fibers against damaging vibration and other mechanical problems.

ALIGNMENT MECHANICS

The mechanics of alignment are an exacting task in that perfect or at least optimum alignment is necessary to avoid loss. This philosophy must extend to the retaining of optimum alignment in every instance that a connection is made or remade. Furthermore, alignment must be of a method that can be accomplished in the field with a minimum of special equipment. Alignment should also be within the technical capabilities of available personnel who are already maintenance technicians.

Under the next several headings, we shall study several of the mechanical concepts used in the fiber industry today. Many of these were invented and patented by individuals and then were developed by major corporations for use in coupling devices.

Opposed Lens Alignment Concept

A pair of highly polished fiber ends are inserted into opposing cavities within a transparent medium. An objective lens and an immersion lens are provided by the opposed cavities in the transparent medium. The viscous fluid has an index of refraction that allows alignment of the opposing fibers. In Figure 7–45, light rays are shown refracted from the viscous indices, through the transparent medium aligning opposing fibers. The inventor of the opposed lens alignment concept was Melvyn A. Holzman.

Transfer Molded Alignment Concept

Pairs of fiber end faces are highly polished to prevent Fresnel reflections, then insert molded on the axis of the resulting ferrule utilizing a fabricated and very precise tool (see Figure 7–46). The two mating ferrules (plugs) insert within a biconical socket. Within the socket are cushions that are index matched to accommodate the plugs. This concept was defined by P. K. Runge and S. S. Cheng in the *Bell System Technical Journal,* July–August 1978.

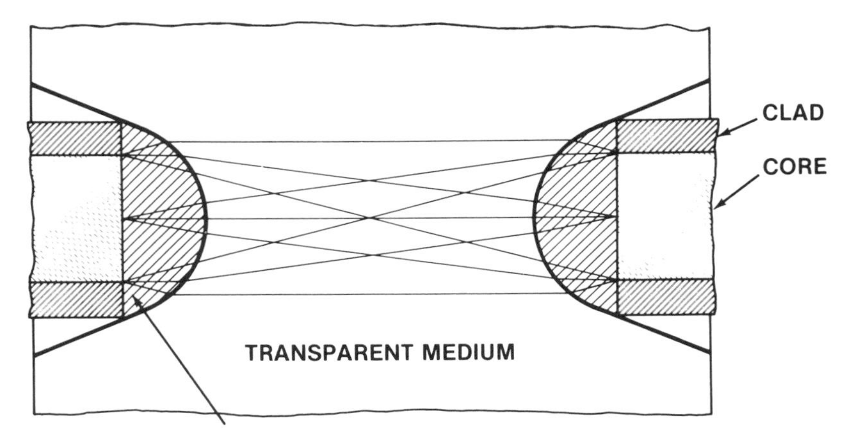

Figure 7–45 Opposed-lens alignment concept (*courtesy of ITT Cannon Electric*).

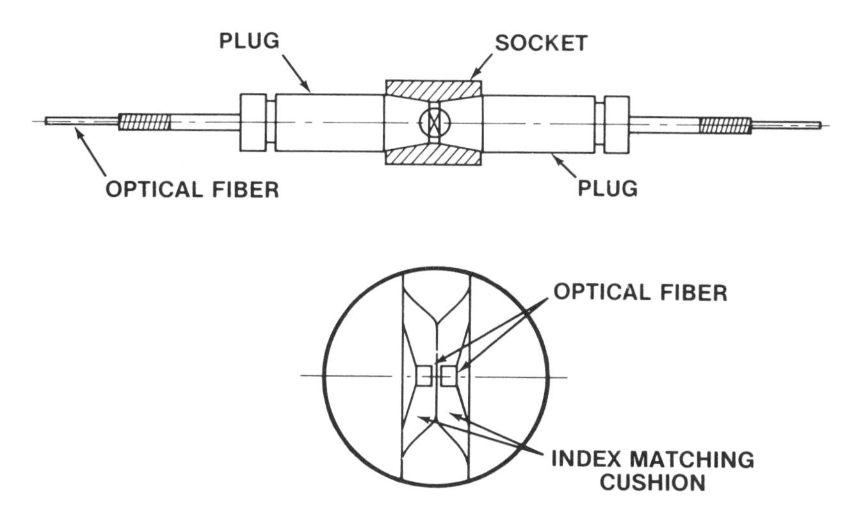

Figure 7–46 Transfer-molded alignment concept (*courtesy of ITT Cannon Electric*).

Multirod Alignment Concept

In Figure 7–47, six precision rods are located around the inside perimeter of a guide sleeve. Three extremely precise ferrule rods provide a mounting for the fiber. An end view of the assembled ferrule and guide sleeve shows the fiber inserted at the exact geometric center within the three-rod ferrule triangle. All rods are calculated for tangency using needle bearings or cam rollers. Tolerances using this method can be held to millionths of an inch. The inventor of the multirod alignment concept was Carl R. Sandahl.

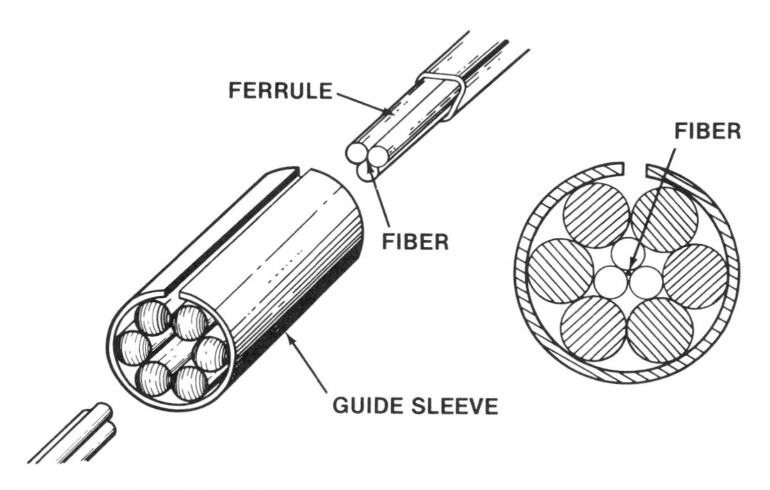

Figure 7–47 Multirod alignment concept (*courtesy of ITT Cannon Electric*).

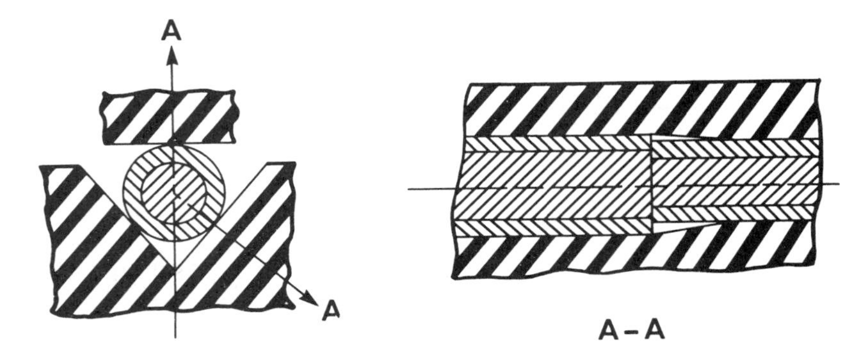

Figure 7-48 Resilient self-centering alignment concept (*courtesy of ITT Cannon Electric*).

Resilient Self-Centering Alignment Concept

Two fibers of slightly different diameters may be aligned using the resilient self-centering alignment concept. In Figure 7-48, the end view illustrates the fiber installed in a V-groove that establishes three planes of force on the fiber. Force is transferred equally to three arcs on the perimeters of the fibers. The material surrounding the fibers is resilient and will remember its original shape if the fibers are removed.

In the cross-sectional view A-A, the slightly different-diameter fibers are shown center aligned. Grinding, polishing, and bonding are accomplished using several standard procedures. The resilient self-centering alignment concept was invented by Robert M. Hawk.

Double Eccentric Alignment

The double eccentric method of fiber alignment is one of the standard alignment procedures. The procedure deals with an eccentric motion similar to X-Y mechanisms. Alignment is made in one direction or the other until fibers align. In Figure 7-49

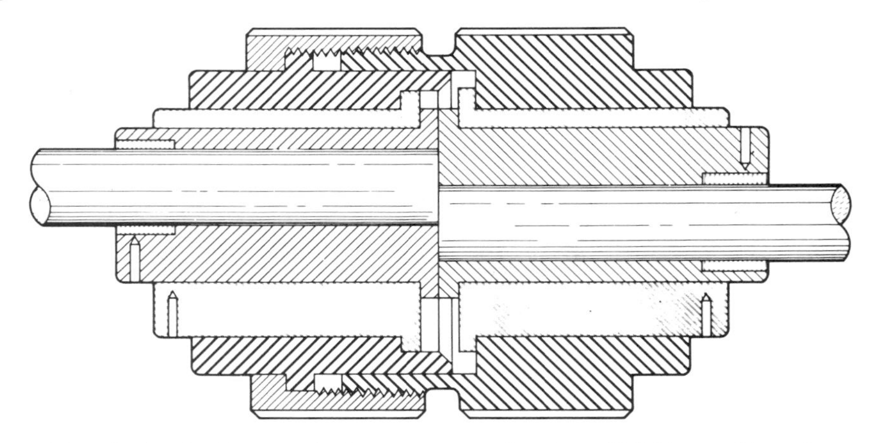

Figure 7-49 Double-eccentric alignment concept (*courtesy of ITT Cannon Electric*).

a pair of opposing ferrules is inserted into a guide cylinder and/or a male–female arrangement. The ferrules are utilized to adjust the fiber core to the axis of the outer cylinder surface. Adjustment is made under a microscope. When aligned, the ferrules are locked into place. The fiber ends are ground and polished. Epoxy is used to bond the fiber into the center sleeve.

A second method used with the double eccentric concept is to mate the fibers into a system, then adjust the eccentric until optimum power is realized at the receiver. The double eccentric concept was invented by Setsuo Sato.

Watch Jewel Concept

Readily available watch jewels are utilized to precisely align fibers as in the arrangement shown in Figure 7–50. The jewels are produced at reasonable prices, with 3-micron concentricity inside or outside diameter. The alignment process involves measuring the diameter of the fiber with accuracy, then choosing the jewel to fit that

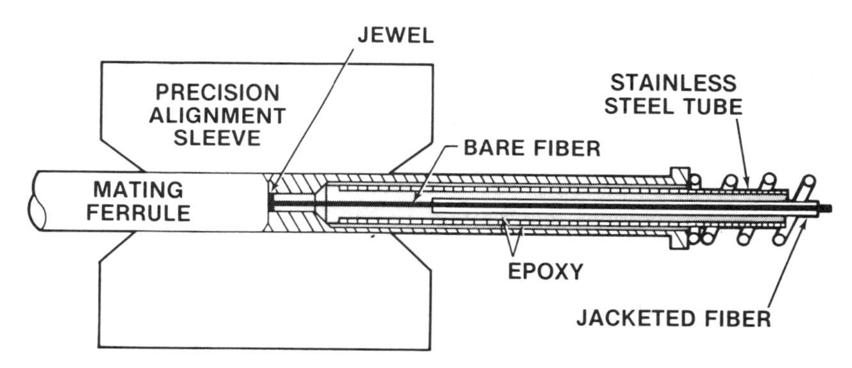

Figure 7–50 Watch jewel alignment concept (*courtesy of ITT Cannon Electric*).

is just larger in size than the fiber. The fiber is then expoxied in place and ground and polished to tolerance. The second fiber end is treated in the same manner. Finally, the two opposing jewel/ferrules are mated in a guide sleeve as pistons in a cylinder.

Three Sphere Concept

Three spheres of exact diameter are placed in a pair of ferrule connectors (see Figure 7–51). Each connector utilizes the three spheres in a plane of 120° increments with the spheres retained in a spherical race. In the center of the three spheres, a defined space for the fiber end is positioned. When the two pairs of ferrules are brought together, the two sets of spheres nest with respect to each other at 60° increments, precisely aligning the optical fibers. Alignment accuracy is influenced only by the variation in sphere diameter. Spheres are available with ten-millionths of an inch tolerance.

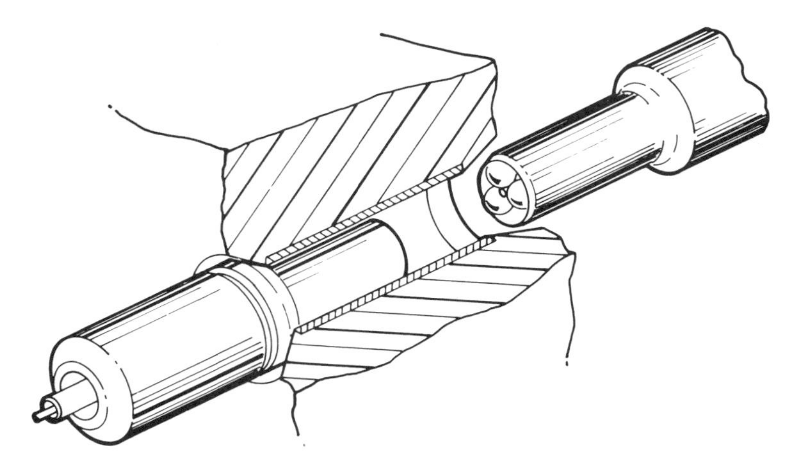

Figure 7–51 Three-sphere alignment concept (*courtesy of ITT Cannon Electric*).

FIBER END PREPARATION

Before placing a fiber into a connector, it must be properly prepared. The jacket of the fiber must be removed, the fiber is then broken and the fiber end is ground and polished. Final procedures also involve cleaning and coating the fiber with a suitable refractive-index-matching compound. Procedures below are utilized for both step-index and graded-index glass fibers.

Jacket Removal

The initial step in fiber end preparation is the removal of the fiber jacket. In Figure 7–52, the fiber is taped to a flat surface and is placed under tension by pulling. The plastic is carefully stripped away using a razorlike knife (X-Acto) or a razor. The knife is placed on a shallow angle ensuring that there is little or no abrasion of the fiber. It is also possible to remove the jacket using a wire stripper of the correct size. Wire strippers are used only as an alternative. Once the fiber jacket is removed, the fiber is cleaned (see Figure 7–53). Soft tissue soaked in isopropyl alcohol or Freon is usually the agent for cleaning. The bare fiber (see Figure 7–54) is now ready for end cutting and finishing.

Scribing and Breaking the Fiber

Before placing a fiber into a connector or splice it must be cut to size. This is accomplished using one of several techniques. Most are simple and can be easily learned.

Quartz and glass fibers are cut by scribing. The first involves scribing a line around the fiber with a fine file. Then by application of a constant pressure pull on both ends of the fiber and it will snap in two. This method is not very efficient when dealing with small-diameter fibers. It will, however, provide large-diameter fibers

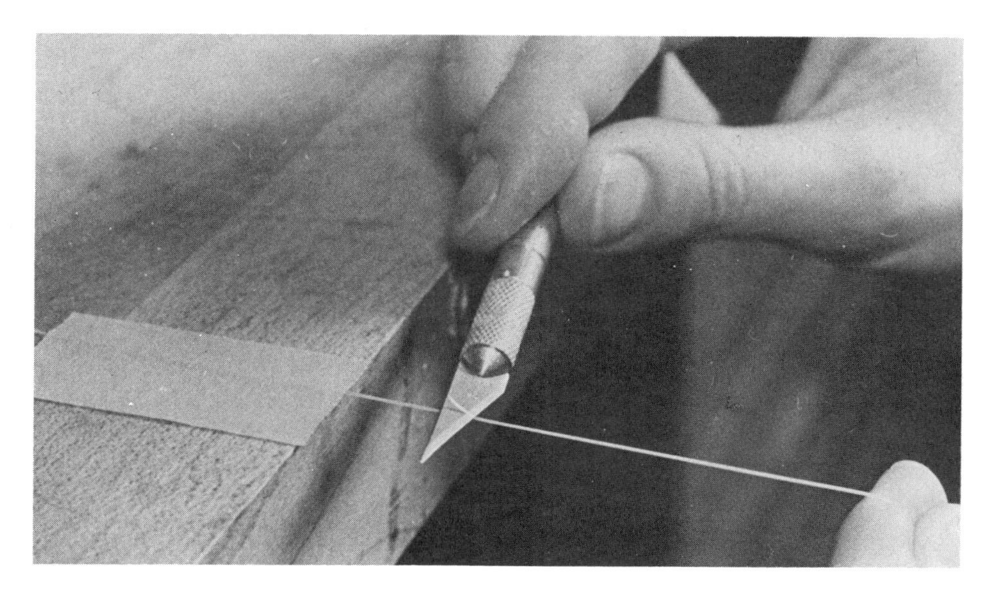

Figure 7–52 Removing the fiber jacket (*courtesy of Alcatel Cable Systems Group*).

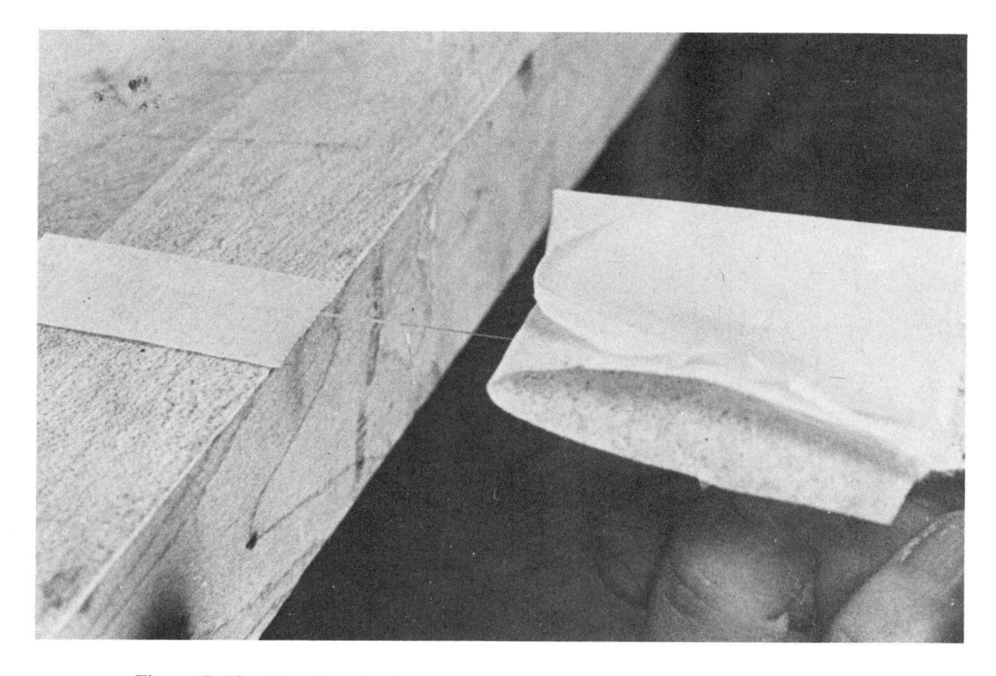

Figure 7–53 Cleaning the fiber (*courtesy of Alcatel Cable Systems Group*).

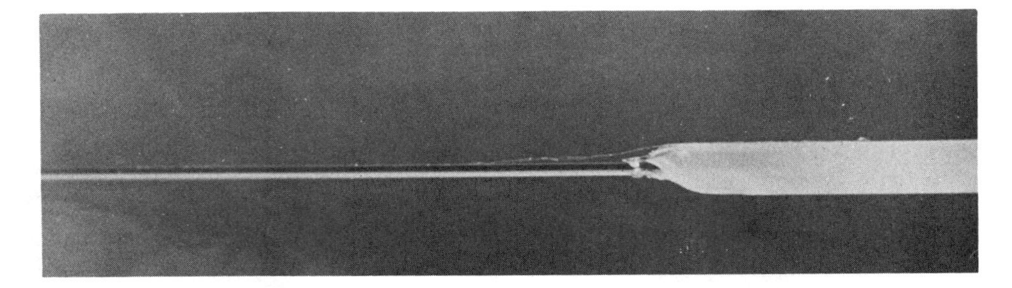

Figure 7–54 Bare fiber ready for finishing (enlarged) (*courtesy of Alcatel Cable Systems Group*).

with a clean break where the line was scribed. In place of the fine file, a sapphire or a diamond cutter may be used for smaller fibers (see Figure 7–55). A steady tension on the fiber will snap them. This tension is not large (around 0.25 pound-foot). The force, though hand-applied, will snap the fiber end smooth (see Figure 7–56). There are other methods for cutting. These are commercial setups with special tooling to be purchased. Most follow a scribing philosophy of some sort. Plastic fibers may be cut with a razor blade.

Inserting and Sealing Fiber into Ferrule

If the fiber is to be placed into a ferrule for further installation in a connector, the end must be lapped and polished. The fiber end is measured using a micrometer (see Figure 7–57) to ensure that the fiber and the ferrule are matched. In this procedure, the fiber end is broken and inserted into the end of the ferrule. Figure 7–58 illustrates the hand alignment necessary to place the fiber into the ferrule. Figure 7–59 shows

Figure 7–55 Scribing the fiber (*courtesy of Alcatel Cable Systems Group*).

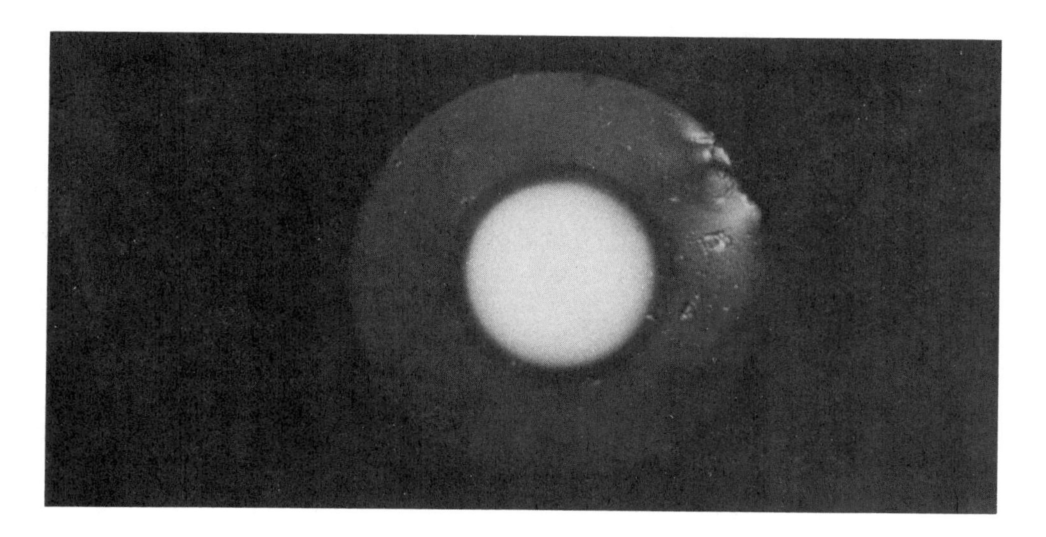

Figure 7–56 Fiber end after breaking at scribe mark (enlarged) (*courtesy of Alcatel Cable Systems Group*).

Figure 7–57 Measuring the fiber diameter. (*courtesy of Alcatel Cable Systems Group*).

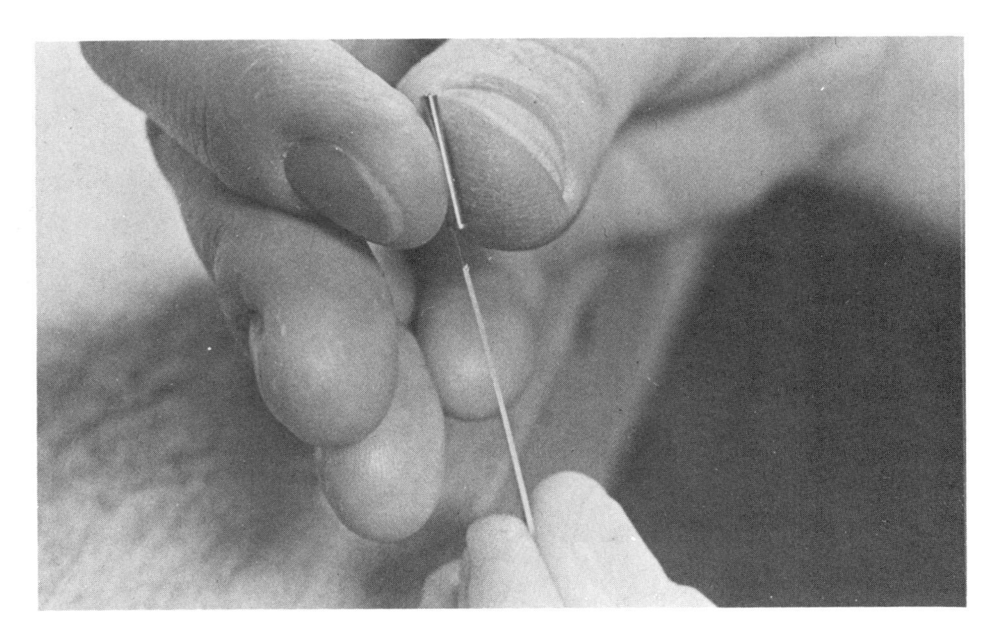

Figure 7–58 Hand aligning of fiber with ferrule (*courtesy of Alcatel Cable Systems Group*).

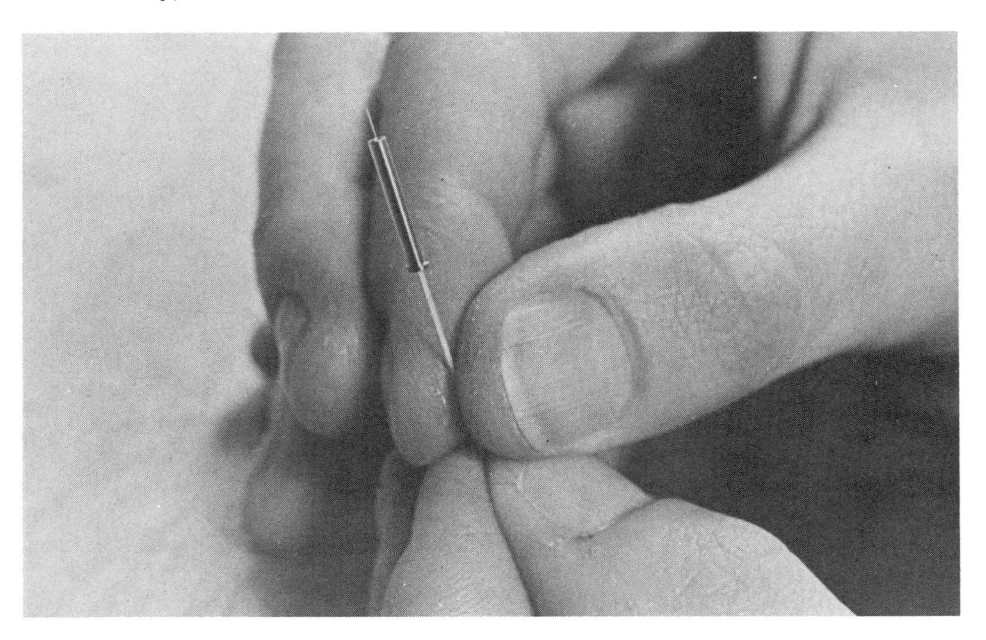

Figure 7–59 Fiber in ferrule (*courtesy of Alcatel Cable Systems Group*).

Figure 7–60 Close-up of fiber in ferrule (enlarged) (*courtesy of Alcatel Cable Systems Group*).

the fiber in the ferrule with part of the fiber end protruding from the ferrule tip. A closeup look at the inserted fiber is provided in Figure 7–60. Epoxy application to the fiber and the ferrule is now to be accomplished. In Figure 7–61, a cross-sectional line drawing of the inserted fiber in the ferrule is presented. The plastic-coated fiber is surrounded in a bed of epoxy resin with the bare end of the fiber aligned in a watch jewel. Figure 7–61 presents the assembly after epoxy application. The next several illustrations describe the process.

Epoxy is now to be applied to the ferrule. Epoxy should have a low viscosity, good adhesion to the stainless steel ferrule and the glass fiber, and reasonable life. The epoxy should also have a short cure time, but not so short that it will prevent the ferrule from being filled prior to hardening. Epoxy recommended by the vendor

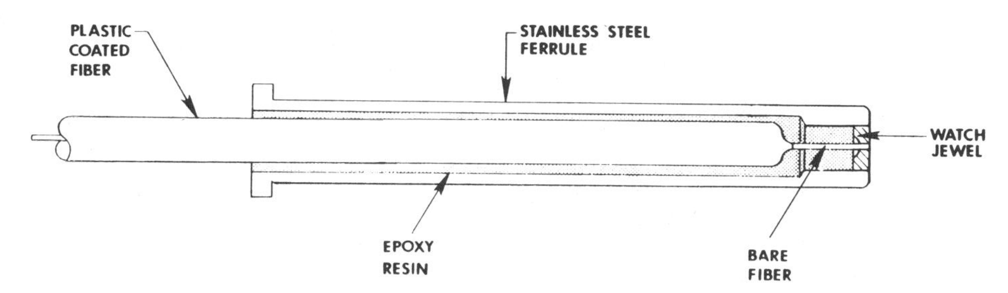

Figure 7–61 Cross section of fiber in a stainless steel ferrule (*courtesy of Alcatel Cable Systems Group*).

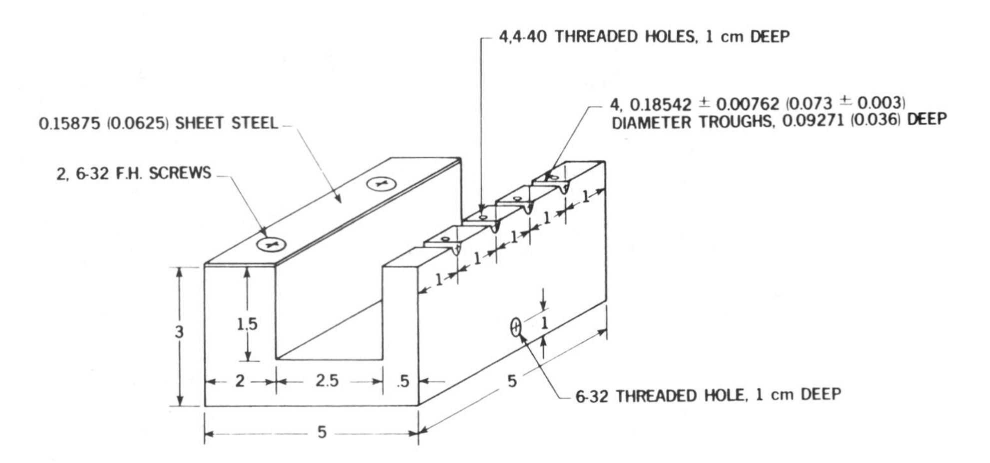

Figure 7–62 Fiber termination station (brass) (*courtesy of Alcatel Cable Systems Group*).

should be used whenever possible, for they have taken the research time to choose the most compatible type.

The fiber is attached to a termination station such as that drawn in Figure 7–62. Epoxy is applied to the front and then the back of the ferrule as shown in Figure 7–63. The ferrule should be heated to approximately 80 °C in the heating fixture at-

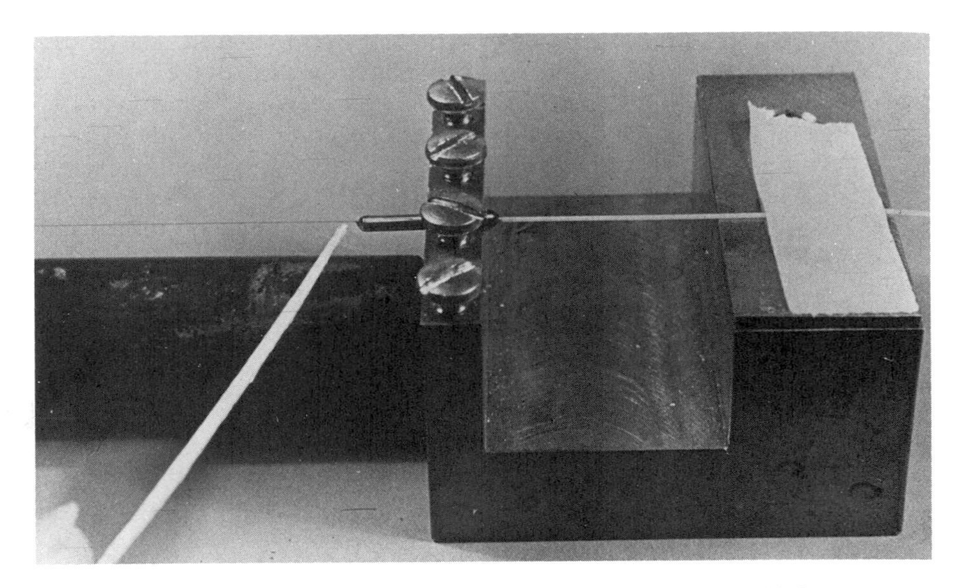

Figure 7–63 Application of epoxy to the ferrule (*courtesy of Alcatel Cable Systems Group*).

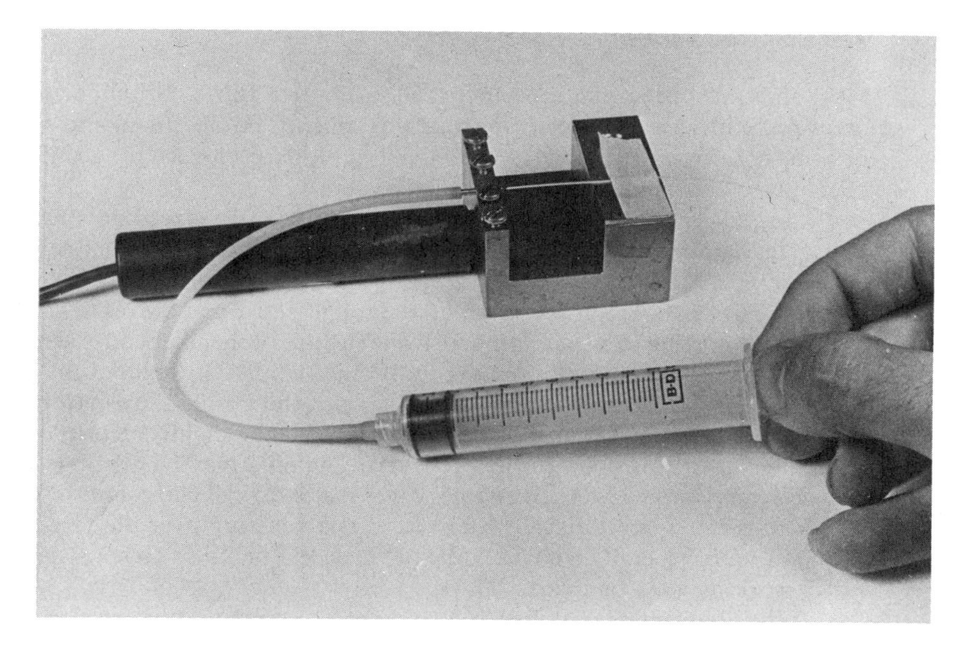

Figure 7–64 Drawing epoxy into ferrule (*courtesy of Alcatel Cable Systems Group*).

tached to the fiber-termination station. The epoxy should wick (flow down thin cracks) completely into the ferrule from the back. If it does not, a syringe may be used to draw epoxy the rest of the way into the ferrule (see Figure 7–64). To speed the curing of the epoxy the assembly is kept in the termination station and heated to 80 °C for about an hour or until the epoxy is cured. The ferrule, as it appears in Figure 7–65, is ready for lapping and polishing.

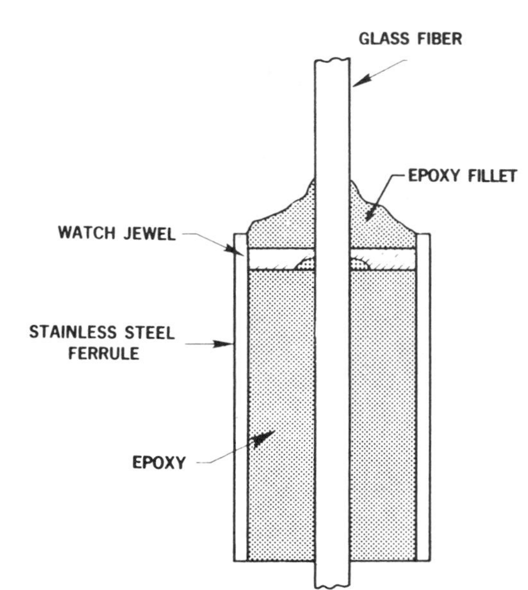

Figure 7–65 Fiber and ferrule just after epoxy is applied (*courtesy of Alcatel Cable Systems Group*).

Lapping and Polishing Fiber Ends

Fiber ends must be ground, then polished to a mirror finish. Grinding and polishing takes place in several stages from rough to finish. The fiber is placed in a pin vise or some other type of holder so as to have freedom of movement. Grinding usually takes place in a mixture of alcohol and water.

Several grades of silicon carbide finishing paper are placed on a firm, flat surface. The lapping and polishing tool is held firmly in the hand and rotated in circular, elliptical, and figure-eight patterns on the surface of the finishing paper (see Figure 7–66). The process is started on the rough paper and moves toward the finer grit. Before moving to a finer-grit paper, the surface of the fiber end is gently brushed off and cleaned in alcohol to remove loose silica chips. Throughout the process the fiber end is inspected and lapping is slowed when the ferrule shows evidence of having been ground flat. The front face of the jewel may be slightly recessed with respect to the ferrule at this point in the procedure. Lapping may be continued with light pressure to bring the jewel and ferrule planes just into or slightly short of coincidence (see Figure 7–67), with a slight film of epoxy remaining over the ferrule endface. This procedure generally requires 1 to 3 minutes. Plastic fibers do not require as much finish work as glass or quartz.

When polishing is finished, a coating of a compound that has near the refrac-

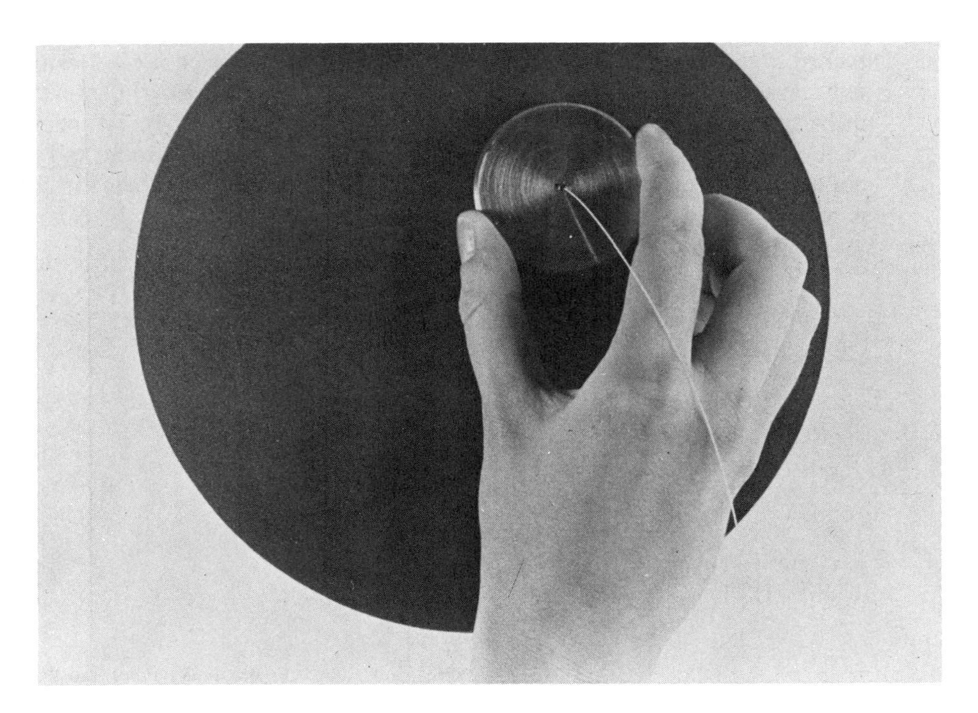

Figure 7–66 Lapping the fiber end (*courtesy of Alcatel Cable Systems Group*).

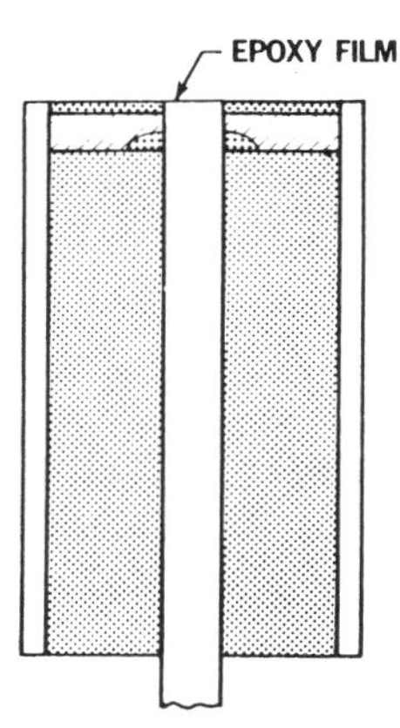

Figure 7–67 Fiber end after lapping (*courtesy of Alcatel Cable Systems Group*).

tive index of the glass is applied to the fiber end. This compound will help to decrease attenuation losses.

Before proceeding with polishing, the ferrule and lapping tool are thoroughly washed to remove residue grit remaining from lapping. A polishing paste which has suspended is used as the polishing compound. This paste is applied to moist adhesive-backed paper which has been mounted on a firm surface (see Figure 7–68).

Depending on the smoothness of the lapped surface, a well-polished surface can be obtained in as little as 1 minute using moderate pressure. The fiber surface

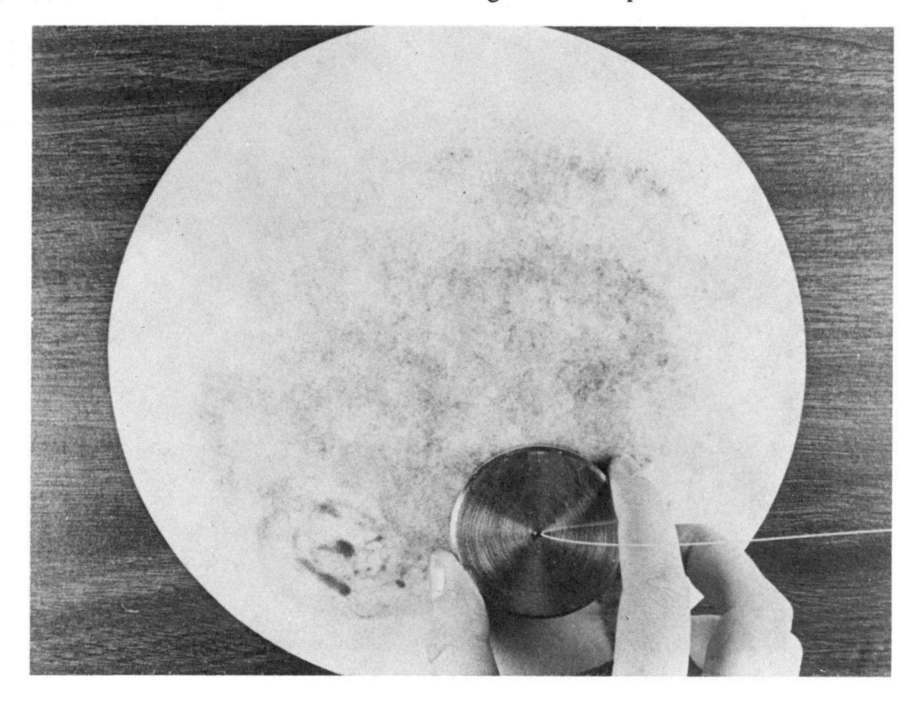

Figure 7–68 Polishing the fiber end (*courtesy of Alcatel Cable Systems Group*).

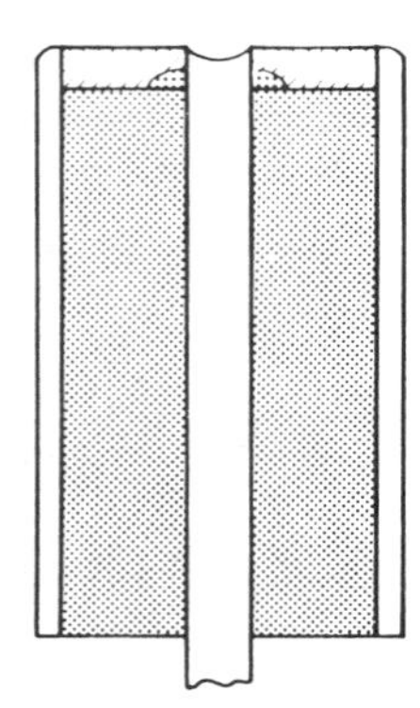

Figure 7–69 Cross section of polished fiber and ferrule (*courtesy of Alcatel Cable Systems Group*).

will become very slightly concave and recessed with respect to the face of the jewel (see Figure 7–69). This characteristic is highly desirable for protection of the fiber during repeated connector matings. A photograph of the polished fiber and ferrule is provided in Figure 7–70.

Inspection Criteria

After polishing and cleaning, the terminated ferrule end should be inspected. Acceptable terminations should look somewhat as the two photographs on the left side of

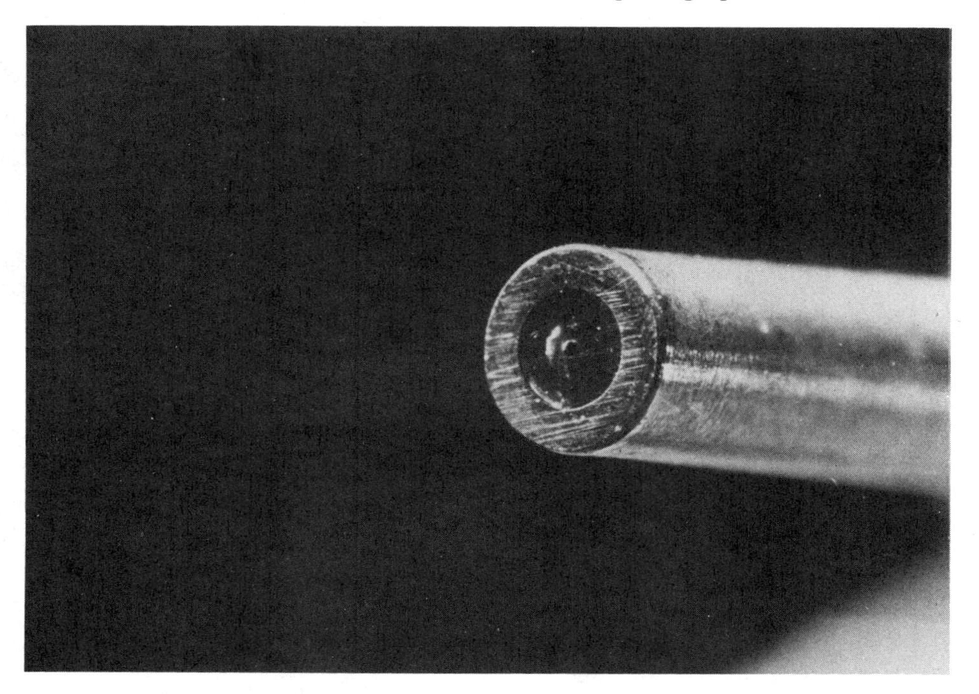

Figure 7–70 Lapped and polished fiber end in ferrule (enlarged) (*courtesy of Alcatel Cable Systems Group*).

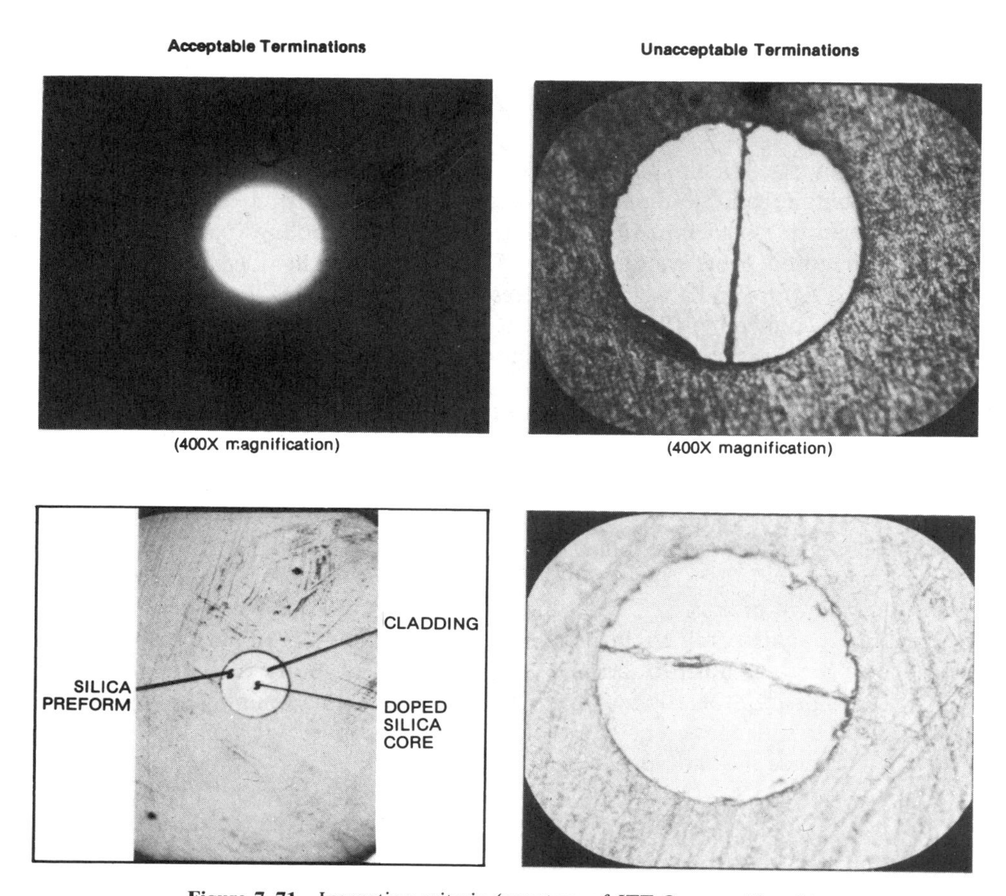

Figure 7–71 Inspection criteria (*courtesy of ITT Cannon Electric*).

the Figure 7–71. Note the mirrorlike finish. The fiber should have no cracks in its core. The surface of the fiber should be flush with the face of the jewel or slightly concave. Under no circumstances can the fiber be convex. The photographs in the right side of Figure 7–71 illustrate unacceptable terminations. As shown, these fibers are under 400× magnification.

SPLICING

There are dozens of different splices and splicing techniques used in the fiber-optic industry. All splices have several aspects in common. They all must have the attribute of low power loss. Attenuation must be held to as near 0.1 decibel as possible. Splices must be quick to install, should be lightweight, extremely strong, and small.

Probably the most critical requirement of splicing is alignment of the fiber. Alignment philosophy and mechanics were covered in detail in previous paragraphs. Although fiber optic manufacturers provide a large variety of splices, most of these utilize several basic techniques.

The Fused Splice

The most difficult splice to make is the fused splice (see Figure 7–72). It is also the splicing technique that provides the lowest attenuation. Economics and efficiency are a trade-off. Two fiber ends are placed in a tooling fixture that aligns the ends to be fused. Fiber ends are polished and prepared. The fibers usually are set in grooves in the fixture and are clamped in resilient vise jaws. Heating elements are then brought to the connecting fiber ends. Heat is applied and the fusing takes place. Several fusion heaters used are microtorches, nichrome wire heaters, and electric arc fusers.

The Siecor Model 68 Fusion Splicer

This method employs the recognized method of permanently fusing two butted optical fibers together by heating with a tiny well-regulated electric arc (see figure 7–73).

The field-usable splicer consists of fiber-holding and alignment devices, a viewing microscope, fiber positioning and arc intensity controls, a built-in arc power supply and a high-precision fiber cutting tool. A built in LID-System (R) unit is also provided. The tool also emits light into one fiber and attempts to detect it in the other fiber. The unit's built-in microprocessor uses the resulting data to maximize the light transmission, thereby maximizing core alignment and minimizing splice loss. Typical average splice losses between telecommunications-grade optical fibers achieved with this device are 0.03 decibels.

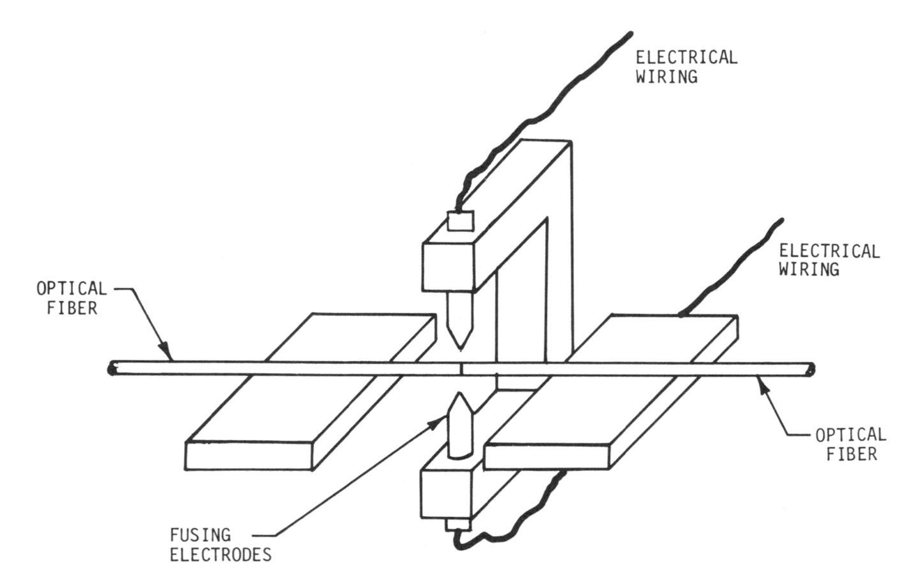

Figure 7–72 Fused splice (*Mike Anaya*).

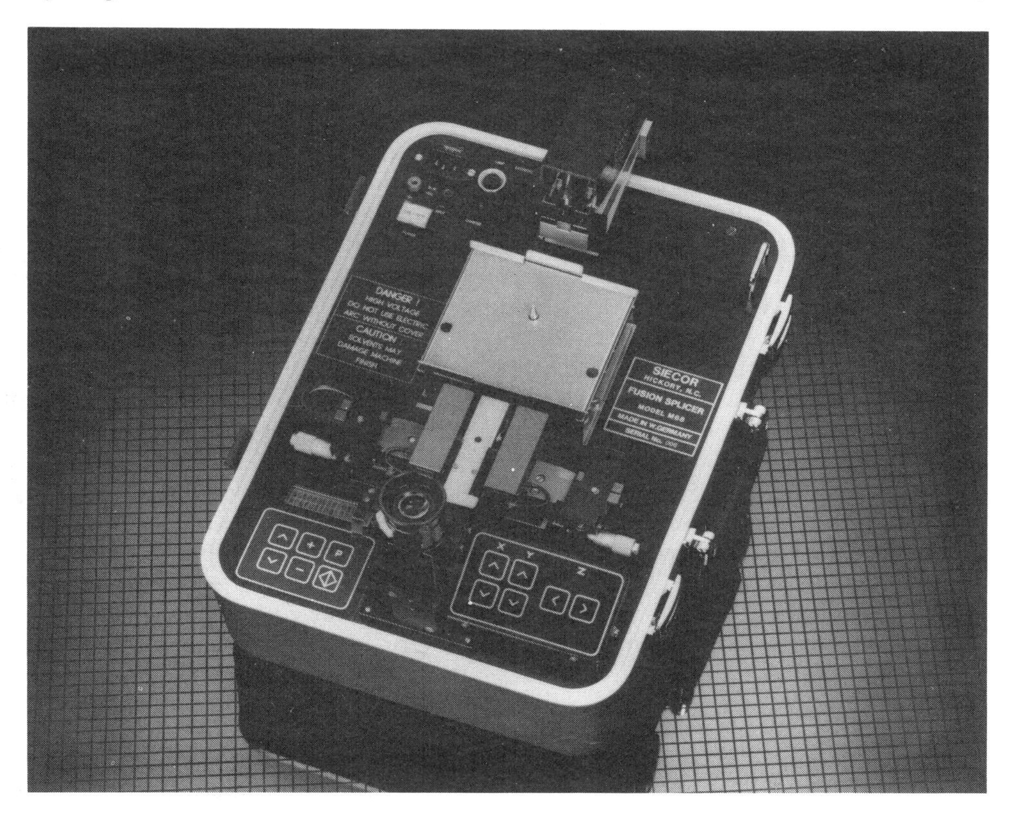

Figure 7–23 The Siecor® model 68 fusion splicer (*courtesy Siecor Corporation.*)

Field experience has shown that the Model 68 requires minimal operator training or experience for the production of low-loss splices.

After the fibers are cleaned and cut, they are clamped in the holders and the built and roughly positioned with the aid of the microscope. After rough alignment, the splicers operation is completely automatic. A button is pushed and the fusing process begins. A brief arc is used to clean fibers and slightly round the fibers ends. The fibers are moved automatically in a spiral search until optimal alignment is achieved. With a higher arc power, the fibers are then fused together. The M68 shows the splice loss readings on its own LCD thereby eliminating the need to verify splice loss readings after each splice with an Optical Time Domain Reflectometer (OTDR). The unit is compact, lightweight and self contained. Splice time is less than one minute.

After splicing, the joint is placed in a splice organizer to protect it. Once all the fibers in a cable are spliced, the organizers are placed in a splice closure. The closure is then sealed for environmental protection.

CONNECTORS AND THEIR INSTALLATION

Connector Basics

Connectors are required to join fibers at transmitters (light sources), receivers (detectors), and bulkheads. That is, the connector may be used to join fiber to fiber or device to fiber. Whichever the situation, the connector must fulfill the requirements of alignment so as to avoid the problem of attenuation. The subject of alignment has been brought up many times in this chapter and necessarily so. However, that is just one of the demands placed upon connectors. The connector must provide termination for jackets, strength members, and buffering materials at the fiber's end. It also must ensure that loads applied are not transferred from one fiber to the next or from a fiber to a cable, or vice versa. Other uses of the connector are protection of the fiber during installation and for fiber-end preparation.

In separable connectors (connectors that have repeated matings), it is extremely important that the two fiber ends do not meet or touch. The reason being is that repeated matings due to shock or vibration will chip or scratch the polished fiber end and thereby degrade the optical efficiency and the performance of the fiber, therefore the system. The gap, however, cannot be large due to gap loss attenuation. The standard gap usually does not exceed 10% of the core diameter. An example is a 5-micron (0.0002-inch) gap for a 50-micron (0.002-inch) core.

Connector Types

Connectors are designed to provide connection for single fiber, multichannel, bundle fiber, and so on. Figure 7–74 is an illustration of a variety of fiber-optic connectors. Two optical cable terminations and a threaded sleeve illustrate how an in-line connection can be made that gives insertion loss less than 1 decibel (see Figure 7–75).

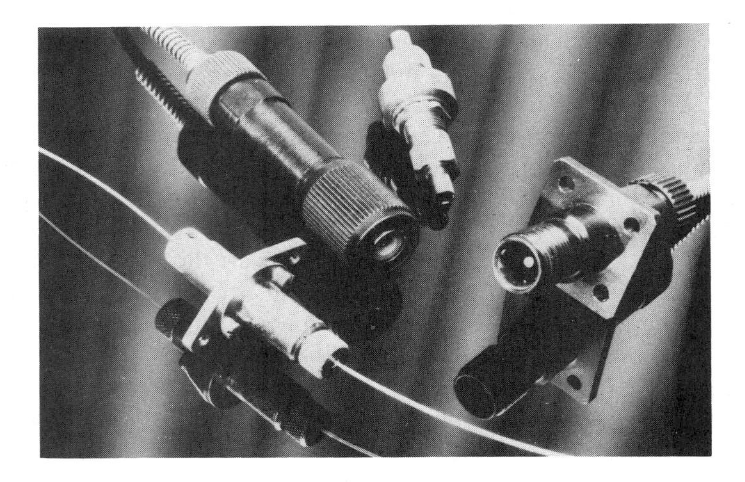

Figure 7–74 Variety of fiber optic connectors (*courtesy of ITT Cannon Electric*).

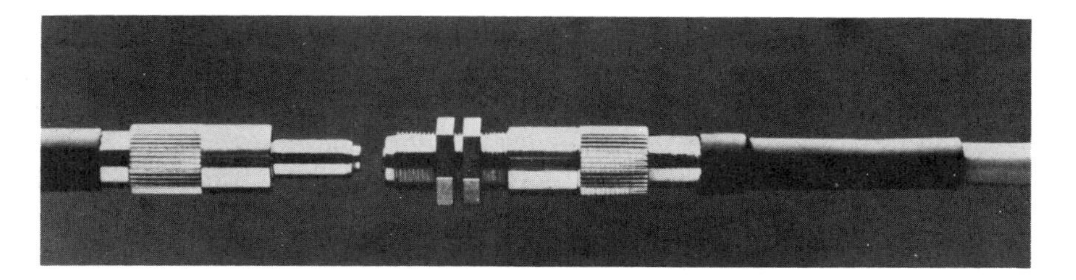

Figure 7-75 In-line connectors (*courtesy of Siecor Corporation.*)

The sleeve's lining compensates for allowable variations in termination dimensions and tends to self-center the assembly. Such a connection is also suitable for a feed-through, bulkhead application. Receptacles containing light-emitting diode (LED) sources and PIN photodetector diodes provide a means of mechanically and optically connecting a terminated optical cable to a transmitter and receiver (see Figure 7-76). The mounting of the receptacles is compatible with normal PC-board practice.

Design Considerations

Alignment considerations are the basic starting points in the design of any fiber-optic connector. For the mating of opposing fibers, each degree of freedom of alignment is responsible for a loss, which is additive. Optical isolation between adjacent ferrules must be provided by sealing off the light from each other.

Heat-dissipation techniques must be designed into receptacles for device mounting. Electrical isolation must also be considered between adjacent channels for the electro-optical devices. Cable strength members must be terminated within the connector such that any axial load on the cable is resisted by the connector hardware and its mounting, and not transferred to the fiber. This precludes the possibility of varying alignment and subsequent coupling efficiency change.

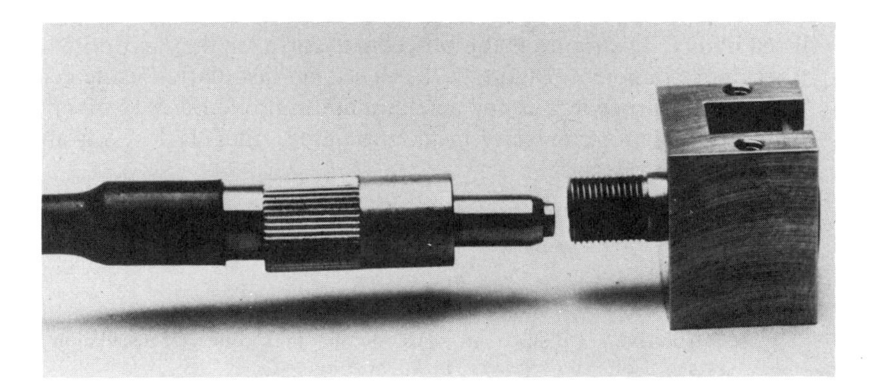

Figure 7-76 Source or detector connector (*courtesy of Siecor Corporation.*)

The effects of environment are different on an optical interconnection. For example, humidity may improve coupling efficiency since the refractive index of water is better matched to the fiber's index than that of the air between mated fibers. Sand and dust are a more severe environmental condition than for electrical interconnections. The sand used in MIL–STD–202 (test methods for electronic and electrical component parts) testing is the same silica as the fiber and, therefore, erodes the fiber's polished endface. Vibration tends to modulate the optical pulses if the ferrules vibrate independently of each other.

Connector Installation Procedures

The process of connecting as a norm follows a rather exacting procedure. The fiber optic connector is a much more critical and refined item of hardware than the standard electrical connector. Procedures for installing a connector on a fiber are usually developed for the peculiar fiber and by the special features of the connector itself. However, most connectors do follow general procedures. These are as follows:

1. The fiber jacket and buffering material are removed from the end of the fiber or cable of fibers.
2. The exposed fiber is broken for the precise length required. Fixtures are available for producing controlled, consistent breaks.
3. The fiber must be ground and polished.
4. Cleaning the fiber is a necessity. Usually, this takes place between any step in the entire procedure.
5. The fiber must be aligned precisely in the connection device. This critical procedure is usually accomplished using special fixtures. However, in the field assembly may simply require a connection that is manufactured to fulfil the philosophy of alignment using good mechanics. Low attentuation due to precisely constructed connectors is the goal.

Figure 7–77 is an installation procedure incorporating the general procedures listed above. The figure is the procedures for installing an optimate single-position fiber optic connector. Figure 7–78 is a second installation using general procedures. This second procedure is for an optimate multiple-position fiber optic connector. All procedures for connector installation are similar. Industry is attempting to standardize the process.

Figure 7–77 Optimate single-position fiber optic connector installation procedures (sheet 1 of 8) (*courtesy of AMP, Incorporated*).

AMP* OPTIMATE* SIMPLEX FIBER-OPTIC CONNECTORS

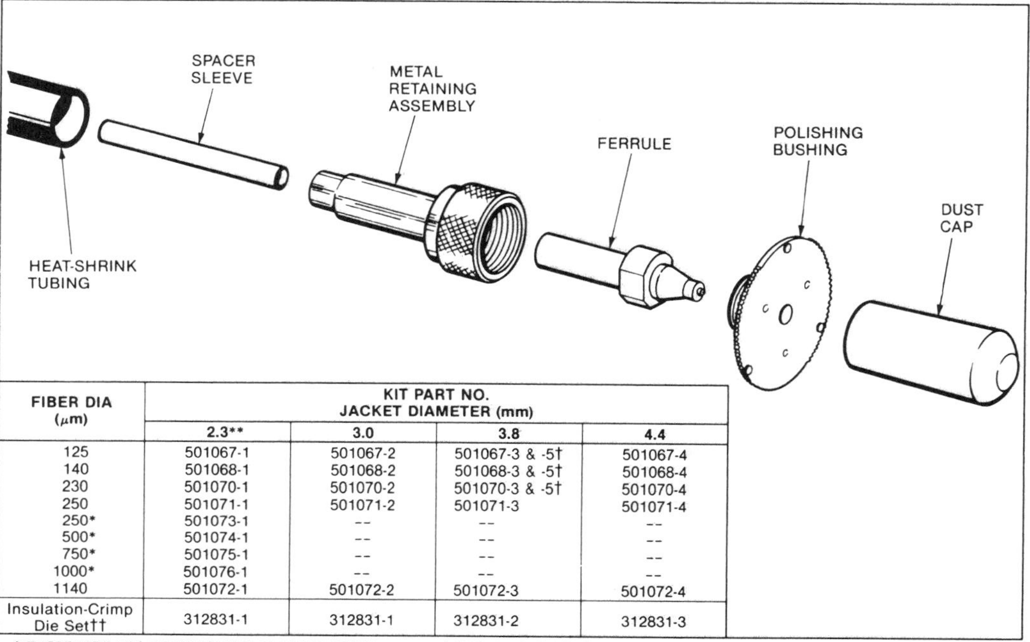

FIBER DIA (μm)	KIT PART NO. JACKET DIAMETER (mm)			
	2.3**	3.0	3.8	4.4
125	501067-1	501067-2	501067-3 & -5†	501067-4
140	501068-1	501068-2	501068-3 & -5†	501068-4
230	501070-1	501070-2	501070-3 & -5†	501070-4
250	501071-1	501071-2	501071-3	501071-4
250*	501073-1	--	--	--
500*	501074-1	--	--	--
750*	501075-1	--	--	--
1000*	501076-1	--	--	--
1140	501072-1	501072-2	501072-3	501072-4
Insulation-Crimp Die Set††	312831-1	312831-1	312831-2	312831-3

* THESE KITS ARE DESIGNED FOR PLASTIC FIBERS.
** SPACER SLEEVE MUST BE USED WITH CONNECTOR.
† THE -5 KIT IS DESIGNED FOR CABLES THAT HAVE EXTRA STRENGTH MEMBERS.
†† USED IN AMP HAND CRIMPING TOOL 58190-1 (IS 9047).

Fig. 1

1. INTRODUCTION

This instruction sheet (IS) covers the application of AMP OPTIMATE simplex fiber-optic connectors to optical fibers. The connectors can be used with glass or plastic fibers and they mate with AMP simplex active device mounts (ADMs) or coupling bushings. See IS 9101 for information on these products.

Also available are installation and designer kits to aid in the application procedure. Installation Kits 501258-1 through -4 contain all tools and materials required. Designer Kit 501220-2 contains an assortment of connector kits, ADMs, splice bushings, and tools.

Read this material thoroughly before starting assembly.

NOTE *All dimensions on this instruction sheet are given in millimeters, followed by inch equivalents in parentheses.*

2. DESCRIPTION

Figure 1 lists connector kits for various fiber sizes and jacket diameters. Besides the ferrule, into which the fiber is epoxied, each kit contains a metal retaining assembly (a coupling nut and eyelet), a polishing bushing, a length of heat-shrink tubing, and a dust cap.

A spacer sleeve, used with cables having a jacket diameter of 2.3 mm (.09 in.) or less, is also included with appropriate kits.

Epoxy is also required. AMP Epoxy 501195-1 cures in 4 hr at 65°C or 24 hr at 25°C. Epoxy 501195-3 has a cure time of 45 minutes at 25°C.

The buffer coating of the fiber can be removed by using a fiber stripper or by dipping the fiber into a heavy-duty paint remover.

The connectors are crimped with AMP Hand Crimping Tool 58190-1, which must be equipped with the proper insulation-crimp die set (available

* Trademark of AMP Incorporated

Figure 7–77A

separately from tool), as listed in Figure 1. See IS 9047, packaged with the tool, for information on the hand crimping tool.

3. ASSEMBLY PROCEDURE

WARNING *Be very careful to dispose of fiber ends properly. The fibers create slivers that can easily puncture the skin and cause irritation. Also, always wear safety glasses when working with optical fibers.*

A. Preparing Fibers

1. Slip heat-shrink tubing over the fiber jacket.

CAUTION *Do NOT use heat-shrink tubing with plastic fiber cables. The heat required to shrink the tubing will damage the fiber.*

2. If the fiber jacket has a diameter of less than 2.3 mm (.09 in), slip the spacer sleeve over the jacket.
3. Slide the metal retaining assembly onto the cable.
4. Strip the cable to the dimensions shown in Figure 2.

CAUTION *Do not touch bare fibers with your fingers; oil from the skin weakens the fiber by making it brittle.*

5. Strip the outer jacket using AMP Cable Stripper 501198-1.
6. Cut the strength members to length using sharp scissors.
7. Remove the buffer from the fiber. Most buffers can be stripped mechanically using AMP Fiber Stripper 501013-1. Some very tight buffers and plastic-clad silica (PCS) cables may have to be soaked in paint remover for 2 minutes to remove the buffer.

CAUTION *When using paint remover, soak only the length of buffer to be removed.*

8. Wipe the fiber clean with a soft tissue.

B. Terminating Fibers

NOTE *When applying epoxy to the ferrule, keep the ferrule vertical with the nose pointing down. This ensures the proper spreading of the epoxy into the front of the nose.*

1. Remove the epoxy from the envelope; then remove the separating clip from the epoxy packet and mix the epoxy thoroughly. We recommend using AMP Epoxy Mixer 501202-1 and stroking the packet vigorously for 20 to 30 seconds. Cut the packet open and pour the epoxy onto the envelope.
2. Apply six drops of epoxy deep inside the ferrule. Figure 3 shows the points at which to apply the epoxy.
3. Gather the strength members to one side, and dip the fiber into the epoxy until the fiber is lightly coated.
4. Insert the fiber into the ferrule using a semi-circular rotating motion. *Do not force the fiber into the ferrule.* The fiber must extend out the front of the ferrule. The strength members must be outside the ferrule.

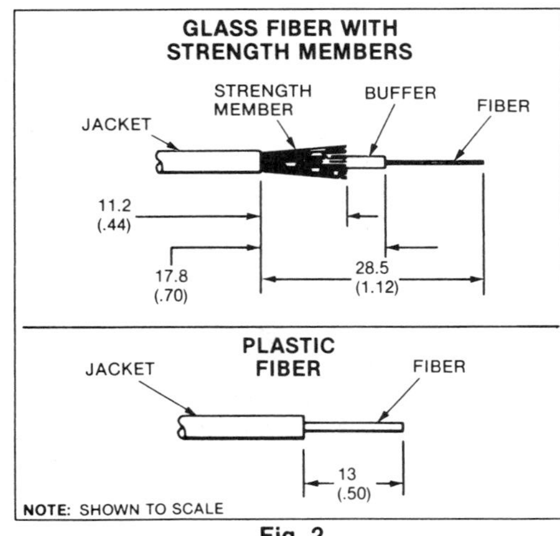

Fig. 2

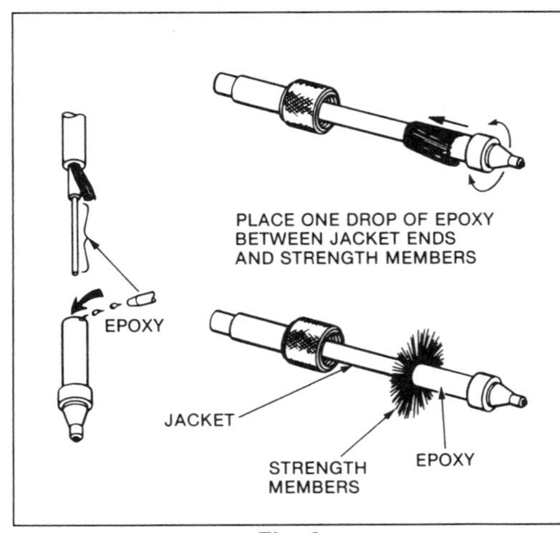

Fig. 3

Figure 7–77B

5. Press the polishing bushing onto the nose of the ferrule.

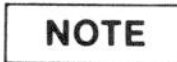

In the following steps, use only a small amount of epoxy. Excess epoxy may bleed out during curing and cause the coupling nut to bind.

6. Slide ferrule onto fiber until strength members flare, then spread small dabs of epoxy on the outside of the ferrule barrel about 6.3 mm (1/4 in.) from the cable entry end. See Figure 3.

7. Apply a drop of epoxy to the end of the fiber jacket.

8. While holding the ferrule back against the jacket, slide the retaining assembly over the ferrule, trapping the strength members between eyelet and ferrule. See Figure 3. Thread the coupling nut onto the polishing bushing.

C. Crimping

1. Squeeze the handles of the hand tool until the ratchet releases. Open the tool fully.

2. Place the connector into the tool so that the coupling nut rests against the locator. See Figure 4.

3. Close the handles until the ratchet releases, then allow the tool handles to open and remove the crimped assembly.

4. Slide the heat shrink tubing over the eyelet 1.5 mm (.06 in.) from nut and apply heat evenly until the tubing shrinks tightly on the eyelet and jacket.

5. Set the assembly aside until the epoxy cures. The cables should be hung, with the connector down, until the epoxy cures.

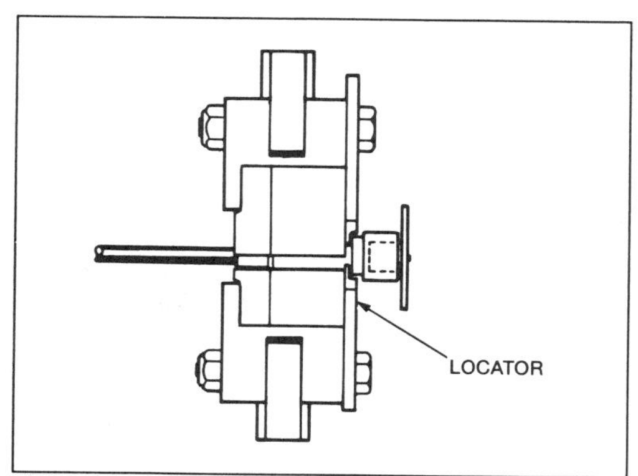

Fig. 4

D. Polishing the Fiber

Polishing may be done by hand or with AMP Polishing Machine 501186. This sheet describes hand polishing. For information on using the polishing machine, see IS 9107, supplied with the machine.

Safely dispose of the excess fiber after it is removed. A handy method is to put a small piece of masking tape on the fiber before scribing it. The fiber is then easily retrieved and disposed of.

1. With the beveled edge of the scribe tool facing up, lightly scribe the fiber, then pull the fiber straight away from the polishing bushing (see Figure 5).

NOTE *Plastic fiber can be cut with a sharp knife. When doing so, do NOT cut the fiber flush with the polishing bushing — leave a small amount protruding so that it can be removed during polishing.*

2. Make sure you have a clean surface on which to polish the fiber. The polishing surface must be flooded with water, preferably running water.

During polishing, the water and lapping film must be kept clean to prevent abrasive particles from scratching or chipping the fiber surface.

3. Place the 5-μm lapping film, dull side up, on the polishing surface.

4. Fine-grind the tip with an elongated figure-8 motion. Begin with light pressure; too much pressure could cause the protruding fiber to fracture or splinter. See Figure 6.

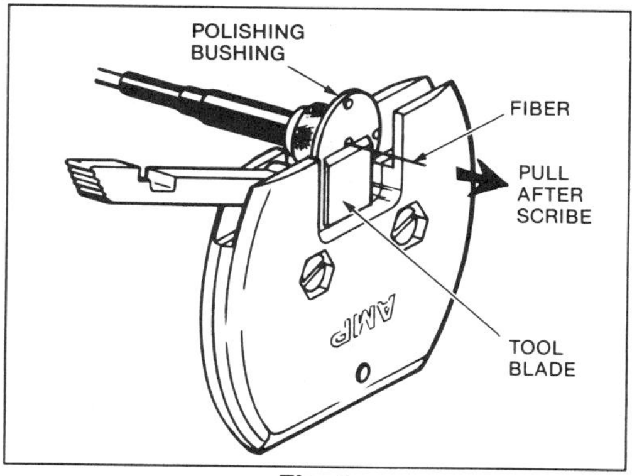

Fig. 5

Figure 7–77C

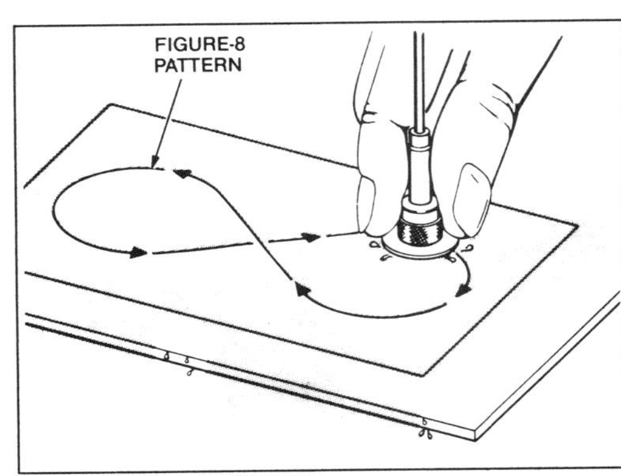

Fig. 6

NOTE *Occasionally rotate the connector in your grip as you are polishing it.*

5. Fine grind the tip until the following conditions are met:
   — The outer pads on the bushing are almost flush with the flat of the bushing.
   — Polishing may be apparent on the face of the bushing where the high spots have been removed.
   — About 90% of the ferrule tip is flat.

6. Replace the 5-μm film with the 1-μm film.

7. Continue polishing on 1-μm film until the following conditions are met:
   — The ferrule is flat and has a high-gloss finish.
   — Signs of polishing are apparent in a circle around the face of the bushing.
   — All pads are flush with the face of the bushing.

WARNING *Never inspect or look into the end of a fiber when optical power is applied to the fiber. The infrared light that is used, although it cannot be seen, can injure eyes and cause blindness.*

8. Inspect the ferrule and fiber under the magnifier or microscope. See Figure 7. To view the ferrule properly through the magnifier, light must be reflected off the ferrule and through the magnifier. Check for the following:
   — Deep scratches on the ferrule indicate the need for further polishing with 1-μm film. (Fine polishing lines are acceptable.)
   — *Small* chips in the outer rim of the fiber are permissible. Large chips, or chips in the center of the fiber, mean either that further polishing is needed or that the termination is unacceptable and the fiber must be reterminated.

NOTE *See IS 9111 for information on the use of AMP Microscope 501196.*

9. Remove the polishing bushing. Carefully dry the connector with a soft cloth or tissue.

10. Place the dust cover over the connector if the fiber is not going to be installed in the housing immediately.

The assembly is now ready for use. See Figure 8.

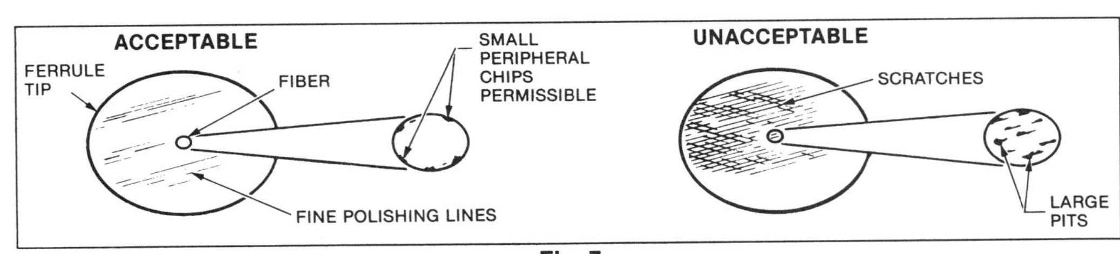

Fig. 7

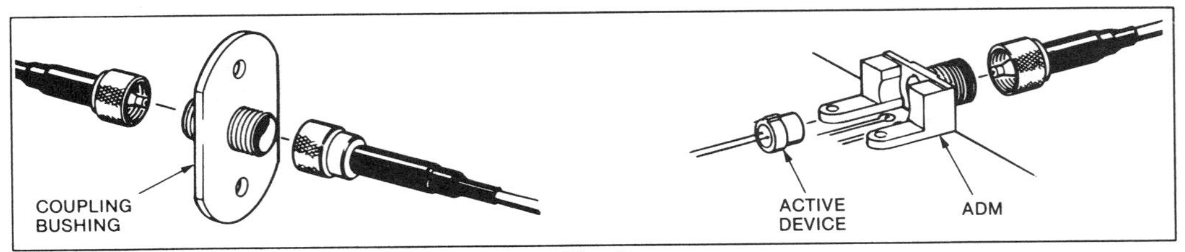

Fig. 8

Figure 7–77D

AMP★ OPTIMATE★ ACTIVE DEVICE MOUNT KITS AND COUPLING BUSHING FOR AMP SIMPLEX FIBER-OPTIC CONNECTORS

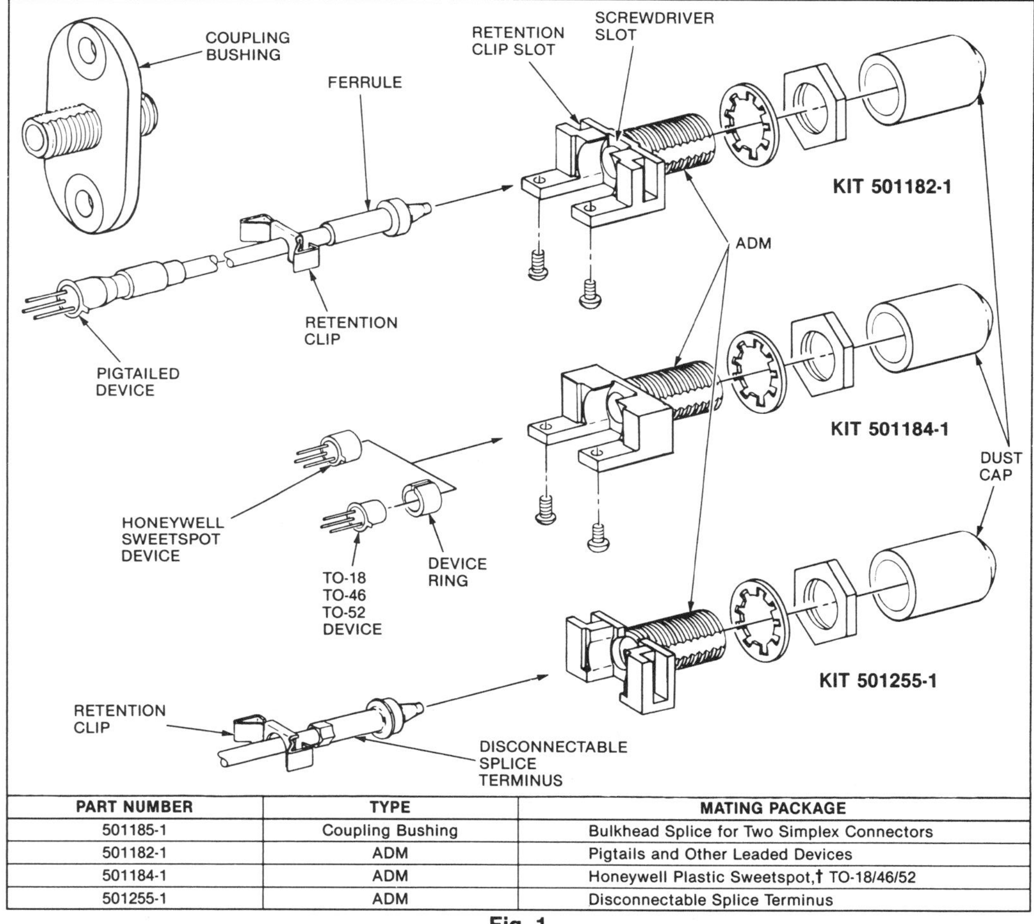

PART NUMBER	TYPE	MATING PACKAGE
501185-1	Coupling Bushing	Bulkhead Splice for Two Simplex Connectors
501182-1	ADM	Pigtails and Other Leaded Devices
501184-1	ADM	Honeywell Plastic Sweetspot,† TO-18/46/52
501255-1	ADM	Disconnectable Splice Terminus

Fig. 1

1. INTRODUCTION

This instruction sheet (IS) covers AMP OPTIMATE active device mount (ADM) kits and coupling bushing used with AMP Simplex fiber-optic connectors.

See Instruction Sheet IS 9100 for information on AMP Simplex Connectors. See IS 9102 for information on disconnectable splices.

NOTE *All dimensions on this sheet are in millimeters first, with inch equivalents in parentheses.*

2. DESCRIPTION (Figure 1)

The coupling bushing is used for bulkhead-mounted fiber-to-fiber splices.

ADMs connect simplex connectors to active devices or other fibers. Three versions are available. One version accepts Honeywell low-cost plastic Sweetspot devices, as well as devices meeting JEDEC specifications for TO-18, TO-46, and TO-52 devices, and having an O.D. of 4.67 ± .025 mm (.184 ± .001 in.) × 3.12 mm (.123 in.) (min) long

★ Trademark of AMP Incorporated
† Trademark of Honeywell Incorporated

Figure 7–77E

for optimum fit and performance. The Honeywell devices are held in the ADM by a press-fit. Other devices require a device ring that fits around the device to center it in the bore. The alignment tab of the device must be removed to use the device in the ADM.

The second type of ADM accepts pigtailed devices with the fiber terminated in an AMP fiber-optic ferrule.

The third type is used for fiber-to-fiber connections, such as those found in AMP undercarpet floor fittings. It accepts fiber terminated with a disconnectable splice terminus.

ADM Kits 501182-1 and 501184-1 can be board mounted and are supplied with screws; Kit 501255-1 can only be panel mounted. All are supplied with a lockwasher and a jam nut for panel mounting, and a protective dust cap. A device ring is supplied with ADM Kit 501184-1. A retention clip to hold the ferruled pigtail or terminus in the ADM is provided with ADM Kits 501182-1 and 501255-1.

Maximum recommended panel thickness for panel mounting an ADM is 1.57 mm (.062 in.). Nominal board thickness recommended for the screws provided is also 1.57 mm (.062 in.).

The ADMs for pigtailed devices or termini have a slot for a screwdriver or similar tool to aid removal of the ferrule. Figure 2 shows the use of the screwdriver.

Figure 3 provides the dimensional information for the coupling bushings and ADMs that is required for planning pc board layouts or panel cutouts.

3. ASSEMBLY AND INSTALLATION

The following are suggested methods for installing bushings and ADMs. The exact procedure to be used depends on your specific manufacturing and assembly requirements.

A. Coupling Bushing

The bushing is simply mounted in a properly dimensioned cutout; then the simplex connectors are connected.

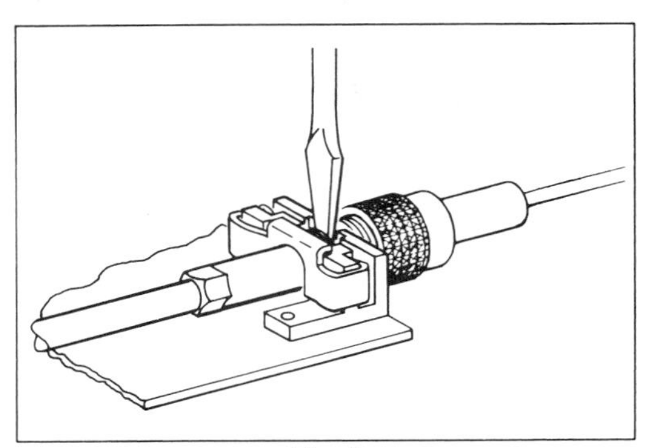

Fig. 2

B. ADM Kit 501182-1 (For Pigtails) or 501255-1 (For Termini)

1. As necessary, secure the ADM to the board with the screws provided or in a panel with the washer and nut.

NOTE *ADM Kit 501255-1 can be panel mounted only.*

2. Press-fit the ferruled pigtail or terminus into the rear of the ADM.
3. Install the retention clip in the slots of the ADM to secure the ferrule.
4. Install the dust cap if a simplex connector is not to be mated with the ADM immediately.

C. ADM 501184-1 (For Discrete Devices)

1. As necessary, secure the ADM to the board with the screws provided or in a panel with the washer and nut.
2. For TO cans, slip the device ring around the can. A Honeywell plastic Sweetspot device does not require the ring. Be sure to remove the tab from the device.
3. Press-fit the device into the rear of the ADM.
4. Insert the device leads into the pc board and solder them.
5. Install the dust cap if a simplex connector is not to be mated with the ADM immediately.

Figure 7–77F

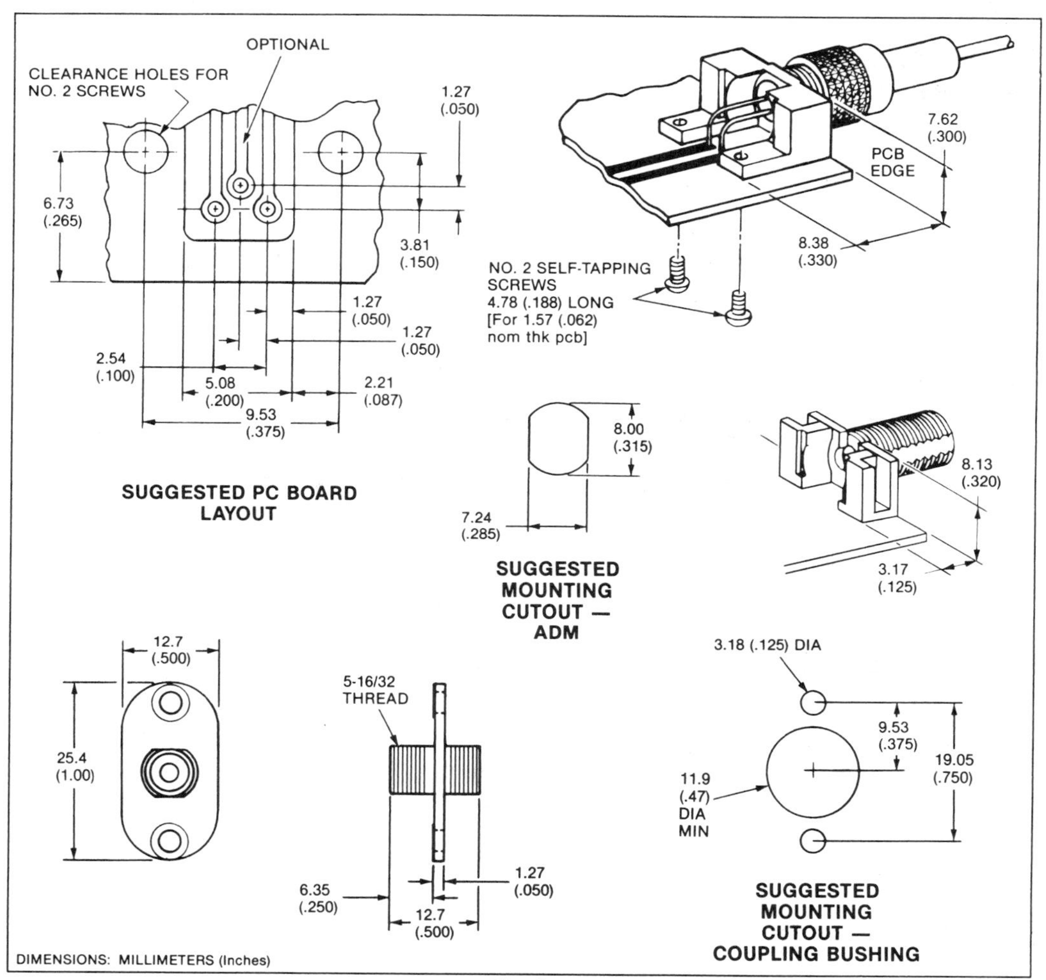

Fig. 3

Figure 7–77G

AMP* OPTIMATE* DUPLEX
FIBER-OPTIC CONNECTORS

Instruction Sheet
IS 6856
8/25/86
RELEASED

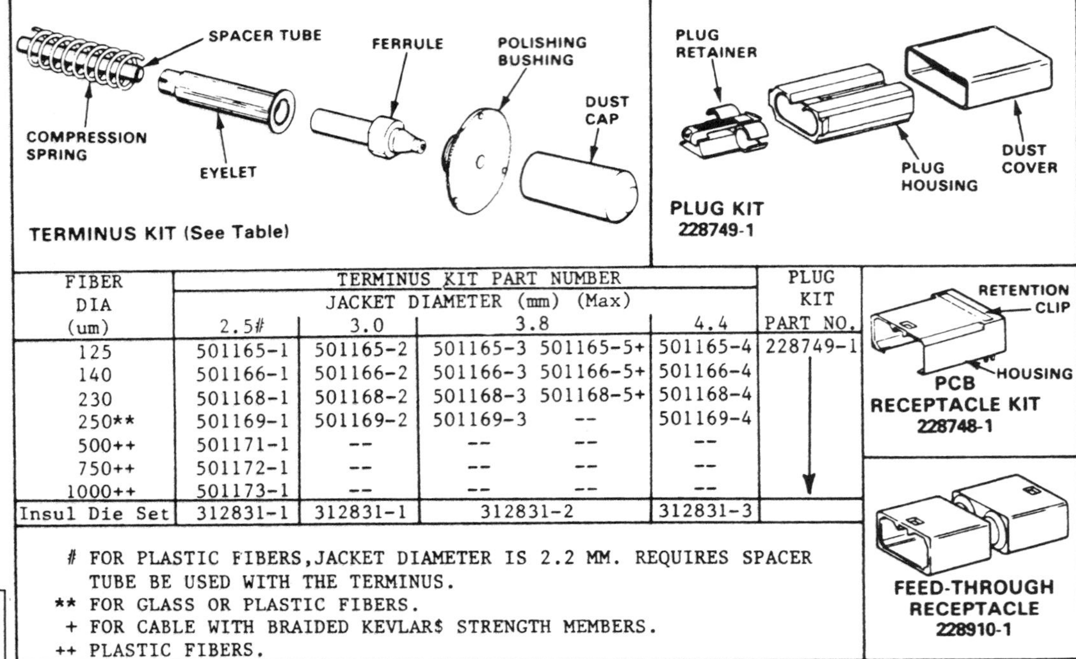

FIBER DIA (um)	TERMINUS KIT PART NUMBER — JACKET DIAMETER (mm) (Max) 2.5#	3.0	3.8		4.4	PLUG KIT PART NO.
125	501165-1	501165-2	501165-3	501165-5+	501165-4	228749-1
140	501166-1	501166-2	501166-3	501166-5+	501166-4	↓
230	501168-1	501168-2	501168-3	501168-5+	501168-4	
250**	501169-1	501169-2	501169-3	--	501169-4	
500++	501171-1	--	--	--	--	
750++	501172-1	--	--	--	--	
1000++	501173-1	--	--	--	--	
Insul Die Set	312831-1	312831-1	312831-2		312831-3	

FOR PLASTIC FIBERS,JACKET DIAMETER IS 2.2 MM. REQUIRES SPACER TUBE BE USED WITH THE TERMINUS.
** FOR GLASS OR PLASTIC FIBERS.
+ FOR CABLE WITH BRAIDED KEVLAR$ STRENGTH MEMBERS.
++ PLASTIC FIBERS.

Fig. 1

1. INTRODUCTION

This instruction sheet (IS) covers the assembly and use of AMP OPTIMATE duplex fiber-optic connectors. The duplex connectors include the following:

- Terminus kits, as listed in Figure 1, which terminate optical fibers. Two kits must be ordered for each duplex connector.

- Plug Kit 228749-1, which holds two termini.

- PCB Receptacle Kit 228748-1, which houses two TO-52 active devices and mates with the plug.

- Feed-Through Receptacle 228910-1, which accepts two plugs to form a fiber-to fiber splice.

This sheet is organized as follows:

- Paragraph 2 describes the connector components.

* Trademark of AMP Incorporated
$ Trademark of E.I. Dupont De Nemours & Co.

PAGE 1

Figure 7–78 Optimate multiple-position fiber optic connector installation procedures (sheet 1 of 8) (*courtesy of AMP, Incorporated*).

- Paragraph 3 provides instructions for applying a terminus kit to an optical fiber.

- Paragraph 4 covers the assembly and application of plugs and receptacles.

Also available are installation and designer kits to aid in the application procedure. Installation Kits 501258-1 through -4 contain all tools and materials required (see IS 9149 for kit contents). Designer Kit 501219-2 contains an assortment of terminus kits, plugs and receptacles, and tools.

Read this material thoroughly before starting assembly.

2. DESCRIPTION

Figure 1 lists the terminus kits for various fiber sizes and jacket diameters. Besides the ferrule, into which the fiber is epoxied, each kit contains one each of the following: an eyelet, a compression spring, a polishing bushing, and a dust cap. A spacer tube, used with cables having a jacket diameter of less than .1 in. (2.5 mm), is also included with appropriate kits.

The PCB receptacle consists of the housing and retention clip. The housing accepts active devices having a TO-52 outline. The retention clip secures the housing to the printed-circuit (pc) board. The feed-through receptacle is a single piece that accepts a plug at both ends.

The plug assembly consists of a plug housing, a plug retainer, and a dust cover. The plug retainer features locking latches that secure internal components in the plug housing when mated.

Both plugs and receptacles are polarized so that a plug mates with a receptacle in only one way.

Epoxy is also required. AMP Epoxy 501195-1 has a cure time of 4 hr at 65°C or 24 hr at 25°C. Epoxy 501195-3 has a cure time of 30 minutes at 25°C.

The buffer coating of the fiber can be removed by using the fiber stripper or by dipping the fiber into a heavy-duty paint remover.

The termini are crimped with AMP Hand Tool 58190-1, which must be equipped with the proper insulation crimp die set listed in Figure 1. See IS 9047, packaged with the tool, for further information on the hand crimping tool.

3. ASSEMBLY PROCEDURE

WARNING: Be very careful to dispose of fiber ends properly. The fibers create slivers that can easily puncture skin and cause irritation.

WARNING: Always wear safety glasses when working with optical fibers.

You can terminate more than one fiber at a time. Do not, however, terminate more than what is practical at a time. Generally, two or three fibers can be terminated before the epoxy begins to gel.

Figure 7-78B

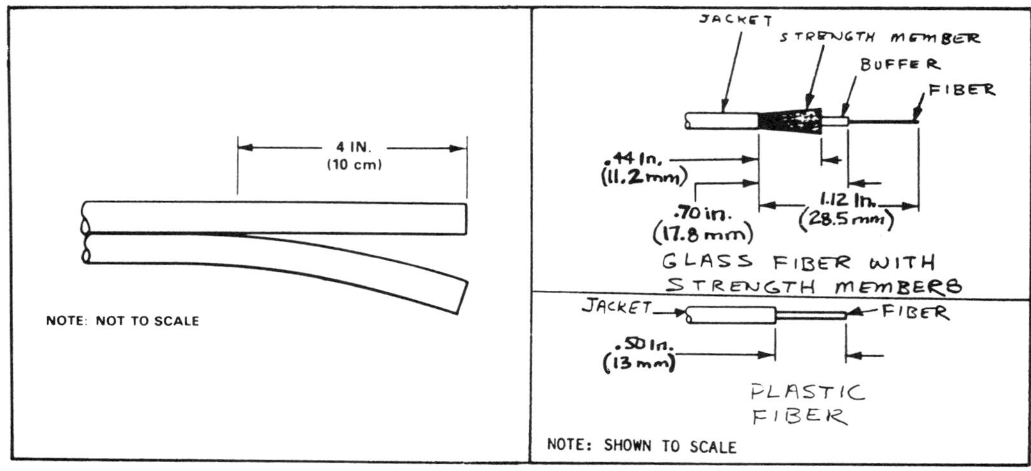

Fig. 2

A. Preparing Fibers

1. For a dual-channel cable, separate the channels at least 4 in. (10 cm). See Figure 2.

NOTE: If the fiber jacket has a diameter over .15 in. (3.8 mm), slip the plug retainer over the jackets before stripping the cable.

2. If the fiber jacket has a diameter of .1 in. (2.5 mm) or less, slip the spacer tube over the jacket.

3. Slide the compression spring onto the cable.

4. Slide the eyelet onto the cable.

5. Strip the cable to the dimensions shown in Figure 2.

CAUTION: Do not touch bare fibers with your fingers, because the oil from the skin weakens the fiber by making it brittle.

6. Remove the primary buffer coating from the fiber. This can be done either by soaking the end of the fiber in paint stripper for about 2 minutes, or by using the fiber stripper.

7. Wipe the fiber clean with a soft cloth or tissue.

B. Terminating Fibers

NOTE: When applying epoxy to the ferrule, keep the ferrule vertical with the nose pointing down. This ensures the proper spreading of the epoxy into the front of the nose.

1. Remove the separating clip from the epoxy packet and mix the epoxy thoroughly for 20 to 30 seconds. Use of Epoxy Mixer 501202-1 is recommended. Open the tube and pour the epoxy into the foil.

Figure 7–78C

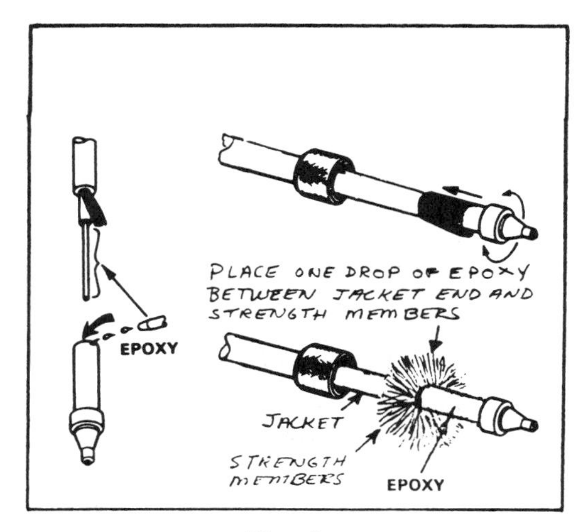

Fig. 3

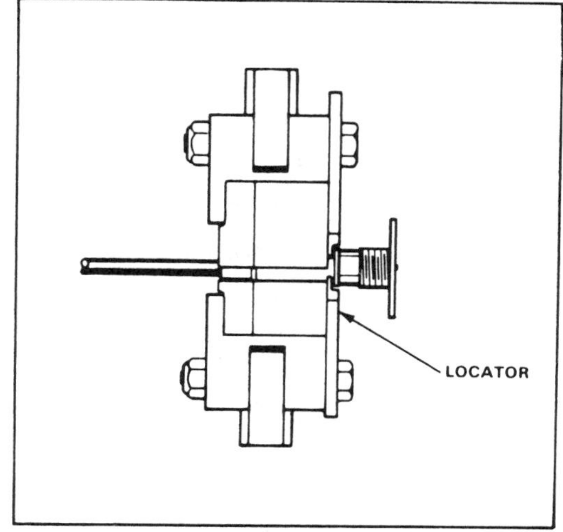

Fig. 4

2. Apply six drops of epoxy deep inside the ferrule. Figure 3 shows you where to apply the epoxy.

3. Gather the strength members to one side, and drag the fiber through the epoxy until the fiber is coated.

4. Insert the fiber into the ferrule with a rotating motion. Do not force the fiber into the ferrule. The fiber must extend out the front of the ferrule. The strength members must be outside the ferrule.

5. Press the polishing bushing onto the nose of the ferrule.

6. Gather the strength members to one side, and apply small smears of epoxy to the outside of the ferrule barrel about 1/4 in. (6.3 mm) from the cable entry end.

7. Apply a drop of epoxy to the end of the fiber jacket.

8. Spread the strength members over the ferrule barrel.

9. While holding the ferrule back against the jacket, slide the retaining assembly over the ferrule, trapping the strength members between the eyelet and the ferrule. See Figure 3.

C. Crimping

1. Squeeze the handles of the hand tool until the ratchet releases. Open the tool fully.

2. Place the terminus in the tool so that the flange of the eyelet rests in the flange of the locator. See Figure 4.

Figure 7–78D

3. Squeeze the handles shut to crimp the eyelet.

4. Set the assembly aside until the epoxy cures. We recommend that the cable be hung, with the terminus down, until the epoxy cures.

D. Polishing the Fiber

Polishing may be done by hand or with AMP Polishing Machine 501186. This sheet describes hand polishing. For information on using the polishing machine, see IS 9107, supplied with the machine.

WARNING: SAFELY DISPOSE OF THE EXCESS FIBER after it is removed. A handy method is to put a small piece of masking tape on the fiber before scribing it. The fiber is then easily retrieved and disposed of.

1. With the beveled edge of the scribe tool facing up, lightly scribe the fiber. The pull the fiber straight away from the polishing bushing. See Figure 5.

NOTE: Plastic fiber can be cut off with a sharp knife. Leave a bit of fiber protruding; do not cut it flush with the bushing.

2. Make sure you have a clean surface to polish the fiber on. The polishing surface should be flooded with water, preferably running water.

CAUTION: During polishing, the water and lapping paper must be kept clean to prevent abrasive particles from scratching or chipping the fiber surface.

3. Place the 5-um lapping film, dull side up, on the polishing surface.

4. Fine-grind the tip with an elongated figure-8 motion. BEGIN WITH LIGHT PRESSURE, since too much pressure could cause the protruding fiber to fracture or splinter. See Figure 6.

NOTE: Rotate the terminus in your grip occasionally as you polish.

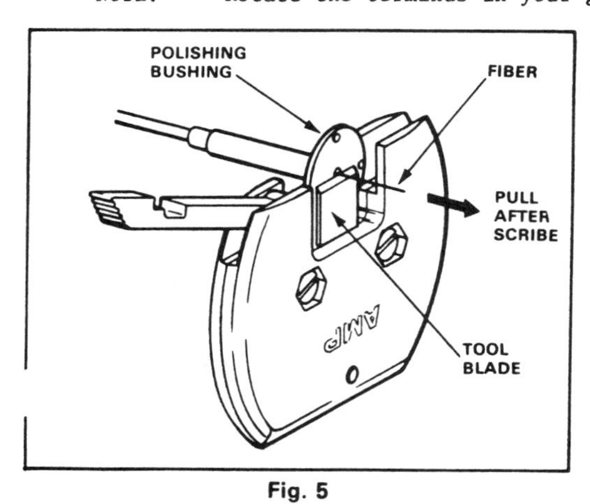

Fig. 5

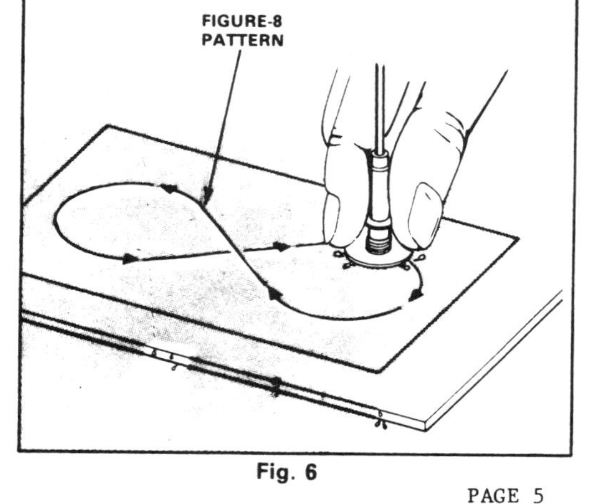

Fig. 6

Figure 7-78E

5. Fine-grind the tip until the following conditions are met:

-- The outer pads on the bushing are almost flush with the flat of the bushing.

-- Polishing may be apparent on the face of the bushing where the high spots have been removed.

-- About 90% of the ferrule tip is flat.

6. Replace the 5-um film with 1-um film.

7. Continue polishing on the 1-um film until the following conditions are met:

-- The ferrule is flat and has a high-gloss finish.

-- Signs are apparent in a circle around the face of the bushing.

-- All pads are flush with the face of the bushing.

WARNING: Never inspect or look into the end of a fiber when optical power is applied to the fiber. The infrared light used, although it cannot be seen, can injure eyes and cause blindness.

8. Inspect the ferrule and fiber under the magnifier or microscope. See Figure 7. To view the ferrule properly through the magnifier, light must be reflected off the ferrule and through the magnifier. Check for the following:

-- Deep scratches on the ferrule indicate the need for further polishing with 1-um film. (Fine polishing lines are acceptable.)

-- Small chips in the outer rim of the fiber are permissible. Large chips, or chips in the center of the fiber, mean either that further polishing is needed or that the termination is unacceptable and the fiber must be reterminated.

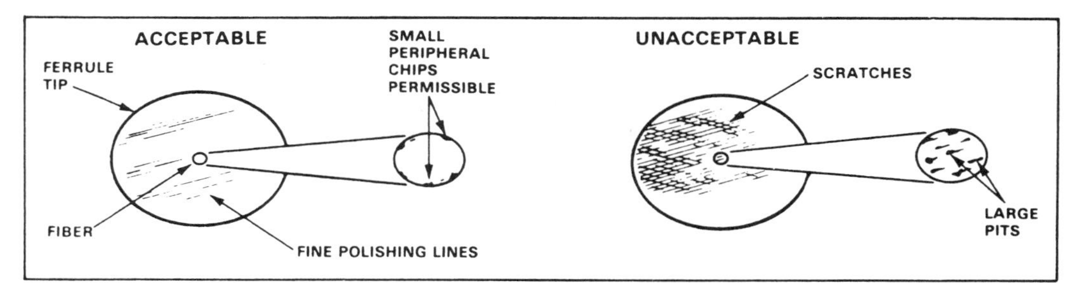

Fig. 7

NOTE: See IS 9111 for information on the use of AMP Microscope 501196.

9. Remove the polishing bushing. Carefully dry the terminus with a soft cloth or tissue.

Figure 7–78F

10. Place the dust cover over the terminus if the fiber is not going to be installed in the housing immediately.

The terminus is now ready for use.

4. PLUGS AND RECEPTACLES

A. Plug Assembly (Figure 8)

1. Make sure the termini to be used are properly applied and polished.

2. Insert both termini into the plug housing.

CAUTION: Ensure that the proper termini are being used and that they are properly located. Make sure each terminus is in the proper side of the retainer at both ends of the cable. Use cable markings or another indicator to ensure this. Once the plug retainer is engaged, it is not removable from the housing.

3. With the plug retainer positioned over the fibers as shown in Figure 8, slide the retainer into the plug housing until the locking latches engage the plug housing.

4. Slide the loaded plug assembly into the receptacle assembly (previously assembled to the pc board) until the latch on the plug assembly engages in the latch hole of the receptacle.

5. To remove the plug assembly, depress the locking latch through the latch hole and withdraw the plug.

6. Place the dust cover over unmated assemblies. The cover should NOT cover the latch of the plug assembly.

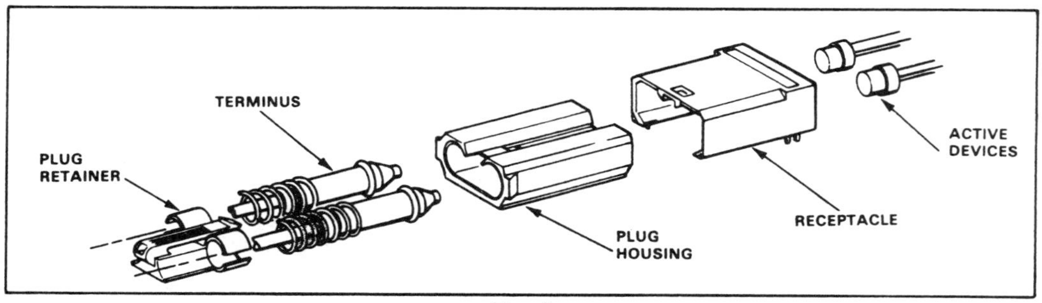

Fig. 8

B. Receptacle Assembly

Figure 9 shows the recommended layout dimensions for the pc board.

1. Install the retention clip through the slots provided in the top of the housing.

Figure 7–78G

IS 6856 AMP OPTIMATE DUPLEX CONNECTOR

2. Install the housing assembly on the pc board by pushing two plastic posts and the retention clip into the appropriate holes in the pc board. See Figure 10.

3. Install the active devices in the housing cavities and insert their leads into the appropriate holes in the board.

4. Hand or wave solder the retention clip posts and the leads of the active device to secure the assembly.

C. Feed-Through Receptacle

1. Insert a plug assembly into each plug cavity of the receptacle until the locking latch engages.

2. To remove a plug assembly, depress the locking latch through the latch hole and withdraw the plug.

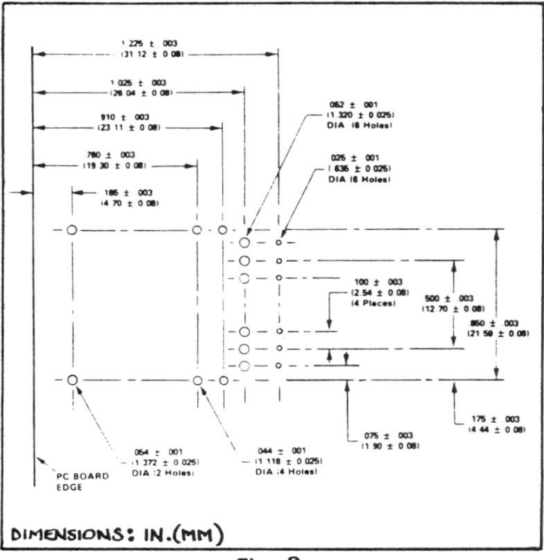

Fig. 9

PC BOARD

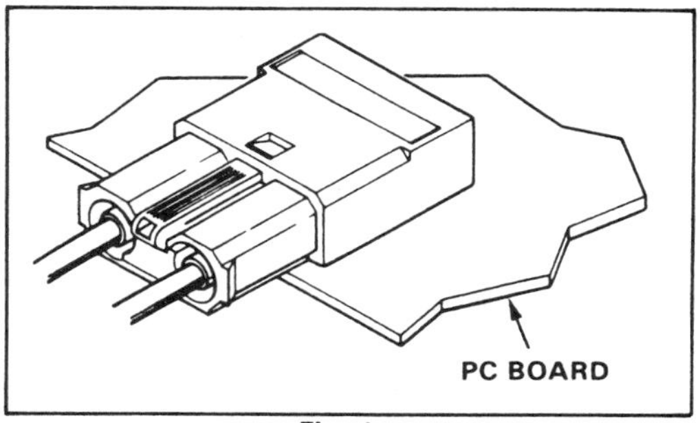

Fig. 10

PAGE 8

Figure 7-78H

FIBER TESTING

As with any industry product, testing of optical fiber is accomplished to ensure that the fiber qualifies under the requirements of specification. Usually tests fall within the major headings of mechanical, optical, electrical, chemical, and physical. This does not preclude the possibility of a company performing a special test peculiar to its own fiber. Tests provide empirical or historical data that help to improve the product or production cost. Tests are performed under controlled temperature, humidity, and atmospheric conditions so as to obtain the same results. Other conditions such as cleanliness, lighting, accuracy of equipment, and personnel proficiency are considerations to be met.

There are literally dozens of tests that are performed on optical fibers. Since there is not space to describe all of these, a sampling of several mechanical and optical tests will be discussed here.

Fiber Tensile Strength

Tensile strength of a fiber is an extremely inportant parameter. It allows the designer of optical fiber systems more flexibility in the choice of components in the system. Further, it provides cable manufacturers with information that may afford the designer a margin of selection for the strength members within the cable. These are opportunities that behoove fiber engineers to develop a method of testing fiber for tensile strength.

Tensile strength of a fiber is governed by stress concentrations along the fiber. It appears from experience that surface laws are principally responsible for high stress concentrations. Failure usually occurs at the deepest surface flaw. Empirical data also have shown that moisture imposed on the fiber surface (and therefore the flaws) tend to enlarge the flaws and cause stress corrosion. Stress corrosion has a fatigue limit that may cause premature failure.

Accurate measurement of tensile strength is complicated by the fact that fibers are usually long. Therefore, testing involves checking of tensile strength for selected or random points on the fiber. By this method, large but rare flaws may be detected and then studied to determine their causes. Once the rare flaws are known, a sophisticated system of testing may be employed.

Proof Test

Each fiber manufacturer has a method of checking rated fiber tensile strength. Figure 7–79 illustrates a test set up whose principle of operation parallels tensile strength testing in the industry. An apparatus is set up with four equal diameter pulleys (A,

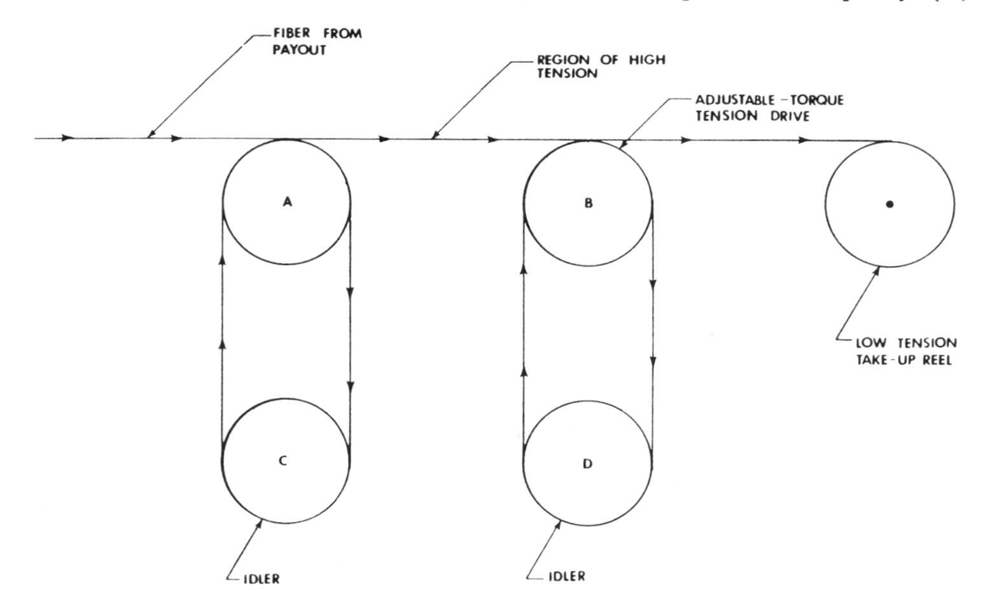

Figure 7–79 Proof test (*courtesy of ITT Cannon Electric*).

B, C, D). The pulleys are mounted on axles on a strong mounting structure. Belt-driven motors provide rotation. Pulley A is motor-driven at a constant speed. A magnetic clutch on pulley B provides for tensile-loading and rapid disengagement. The other two pulleys, C and D, are idlers.

A fiber specimen is strung through the constant-drive pulley A and the idler pulley C, and then the clutch pulley B and idler pulley D. Tension level is set by adjustign torque tension pulley B to a predetermined proof load. A friction wheel within the clutch of pulley B allows it to load-release quickly.

The entire length of the fiber is fed progressively through the setup while subjected to the constant tensile load for a period of time determined by the speed of drive pulley A and the distance between the centerlines of pulleys A and B. On the figure this area is called the region of tension. After passing over pulley B and D, the fiber is wound on a takeup reel with low tension.

Testing parameters are proof load, time under load, length of fiber under load, and number of breaks, if any.

Dynamic Short-Length Strength Test

As you may recall, the proof test called for the testing of rated tension that should not break the fiber. The dynamic short-length strength test has tension applied until the fiber fractures. In Figure 7–80, a length of fiber (say 2 meters) is placed between spools. The top spool is a holding spool that is fixed to a tension gauge. The lower spool is a constant-speed drive consisting of a threaded shaft, drive motor, a power supply, and a pulley attached to the threaded shaft. The power supply drives the motor, which turns the pulley. The pulley traverses the threaded shaft at a constant rate. The entire apparatus is attached to a rail for strength.

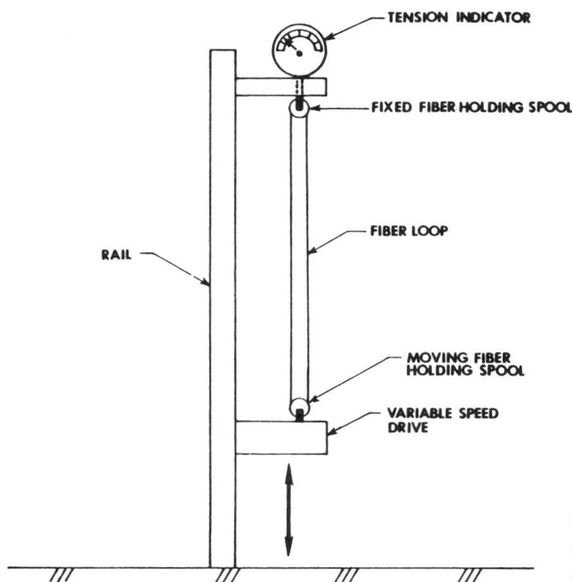

Figure 7–80 Dynamic short-length strength test (*courtesy of ITT Cannon Electric*).

A predetermined tensile load is applied through the lower spool-drive mechanism at a specified speed until the fiber fractures. The fracture load, speed, and diameter of the fiber ends are monitored and recorded. Sample lengths from production runs are tested to accumulate an average for statistical and historical data. Data accumulation includes average tensile strength, load rate, fiber identification, and the number of samples.

Static Short-Length Strength Test

The difference between static and dynamic tests is obvious. The fiber moves in a dynamic test and is static in a static test. As in the dynamic test, loads are applied to the fiber until it fractures. In Figure 7–81, a specimen of fiber is placed between two pulleys under tension to prevent sagging. The pulleys are of equal diameter. A bracket on the right pulley in the figure is mounted to holding posts and/or tension gauges. One or the other pulleys have a ratchet jack for application of tension. The fiber ends are stripped of their jackets. A light source is applied to one end of the fiber and a convenient observation source on its other end. Two gauge markers and a measuring tape are placed on the cable (fiber) between the two pulleys.

Tension is applied to the fiber by operating the ratchet jack. Tension is applied at predetermined increments and monitored on the tension meter. The fiber length between gauge markers is measured at each increment until the fiber fractures. Significant parameters are tensile load, fiber elongation, and load at fracture. Fiber load is 50% of the tension gauge indication.

Attenuation Test

Fiber attenuation is the ratio of power loss in decibels (dB) to the length of the fiber. The test setup for fiber attenuation is illustrated in Figure 7–82. The light source is a standard microscopic tungsten-filament bulb mounted within an illuminator housing. By adjustment of focusing optics a collimated beam is launched through a filter.

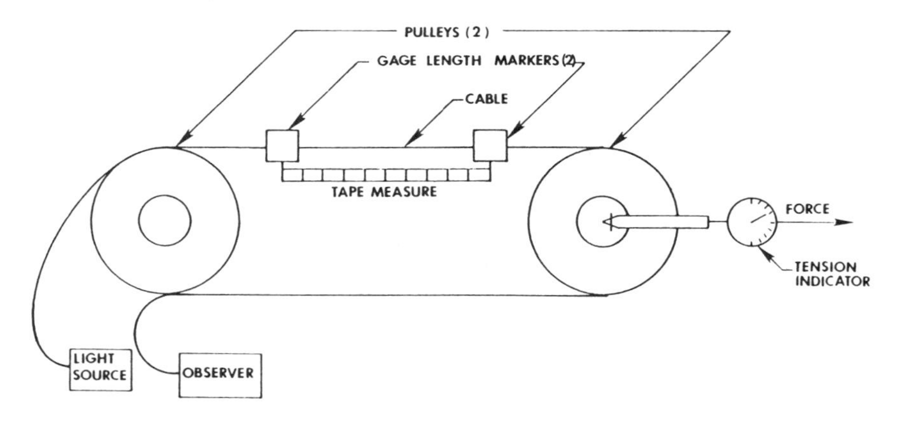

Figure 7–81 Static short-length strength test (*courtesy of ITT Cannon Electric*).

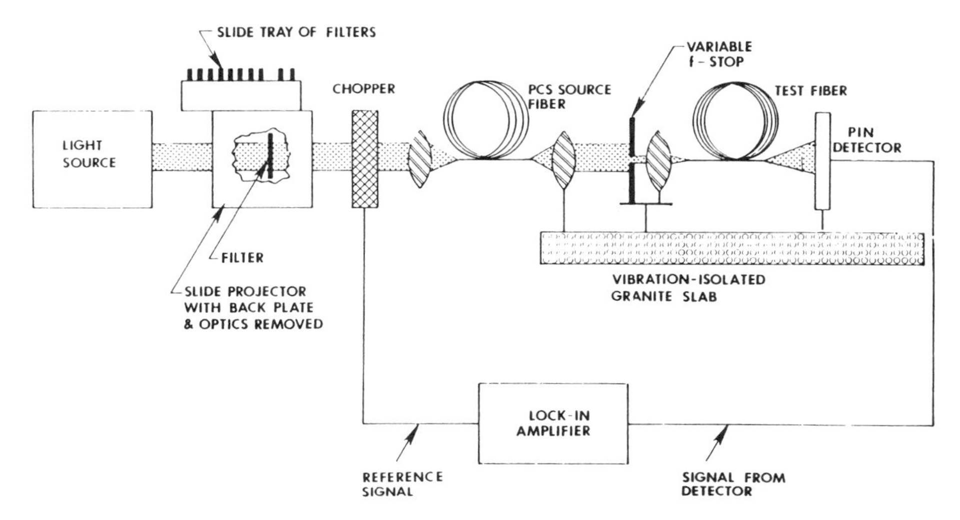

Figure 7–82 Attenuation test (*courtesy of Alcatel Cable Systems Group*).

Collimating is making beams parallel or straight. A slide tray of filters covering a variety of wavelengths from say 40 to 100 microns in length and in 0.01-, 0.02- or 0.04-micron steps drops filters into a slide projector for filtering action. The filtered-collimated beam is directed into an optical chopper, where it is mechanically chopped to a reference signal of, say, 84 hertz. This chopped reference signal is coupled into a lock-in amplifier. The purpose of the reference signal is to screen out unwanted background light and to improve test sensitivity.

The chopped, filtered, and collimated beam is focused into a large-diameter plastic-clad silica fiber that is used as a point source for the fiber under test. The source fiber is attached to optical apparatus mounted on a vibration-free granite slab. The signal is again collimated through a set of television cameras with variable f-stops. The variable f-stops of the cameras allow the fiber under test to be illuminated by specific numerical apertures (NAs). The fiber end is positioned (aligned) with the light beam by a five-axis micropositioner. This establishes maximum light launching into the fiber.

The test fiber output is adjusted by a three-axis micropositioner to ensure maximum light beam to fall on light-shielded, high-quality photodiode (PIN) detector. Leads from the PIN diode are connected to the lock-in amplifier. The test fiber is prepared by stripping its jacket, scribing and breaking the fiber perpendicular to its core, and cleaning. Long-length attenuation testing now takes place.

The fiber input end is placed into the five-axis micropositioner. Bandpass filters are selected for the fiber type and placed into the source optics. Focusing optics adjust for maximum NA and maximum output. The signal output is then compared with a reference signal and the difference voltage signal read from the lock-in amplifier display. Note that the signal readout is a voltage signal that represents fiber attenuation.

Data are collected at each specific filter and NA setting by reading voltage value levels on the lock-in amplifier display. A baseline may be established by reading the lock-in amplifier display with the light beam blocked. Short-length testing of fiber may be compared by cutting a length of fiber off the injection end and preparing it, then installing it back in the test setup. No adjustment of the test setup should be made. Readings from the lock-in amplifier are again taken. Long-length and short-length fiber tests results are then used to establish the attenuation coefficient.

Dispersion Test

Dispersion, as you may recall, is the spreading of ray arrival times at a receiver. Bandwidth reduction increases with length and is usually expressed in megahertz per kilometer. For pulse signals the effect results in pulse spreading which is generally expressed in nanoseconds per kilometer. Testing of pulse dispersion is readily explained. In Figure 7–83, a timing control unit provides a trigger pulse to the external trigger (horizontal) of a sampling oscilloscope. The timing control unit consists of an integrated-circuit clock triggered by a low-frequency oscillator. A second pulse from the timing control unit triggers a laser driver. A step time delay in, say, 100-nanosecond steps is transmitted to the oscilloscope.

A laser diode is driven by the laser driver. A laser trigger pulse from the timing control pulse initiates the operation. The laser produces a narrow light bandwidth into the optical setup. The optical system (in the center of the illustration) consists of four microscope-objective lenses arranged on a square plate. A beamsplitter is

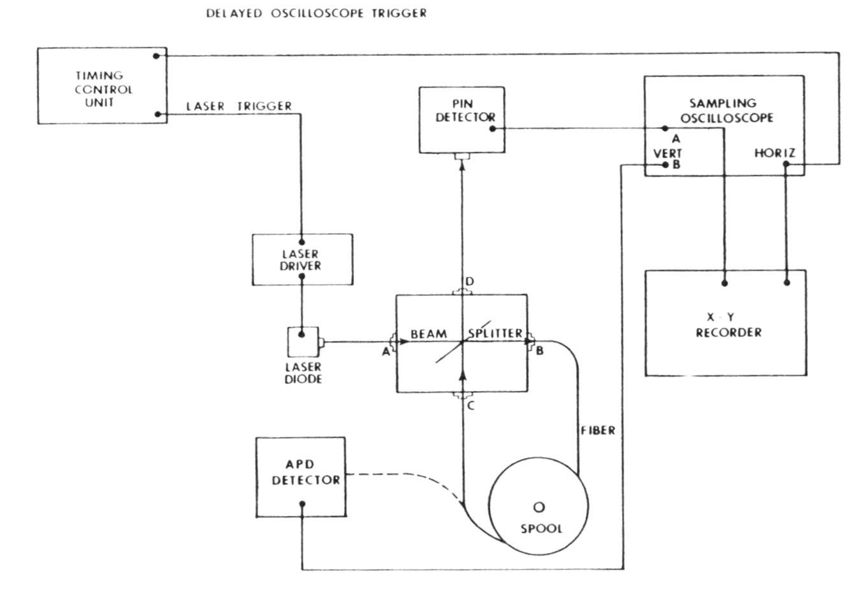

Figure 7–83 Dispersion test (*courtesy of Alcatel Cable Systems Group*).

aligned at a 45° angle with respect to the optical axis on the square plate as shown in the figure. Light directed from the laser diode is collimated at objective lens A. The beamsplitter reflects part of the beam into objective lens D where it is focused onto the PIN detector (photodiode) as a reference level. The remainder of the light is transmitted through the beamsplitter and focused through objective lens B and into the fiber under test. The opposite end of the fiber under test is attached at objective lens C, where the output light is recollimated and directed through the beamsplitter and focused by objective lens D into the PIN detector.

An APD (avalanche photodiode) detector is mounted external to the test setup. Before the test, the fiber end is installed on the APD detector. The APD detector output is displayed on sampling oscilloscope for initial alignment, by manipulation of the input end micropositioner. Long-length measurements are begun. The fiber is removed from the APD detector and placed on the objective lens C. The timing control unit is originally set to 0.0 delay, then in successive increments of 0.1 microsecond until the pulse is observed on the sampling oscilloscope. The fiber input micropositioner is fine-adjusted to get a maximum output picture on the oscilloscope.

Laser driver current is adjusted just above threshold and then at threshold. The traces of the above-threshold and at-threshold laser operation are recorded on the x-y chart recorder.

Short-term measurements are then taken. The fiber is shortened by several meters and reprepared. Alignment is made and traces obtained in the same manner as long-length measurements. Peak signal current is determined by measuring the maximum difference between laser threshold and the pulse waveform on the x-y chart recording. This current difference is then multiplied by 0.5 to determine the −3-decibel pulse height. You may recall from past electronic experience that the −3-decibel point is that point where power loss has fallen to half or actually 0.501.

The half-power pulse height is then marked on the leading and trailing edges of the pulse waveform, using the baseline voltage as a reference. Pulse-width is measured between the leading- and trailing-edge cross marks.

Numerical Aperture (NA) Measurements Using the Scanning Method

A shelf is installed on a rotating stage. The fiber output end is taken from the attenuation test illustrated in Figure 7–84 and attached to the shelf. The shelf is marked so as to be able to bring the fiber endface parallel to the detector at 0° rotation angle of the rotating stage. A large-area detector is positioned about 40 centimeters from the fiber output face. This test setup is identical to the attenuation test setup just discussed and illustrated in Figure 7–82, with minor variations. The rotating stage is placed between the output end of the fiber under test and the detector.

A reference signal is placed into the lock-in amplifier and the detector leads are attached to the lock-in amplifier. The detector is set so that it collects maximum signal at 0° rotation of the stage. To measure the fiber output pattern the rotating stage is adjusted to an angle which produces zero signal (less than 10%) at maximum

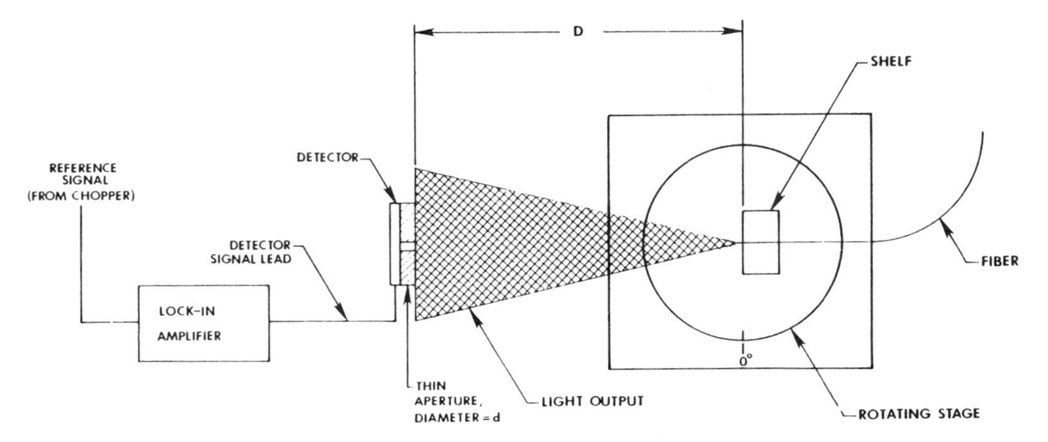

Figure 7–84 Numerical aperture (NA) test setup (scanning method) (*courtesy of Alcatel Cable Systems Group*).

full-degree angle (example 16° rotation). A horizontal line is drawn on graph paper. This line represents the distance (40 centimeters) between the fiber output face and the detector. It also represents the maximum full-degree angle which just produced a zero signal (less than 10%).

The rotating stage is then turned in the opposite direction in 1° increments and detector voltage readings are taken at each stop until the beam limit is reached on the opposite side of the beam pattern. That is the maximum full-degree angle which produces a zero signal (less than 10%). A line is drawn on the graph paper from the fiber endface (the right side of the horizontal line) at an angle equal to the difference between the two maximum zero-signal angles. These are determined by the rotation of the shelf. The fiber NA can then be calculated simply by taking the sine of half the angle between the two zero signals (less than 10% points).

A second method for graphing is to draw a horizontal line that represents the 40-centimeter distance between the fiber face and the detector. Draw a second line vertically in each direction on the detector end of the horizontal line. Indices on the vertical line should represent lock-in amplifier voltage output as related to angle of shelf rotation. Maximum full-degree angle voltages are plotted on the graph vertical as read on the lock-in amplifier. These zero-signal points are connected to the fiber end of the horizontal line. One-half the angle made by the two extremes is used to calculate the NA. That is, NA $= \sin \frac{1}{2}$ the angle between the two extreme zero-signal points.

Numerical Aperture (NA) Measurements Using the Trigonometric Method

The test setup is mounted on a vibration-isolated granite slab (see Figure 7–85). The fiber under test is prepared in the normal manner. The input end of the test fiber is placed on a five-axis micropositioner. A wide-angle light source launches a beam

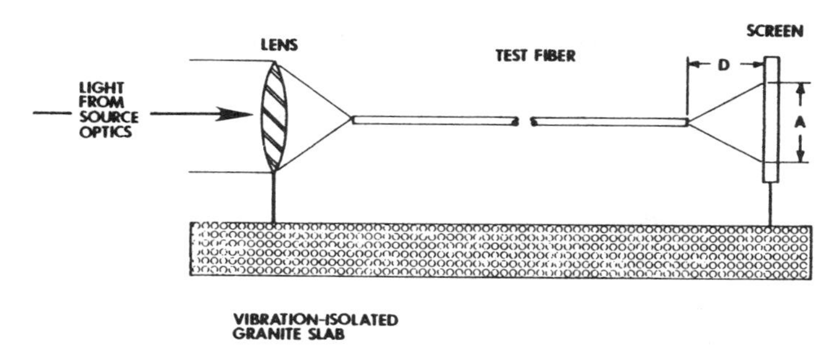

Figure 7-85 Numerical aperture (NA) test setup (trigonometric method) (*courtesy of Alcatel Cable Systems Group*).

of light through optics into the test fiber. The opposite end (output) is placed in front of an image screen. The screen is positioned a known distance (D) from the fiber endface so that the output light spot falls on the screen. Alignment of the fiber and the input optics is made so that the output spot is maximum size and intensity. A vernier caliper is used to measure the spot size.

Calculation of the numerical aperture (NA) is made with the results using the following formula:

$$\text{NA} = \frac{A}{\sqrt{A^2 + 4D^2}}$$

where A = spot diameter
D = distance from fiber to screen

Some General Test Procedures Used in the Fiber Optics Industry

Test procedures are varied and range from simple to complex. The test procedures listed here are just a small number of ones that are in actual use. Each manufacturer will also have their own method or technique to perform the test for best results and speed of accomplishment.

Fiber-size measurement. This test measures the core and cladding of an optical fiber. The test is accomplished using a microscope, a prepared fiber, and a predetermined calibration index.

Fiber bundle measurement. The measurement checks bundle size by dipping the bundle into wetting liquid, then placing the bundle into sizing gauges beginning with the largest gauge.

Number of fibers. Determination of the total number of fibers in a fiber bundle where the number of fibers is large. The fiber bundle end is prepared and photographed. The photo is enlarged to applicable size and the fibers are counted.

Number of transmitting fibers. Test is used to determine the number of fibers in a bundle that are transmitting light. Several photographs are taken using different intensity settings. Photos are enlarged as necessary and the fibers transmitting are counted.

Cyclic flexing. A procedure used to determine the ability of a fiber optic cable to withstand cycle flexing. The fiber is placed in a cyclic arm driven by an electric motor. The cyclic arm flexes the fiber a specified count.

Low-temperature flexibility. A test which determines the ability of a fiber optic cable to withstand bending around an obstacle at low temperature by measuring fiber breakage or transmitted power. The fiber is bent around mandrels, then straightened while under cold temperatures.

Impact testing. A procedure used to determine the ability of a fiber optic cable to withstand impact loads. A drop hammer of specified weight free falls on a length of fiber during this test to determine damage.

Compressive strength. A test to check the ability of a fiber optic cable to withstand slow compression or crushing. A fixture that applies compression to the fiber is used and an inspection is made to determine damage.

Cable twist and bend. Test is used to determine the ability of a fiber optic cable to withstand twisting checks are made for broken fibers and/or attenuation. A pair of gripping blocks hold a length of fiber. One of the blocks is fitted with a rotating clamp to twist the fiber. Fiber is then inspected for damage.

Cable tension load. This test is used to determine percent increase in length under tension, broken fibers, changes in radiant power, and the ability of a strength member to withstand tensile loads. Load is placed on the fiber in a form such as gripping blocks. The fiber is placed under tension. Measuring gauges check tension and length.

Power transmission versus temperature. The effects of temperature upon the transmitted power of an optical cable are defined by this test. This is a relation between two temperatures. A light source and detector are placed on either side of a test chamber. Transmission is made while temperature is varied.

Power transmission versus temperature cycling. A test to measure the effects of temperature upon the transmitted power of an optical cable while cycling temperatures. Fiber is placed in high then low test chambers. The light source and detector are on the outside of the test chamber. The fiber is exposed in cycles to high and low temperatures.

Power transmission versus humidity. This test determines the effect of humidity upon transmitted power. This is a ratio of radiant power of initial humidity preconditioning and at a final condition. A fiber specimen is placed in a humidity-controlled test chamber with the light source and detector externally mounted. Control is made to ensure condensation does not fall on the fiber.

Tensile loading versus humidity. This test defines the effect of humidity on the tensile loading of a fiber optic cable. Test is performed within a test chamber. Loads are placed on the fiber under controlled humidity.

Freezing water immersion—ice crush. A test to determine the effect of crush force caused by freezing water upon the transmitted power of an optic cable immersed in freezing water. This is a ratio of radiant power in cold but unfrozen water and in frozen water. Fiber is placed in a test chamber that is able to lower temperatures quickly at a predetermined rate.

Dimensional stability. This test determines the permanent dimensional changes which occur to a fiber jacket or covering when the cable is exposed to elevated temperatures. The test checks expansion and shrinkage usually in a percent of change. The test fiber is dimensionally checked before and after exposure to heat in a test chamber.

Flammability. The test measures the flammability of a subject fiber by application of a flame such as a bunsen burner, then checking for fiber damage.

Far-end crosstalk. The test measures the crosstalk of two neighboring transmitting fibers. Two fiber specimens are placed parallel to each other, radiation is applied to one fiber, and detectors are used to monitor radiation emanating from the illuminated and neighboring fibers.

Refractive index profile. This test determines the refractive index profile of a graded-index fiber by the interferometric process. The fiber end is prepared and placed under a microscope. The microscope is tilted so as to view a tilted wavefront on the fiber face. A photo is taken for empirical data.

Military Specifications and Standards

Test procedures for fiber optics fall into many categories, the main ones are general physical, electrical, and chemical methods. Most of these standards and specifications are referenced to federal and military documents.

Specifications

Federal

L-P-390	Plastic molding material, polyethylene, low and medium density

QQ-W-343	Wire, electrical and nonelectrical, copper (uninsulated)
QQ-W-423	Wire, steel, corrosion-resisting
Military	
MIL-C-17	Cables, radio frequency; coaxial, dual coaxial, twin conductor and twin lead
MIL-I-631	Insulation, electrical, synthetic-resin composition, non-rigid
MIL-W-5086	Wire, electric, hookup and interconnecting, polyvinyl chloride-insulated, copper or copper alloy conductor
MIL-C-12000	Cable, cord, and wire, electric, packaging of
MIL-C-13777F	Cable special purpose, electrical
MIL-M-20693	Molded plastic, polyamide (nylon) rigid

Standards

Federal	
FED-STD-191	Textile test methods
FED-STD-228	Cable and wire, insulated; methods of testing
FED-STD-601	Rubber, sampling and testing
Military	
MIL-STD-104	Limits for electrical insulation color
MIL-STD-109	Quality assurance terms and definitions
MIL-STD-202	Test methods for electrical and electronic components
MIL-STD-810	Environmental test methods

SELF-CHECK QUESTIONS

Self-check questions allow students to evaluate how well they have learned what they are studying.

1. What are optical fibers?
2. Define the terms *core* and *cladding*.
3. Explain in general terms how transmission takes place within an optical fiber.
4. What is a material's index of refraction?
5. What is the optical fiber cone of acceptance?
6. The term *angle of acceptance* is a critical term used in optical fiber's geometry. Define this term.
7. There are six major properties involved when choosing the correct fiber for an optical system. What are the six? Provide a short definition of each.
8. Fiber optics are made from fused mixtures of three materials. What are these materials?
9. What are the three general techniques of manufacturing optical fibers?
10. Describe several processes that utilize the techniques that were explained in Question 9.
11. There are two types of fibers which are manufactured specifically because of their refractive indexes. What are these two types called?

12. Why are single and multiple cables made in cable form?
13. Explain Snell's law.
14. Transmission and power are dependent on specifications and the geometry of transmission angles. There are several light angles that are critical. State and define these angles.
15. Name five causes of coupling power loss.
16. What are some of the causes of fiber misalignment in coupling?
17. Why are flat planes poor alignment mechanisms for optical fibers?
18. What are the advantages and disadvantages of two flat planes to align fibers?
19. Can three flat planes improve the alignment mechanism for optical fibers?
20. What is the philosophy dealing with three round planes for a fiber alignment mechanism?
21. How can four flat planes be used for an alignment mechanism?
22. What part do overlapping surfaces play in fiber alignment?
23. What are the general approaches to fiber alignment?
24. How do you ensure success of the resilient alignment mechanism?
25. Name several of the alignment concepts developed by individuals for the fiber optic connector discipline.
26. What are the basic procedures for preparing a fiber end for connection?
27. What are the requirements of splicing two fiber ends?
28. What are two methods of splicing optical fibers?
29. List four purposes of the optical fiber connector.
30. Name the three basic connector types in terms of the type of work that they perform.
31. How are optical connectors installed on a fiber and its support members?
32. List several performance tests that must be performed on optical fibers.

Eight

Fiber Optic Systems and Applications

This chapter is dedicated to putting the optical fiber to use. In the first part of the chapter we discuss some of the things that are involved in organizing the system. In the rest of the chapter, functional and unique system applications for use of fiber optics are presented.

OBJECTIVES

After studying this chapter and after completing the self-check questions at the end of the chapter, the reader will be able to:

1. Describe a basic fiber optic system to include the following.
   a. Electrical signal transmitter (driver)
   b. Light source
   c. Light detector
   d. Receiver
2. Interpret typical fiber optic specifications to include the following.
   a. System considerations
   b. Transmitter specifications
   c. Light source specifications
   d. Fiber specifications
   e. Cable specifications
   f. Detector specifications
   g. Receiver specifications

3. Follow typical design procedures for fiber optic systems.
4. Calculate fiber optic system attenuation.
5. Determine the fiber optic system transfer functions.
6. Examine typical fiber optic system cabling and connectors.
7. Explain the telephone standardization system called the American Digital Hierachy.
8. Examine large fiber optic system installations.
9. Describe the functions of several coupler and bus types to include the following.
   a. Multiple-access couplers
   b. Star couplers
   c. Directional couplers
   d. Data buses
   e. Loop buses
   f. Star buses
10. Explain the installation of fiber optic systems within data processing industry to include simplex and duplex applications.
11. Present the purpose, and advantages, of in situ (on site) testing.
12. Present the optical time-domain reflectometer as a tool for in situ (on site) testing, and describe its advantage.

A BASIC FIBER OPTIC SYSTEM

A simple fiber optic system is called a *transmission link*. It consists of a transmitter with a light source, a length of fiber, and a receiver with a light detector. The basic operation of a system is to connect a digital or analog signal to a transmitter. Within the transmitter, the input signals are converted from electrical to optical energy by modulating an optical light source, normally achieved by varying the drive current. The modulated light is launched into a length of fiber where it reflects from wall to wall through the fiber core. At the opposite end of the fiber a detector accepts the light and converts it back to an electrical signal. The electrical signal is converted back to its original form in the receiver. A brief discussion of each major component is provided.

Electrical Signal Transmitter (Driver)

The purpose of the transmitter (driver) is to change the electrical signal into the required current to drive a low-impedance light source. The electrical inputs are either digital or analog. The choice of the converter should depend on the current requirement of the light source. If the signal is digital, the transmitter (driver) should consist of a high-speed pulser to turn the light source on and off. If the signal is analog, the transmitter (driver) should be able to supply current to the light source to transmit the positive and negative alternations of the signal.

In Figure 8–1, a typical analog driver is illustrated. R_2 and R_3 provide a voltage divider for the input signal. The potentiometer P_1 amd resistor R_1 serve to set the operating point so that the positive and negative swings of the analog input signal produce only a positive output. The output current never changes polarity, only amplitude. R_{fb} is the feedback resistance. The light-emitting diode (LED) transmits light as current varies. Resistor R_L provides current limiting and load.

In Figure 8–2a, a very basic LED driver consists of an inverter with an LED tied to its output. The resistor R_L serves as a current limiter. Signal pulses at the input direct pulse current to flow through the LED causing it to radiate.

In Figure 8–2b, a variable (analog) signal input is directed through a pulse modulator. The modulator can be one of three types: PRM—pulse-rate modulation, PPM—pulse-position modulation, or PWM—pulse-width modulation. Modulated output directs pulse current to flow through the LED causing it to radiate. The samples in Figures 8–1 and 8–2 are extremely simplified. More details as to transmitter (driver) parameters and operation are discussed later on in this chapter.

Light Source

The purpose of the light source is to launch a light signal into the optical fiber at an angle that provides maximum signal transfer. There are two basic light sources used in fiber optic electronics. These are the light-emitting diode (LED) and the injection laser diode (ILD). Both of these units provide small size, brightness, low drive voltage, and are able to emit signals at desired wavelengths. Each has characteristics that make them desirable or undesirable for a particular application. The LED has a longer life span, greater stability, wider temperature range, and much lower cost. The ILD is capable of producing as much as 10 decibels more power output than the LED. It can launch the light signal at a much narrower numerical aperture (NA), and therefore can couple more power through the optic fiber than the LED. The disadvantage of using an ILD is that its current range is extremely restricted. Since some system operations vary greatly, the ILD must have compensation devices

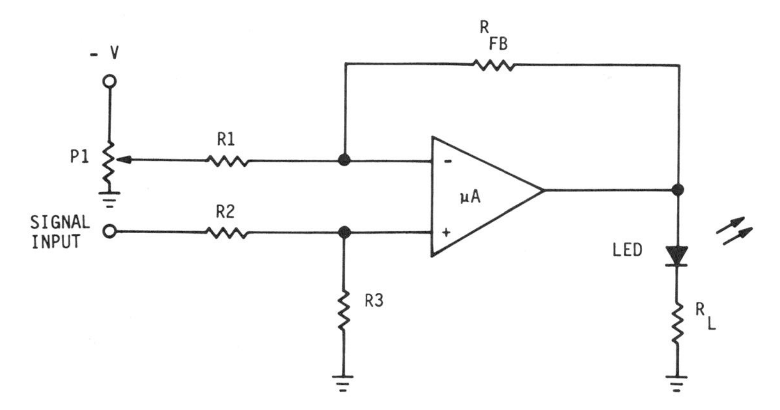

Figure 8–1 Typical analog driver (*Mike Anaya*).

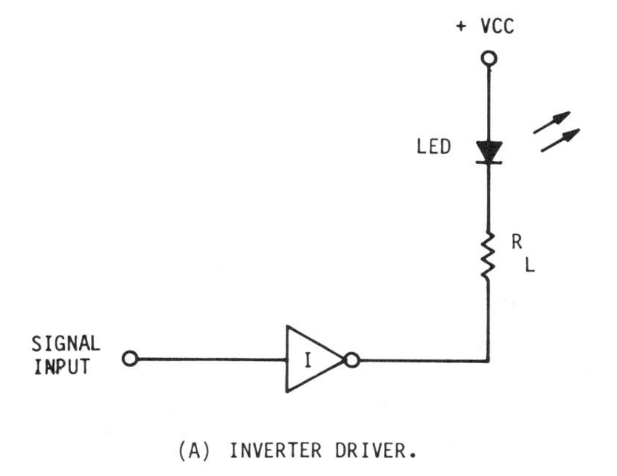

(A) INVERTER DRIVER.

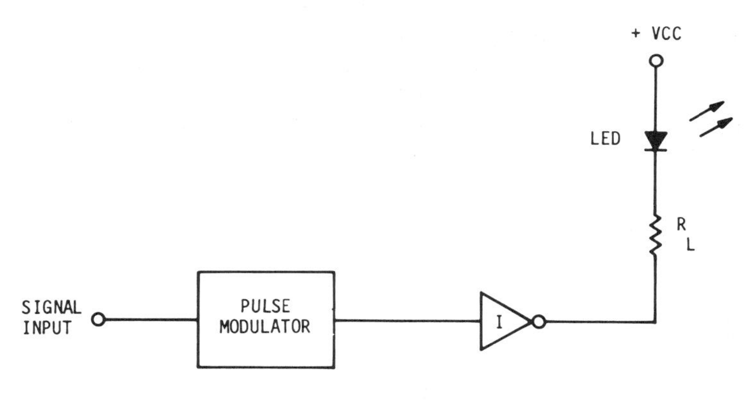

(B) PULSE MODULATED DRIVER.

Figure 8–2 Basic LED drivers (*Mike Anaya*).

added to the electronics. This may make the cost prohibitive. More details as to light-source parameters may be found in Chapter 4.

Light Detector

The light detector accepts a light signal from the optical fiber and converts it into an electrical current. There are three types of detectors in use: the phototransistor, the PIN diode, and the APD.

The phototransistor is inexpensive but has slow rise time and a limited bandwidth capability. Therefore, it is seldom used.

The PIN diode (contains positive, intrinsic, and negative solid-state layers in its construction) exhibits fast rise time and acceptable bandwidth parameters. It is reasonably priced.

The APD diode (avalanche photodiode) exhibits fast rise time and acceptable bandwidth parameters. The APD is more expensive than the PIN diode since it provides greater receiver sensitivity. The APD also requires an auxiliary power supply.

Receiver

The function of the receiver is to accept low-level power from the detector and convert it into a high-voltage output. There are at least two methods of accomplishing this. In Figure 8–3a, detector current produces a voltage drop across a load resistance. The voltage drop is directed into an amplifier. An output voltage representative of the transmitted signal is the result. In Figure 8–3b, the operational amplifier output voltage is the effect of the amplifier driving detector current through the feedback resistance. Again the output voltage is representative of the transmitted signal.

Other electronics may be added to the circuitry to maintain correct response. A gain control may be used on the front end to vary the amplification of the receiver. The operational amplifier is used as a current to voltage converter.

Since signal inputs are generally weak, shielding and power-supply decoupling are a requirement to achieve sensitivity. Sensitivity is set by input noise. Noise, of course, leads to errors in digital systems and restricts signal-to-noise ratio (SNR) in analog systems. Details on optical receivers are discussed later on in this chapter.

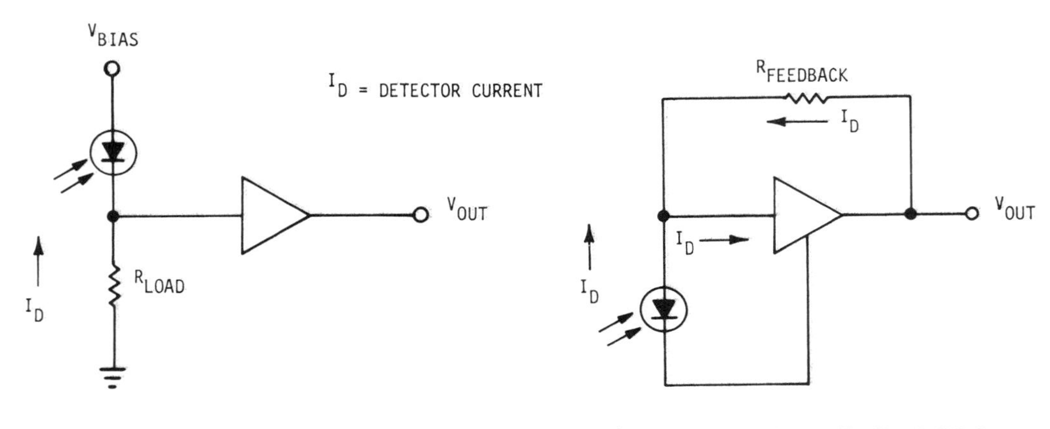

Figure 8–3 Receiver action (*Mike Anaya*).

SYSTEM COMPATIBILITY AND SPECIFICATIONS

It stands to reason that if you acquire all the parts for a fiber optic system from one company, the parts you buy will be compatible. Indeed, it may be intelligent thinking to do just that. Most manufacturers have developed complete systems and have off-the-shelf components available for the asking. Let's consider some of the things that are ultimately important when parts are chosen and a system is being created.

System Considerations

1. Digital
   a. Required bit error rate (BER) in bits per second, upper bit error rate usually in megabits per second (Mb/s), lower bit error rate usually in bits per second (b/s)
   b. Temperature operating range in degrees Celsius (°C)
2. Analog
   a. Bandwidth in hertz (Hz), or megahertz (MHz)
   b. Distortion in decibels (dB) or %
   c. Temperature operating range in degrees Celsius (°C)
3. Audio
   a. Bandwidth in hertz (Hz), or megahertz (MHz)
   b. Distortion in decibels (dB)
   c. Crosstalk in decibels (dB) (for multiple channels)
   d. Temperature operating range in degrees Celsius (°C)
4. Video
   a. Bandwidth in hertz (Hz), or megahertz (MHz)
   b. Distortion in decibels (dB)
   c. Crosstalk in decibels (dB) (for multiple channels)
   d. Temperature operating range in degrees Celsius (°C)

Transmitter Specifications

1. Input impedance in ohms (Ω)
2. Maximum input signal in volts dc (V_{dc}), volts effective (V_{rms} or V_{eff}) volts peak-to-peak (V_{p-p})
3. Optical wavelength in micrometers (μm) or nanometers (nm)
4. Optical output power in microwatts (μW)
5. Optical output rise time in nanoseconds (ns)
6. Required power supply in volts dc, usually $5 \pm 0.25\ V_{dc}$ or $\pm 15 \pm 1\ V_{dc}$

Light-Source Specifications

1. Continuous forward current in milliamps (mA)
2. Pulsed forward current in milliamps (mA)
3. Peak emission wavelength in nanometers (nm)
4. Spectral width in nanometers (nm)
5. Peak forward voltage in volts dc (V_{dc})
6. Reverse voltage in volts dc (V_{dc})
7. Temperature range in degrees Celsius (°C)

8. Total optical power output in microwatts (μW)
9. Rise/fall times in nanoseconds (ns)

Fiber Specifications

1. Mode—single or multimode
2. Index—step or graded
3. Attenuation in decibels per kilometer (dB/km)
4. Numerical aperture (a sine value)
5. Intermodal dispersion in nanoseconds per kilometer (ns/km)
6. Core index of refraction (a ratio)
7. Cladding index of refraction (a ratio)
8. Core diameter in micrometers (μm)
9. Cladding diameter in micrometers (μm)
10. Tensile strength in pounds per square inch (psi)
11. Bend radius in centimeters (cm)

Cable Specifications

1. Numbers of fibers (a unit)
2. Core diameter in micrometers (μm)
3. Cladding diameter in micrometers (μm)
4. Cable diameter in millimeters (mm)
5. Weight in kilograms per kilometer (kg/km)
6. Minimum bend radius in centimeters (cm)

Detector Specifications

1. Continuous forward current in milliamps (mA)
2. Pulsed forward current in milliamps (mA)
3. Peak reverse voltage in volts dc (V_{dc})
4. Temperature range in degrees Celsius (°C)
5. Optical power output in microwatts (μW)
6. Threshold current in milliamps (mA)
7. Rise/fall times in nanoseconds (ns)
8. Radiation pattern in angular degrees

Receiver Specifications

1. Output impedance in ohms (Ω)
2. Output signal level in volts dc (V_{dc}), volts effective (V_{rms} or V_{eff}), volts peak to peak (V_{p-p})
3. Optical sensitivity in microwatts (mW), nanowatts (nW), or decibels (dB), or megabits per second (Mb/s)
4. Optical dynamic range in decibels (dB)
5. Analog output overload (%)
6. Analog output rise time in nanoseconds (ns)
7. Digital output rise time in nanoseconds (ns)
8. Required power supply in volts dc, usually $5 \pm 0.25\ V_{dc}$ or $\pm 15 \pm 1\ V_{dc}$

DESIGN CONSIDERATIONS

Before any thoughts of developing an optical fiber system can be completed, certain factors must be realized. First, the signal information must be known. Is the signal analog or digital? What is the information bandwidth? What power is required? Second, length of the transmission line has to be determined. How far is it between transmitter and receiver? Are there any physical obstacles that must be skirted or must the cable go through anything? Finally, what are the tolerable signal parameters? What is the acceptable signal-to-noise ratio (SNR) if the system is analog? What is the acceptable bit error rate (BER) and rise/fall time if a digital system? Once the basic parameters of the system are set, the system development can take place.

Design Procedures

Procedures for design of an optical fiber system are as follows:

1. Determine the signal bandwidth.
2. Determine the signal-to-noise ratio (SNR) if the signal is analog. This is the ratio of output signal voltage to noise voltage. The ratio is expressed as 10:1, 8:1, 20:1, and so on. The largest signal-to-noise ratio is desirable. The SNR is expressed in decibels (dB). SNR curves are provided on detector data sheets.
3. Determine the tolerable bit error rate (BER) if the signal is digital. BER is the ratio of incorrect bits to total bits of data. A typical good BER is 10^{-8}. BER curves are supplied by the manufacturers of detectors.
4. Determine link distance, that is, the distance between the transmitter and the receiver.
5. Select a fiber based on attenuation.
6. Calculate fiber bandwidth for the system. This is accomplished by dividing the

bandwidth factor in megahertz per kilometer by the link distance. The bandwidth factor is provided on fiber manufacturer's data sheets.

7. Determine power margin. This is the difference between light source output power and receiver sensitivity.
8. Determine total fiber loss by multiplying fiber loss in decibels per kilometer by the length of the link in kilometers.
9. Identify the number of connectors. Multiply the connector loss (provided by the manufacturer) by the number of connectors.
10. Identify the number of splices. Multiply the splice loss (provided by the manufacturer) by the number of optics.
11. Allow 1 decibel for detector coupling loss.
12. Allow 3 decibels for temperature degradation.
13. Allow 3 decibels for time degradation.
14. Sum the fiber loss, connector loss, splice loss, detector coupling loss, temperature degradation loss, and time degradation loss, (add values of steps 8 through 13) to find total system attenuation.
15. Subtract total system attenuation from power margin. If the difference is negative, then the light-source power receiver sensitivity must be changed to create a larger power margin. Second, a fiber with a lower fiber loss may be chosen. Use of fewer connectors and splices may also be an alternative, if it is possible to do so without degrading the system.
16. Determine rise time. To find the total rise time add the rise time of all critical components such as the light source, intermodal dispersion, intramodal dispersion, and detector. Square the rise times. Then take the sum of the total squares. Square this sum and multiply it by some parameter factor such as 110% or 1.1 as in the following example.

$$\text{System rise time} = 1.1 \quad \sqrt{T1^2 + T2^2 + T3^2 + \cdots}$$

Fiber-Optic System Attenuation

The total attenuation of an optical fiber system is the difference between the power leaving the light source and the power entering the detector. In Figure 8–4, power entering the fiber is designated as P_S or source power. L_{C1} is power loss at the source to fiber coupling, usually 1 decibel per coupling. The power is of that signal launched into the fiber from the light source at the fiber coupling. L_{F1} represents the loss in the fiber between the source and the splice. Fiber losses are listed in specifications and have power losses at about 10 decibels per kilometer. L_{sp} represents the power loss at the splice. A representative power loss of a splice is 0.3 to 0.5 decibel. L_{F2} represents power loss in the second length of fiber. L_{C2} is the power loss at the fiber-to-detector coupling. Finally, P_D is the power transmitted into the detector. Other power losses due to temperature and time degradation are generally around 3 decibels loss each. Power at the detectors is then generalized as

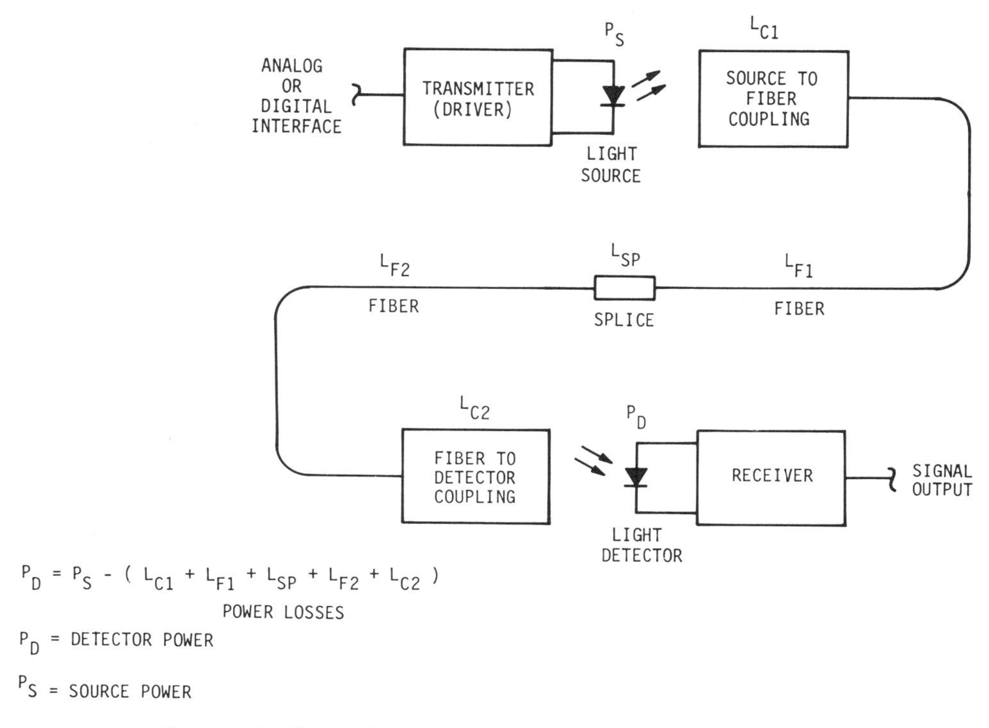

Figure 8–4 Fiber optic system attenuation (power loss) (*Mike Anaya*).

$$P_D = P_S - (L_{C1} + L_{F1} + L_{sp} + L_{F2} + L_{C2})$$

Note: All power and losses must be expressed in decibels (dB).

Transfer Functions

Transfer functions are the power formulas that describe the power at specific points along the transmission line. In Figure 8–5, P_S is representative of the power output from the light source. Power at different points can be represented by a power formula called a *transfer function.* Some typical transfer functions applicable to the fiber optic system in Figure 8–5 are listed in Table 8–1.

TABLE 8–1 TABLE OF TRANSFER FUNCTIONS (SEE FIGURE 8–5)

Parameter	Transfer function
Power out of light source	P_S
Power into fiber at point *W*	$P_W = P_S - L_{C1}$
Power into splice at point *X*	$P_X = P_S - (L_{C1} + L_{F1})$
Power into fiber at point *Y*	$P_Y = P_S - (L_{C1} + L_{F1} + L_{SP})$
Power into fiber at point *Z*	$P_Z = P_S - (L_{C1} + L_{F1} + L_{SP} + L_{F2})$
Power into detector	$P_D = P_S - (L_{C1} + L_{F1} + L_{SP} + L_{F2} + L_{C2})$

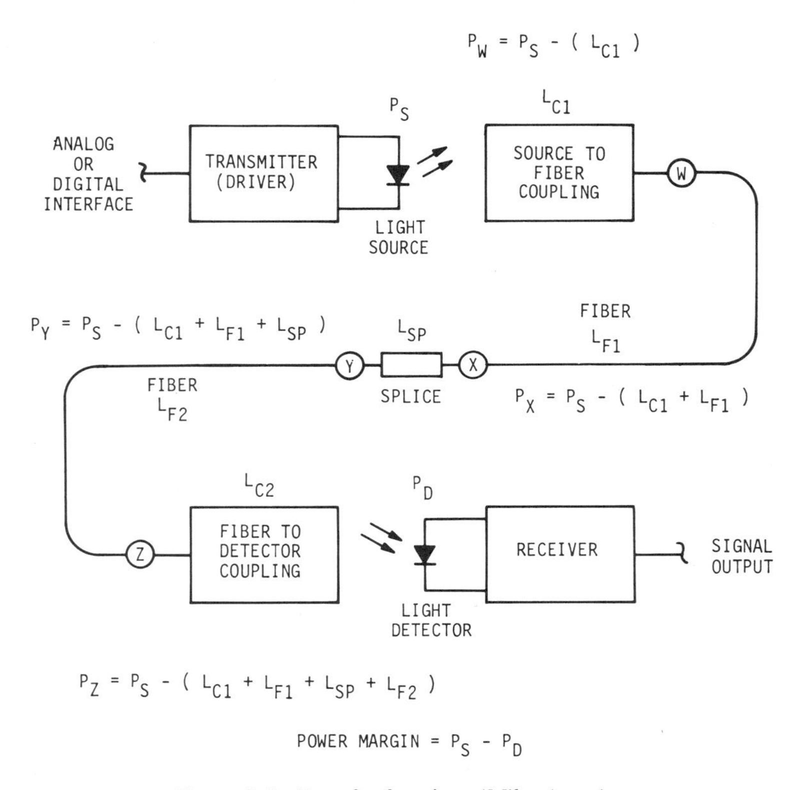

Figure 8–5 Transfer functions (*Mike Anaya*).

SYSTEM CABLING AND CONNECTORS

Although we previously discussed cables and connectors in Chapter 7, it is appropriate to provide some discussion of them in the systems portion of this book. To replace electrical conductors, optical waveguide cables must attain mechanical and environmental capabilities comparable to the metal conductor cables. Acceptable cable structures must achieve thermal and elastic compatibility, and stress fibers to a controlled and uniform degree under both tensile and crush loading. They must prevent fiber-to-fiber contact, especially in a crossover configuration.

Typical Cable

A fiber optic cable that meets the requirements of realistic applications should consist of relatively few fibers, each mechanically protected. A typical fiber cable, illustrated in Figure 8–6, consists of six channels (waveguides), each mechanically and optically isolated from the others. Note that the cable in the illustration is surrounded with a jacket and strength members. This particular cable has a maximum attenua-

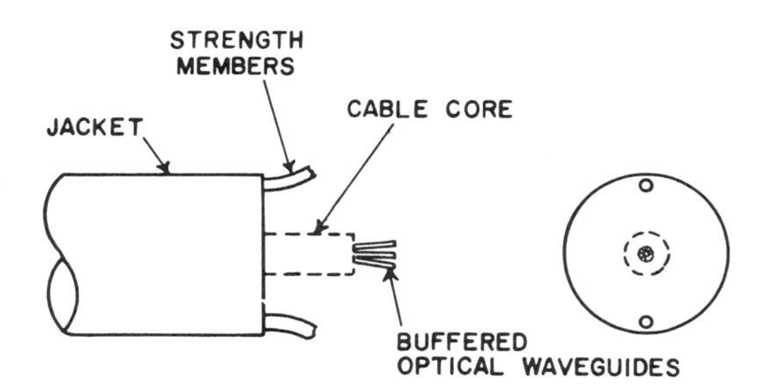

Figure 8-6 Typical fiber cable (*courtesy of Corning Glass Works*).

tion of 20 decibels per kilometer at 820-micron wavelength. The cable is 5 millimeters in diameter and has a tensile strength of 50 kilograms (110 pounds). Reference should be made to Chapter 7 for other cable structures. Typical cables for specific systems will also be illustrated later in this chapter. Other cable designs incorporating various numbers of channels and graded-index waveguides are available.

Cable Connectors

In-line system connectors usually fall into two specific categories. The first has the fiber group butted collectively against the second group of fibers. The second butts individual fibers. In group connections losses in transmission are typically 3 to 5 decibels. The group connectors are easily made. The single-fiber connection has relatively low losses, in the neighborhood of 0.5 decibel, but are more expensive.

The desirable properties of waveguide connectors are limited by the bundle-to-bundle connectors. In addition to the severe insertion loss, grinding and polishing are required, and epoxy or other glue limits the connector performance, especially with regard to temperature, humidity, and chemical attack.

Individual fiber-to-fiber connectors will substantially eliminate the objections to the bundle-to-bundle connectors. However, well-matched fibers and precision connector hardware are required. In addition, care must be taken not to distort the fibers, which can cause significant scattering loss. This effect is encountered especially in connectors using adhesives (epoxies), due to material inhomogeneity or nonuniform curing. Finally, fiber ends must be smooth to avoid another case of scattering losses. Fiber-end losses can be minimized through the use of index-matching fluids, which also reduce Fresnel reflection losses.

In addition to the need to minimize insertion loss in a structure mechanically and environmentally suited for its use and compatible with the cable, a connector should not require adhesive to hold the fiber, or grinding and polishing to finish the fiber end. Simplicity of connector hardware and of the termination process is important relative to future production costs and indispensable for practical field termination.

One connector concept locates the waveguide in the interstices of appropriately dimensioned rods of elastically suitable materials (see Figure 8-7). It is immediate-

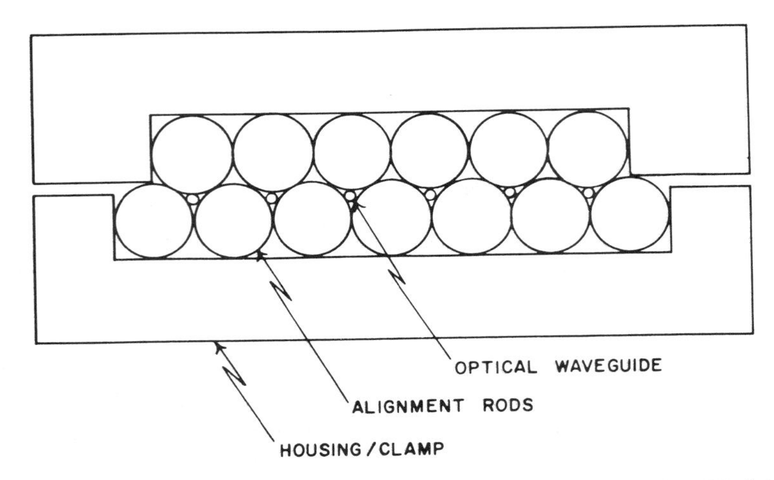

Figure 8–7 Typical cable connector concept (*courtesy of Corning Glass Works*).

ly apparent that this approach has a self-aligning capability minimizing requirements for tight fiber tolerances. The rods are made of a compliant material for fiber retention without the use of adhesives. Due to crosstalk considerations, only every other interstitial space is used.

A second cable-connector concept is the overlap concept. The connector assembly technique allows simultaneous cutting of all waveguides based on controlled scoring of bent and stressed fibers. This technique has consistently viable perpendicular and fault-free ends, eliminating the need for grinding and polishing.

A viable connector concept must not impose the optical dimensional tolerances on the mechanical elements of the assembly. This is accomplished by separation of the mechanical and optical mating planes as shown in Figures 8–8 and 8–9. In addition, the waveguides are made to function as springs, assuring contact of fiber ends. This is likely to prove particularly important under shock and vibration conditions over a broad operating temperature range.

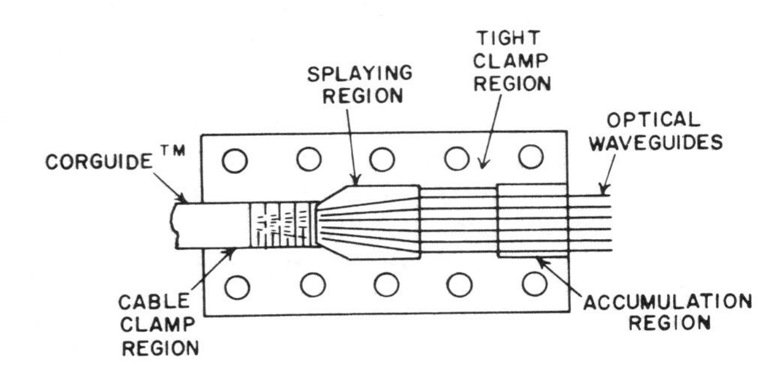

Figure 8–8 Integration of cable end into the overlap connector (*courtesy of Corning Glass Works*).

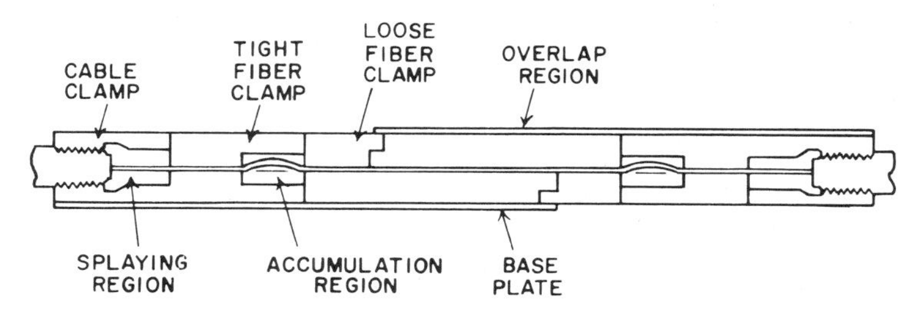

Figure 8–9 Overlap connector (*courtesy of Corning Glass Works*).

TYPICAL SPECIALTY CABLE AND ITS APPLICATION

The type AT optical fiber cable is typical of the cables used throughout the communication industry (see Figure 8–10). The AT cable was developed by General Cable Company and installed by General Telephone of California for use in what was reported to be the first use of optical fiber cable for regular telephone use. The cable was installed in a General Telephone route between the Artesia Central Office and the Long Beach Toll Center. Two regenerators (boosters) were placed in the line.

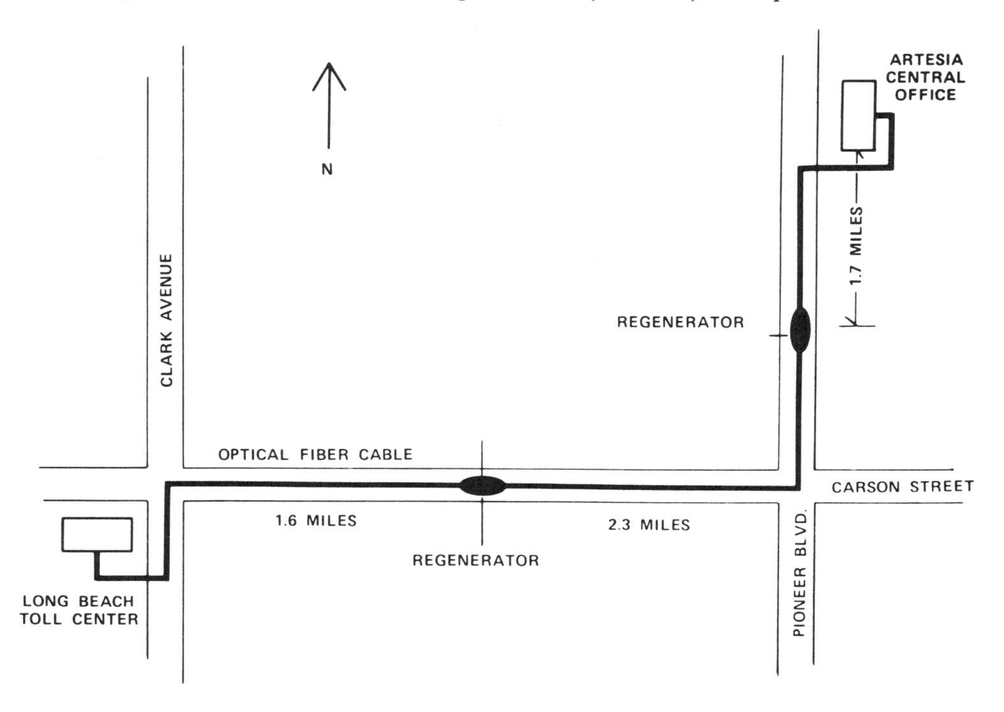

Figure 8–10 Typical communication fiber optic cable layout (*courtesy of GTE Corp.*).

The entire length of the system is 5.6 miles. One regenerator is 1.6 miles from the Long Beach Toll Center and the second regenerator 1.7 miles from the Artesia Central Office. The distance between the regenerators is 2.3 miles.

Outside Cable Construction

A line drawing of the AT outside cable and its parts is illustrated in Figure 8–11. A photograph of the outside cable is shown in Figure 8–12.

Inside Cable Types

Figure 8–13 illustrates typical inside optical fibers used throughout the telephone and communications systems today.

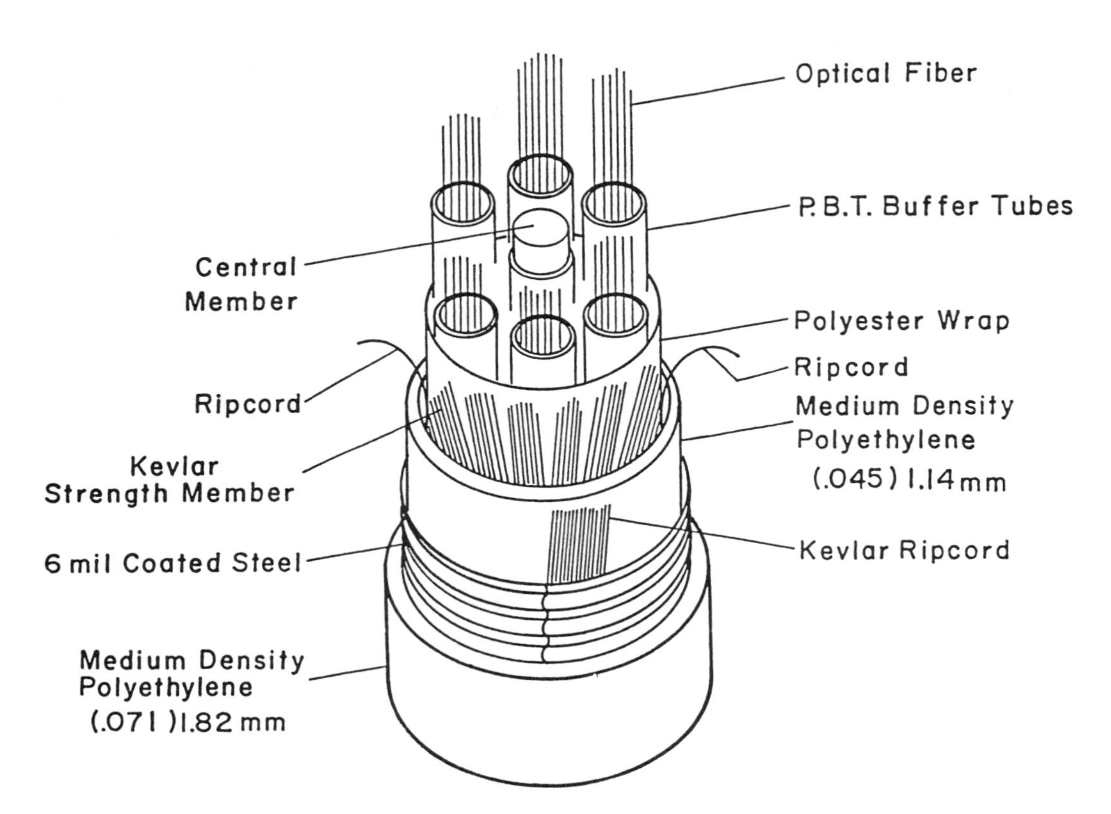

Figure 8–11 Line drawing of outside type fiber optic cable (*courtesy of General Cable Company*).

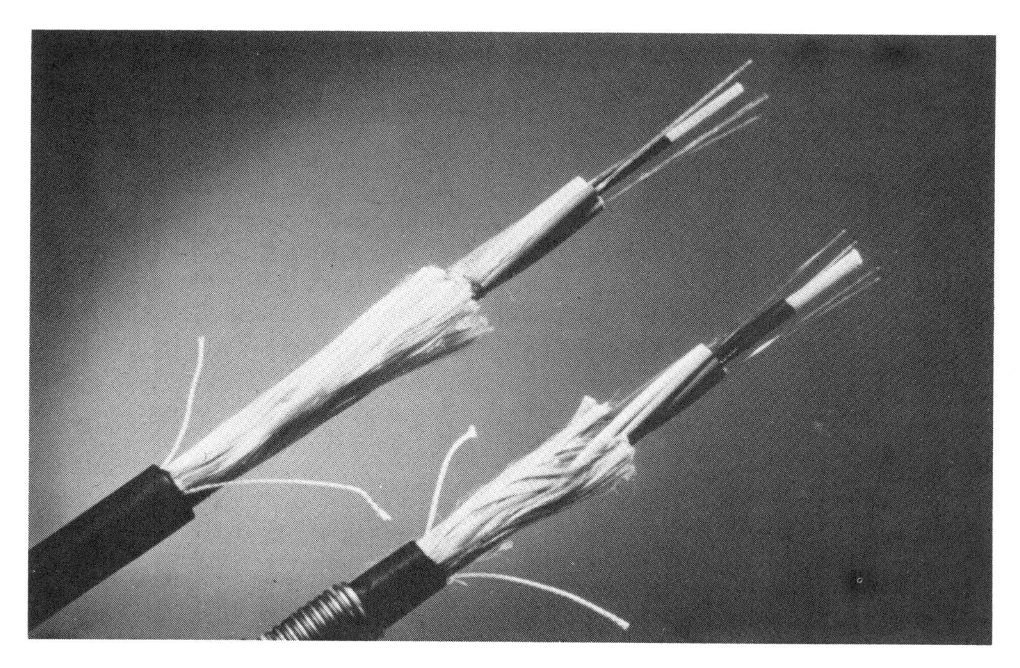

Figure 8–12 Outside-type fiber optic cable (*courtesy of General Cable Company*).

Figure 8–13 Typical inside-type fiber optic cable (*courtesy of General Cable Company*).

SYSTEM APPLICATION

Probably the most significant of the applications of fiber optic systems are in the telephone industry. Substitutions for electrical conductors are being used throughout the country. Standardization within the industry is made possible by the North American Digital Hierarchy, which is the standard in the United States. Essentially the standards are in channels and bit rates. These are listed below. Applications in telephone are intercity trunks, interoffice trunks, and local loops. The key advantages for telephone systems are high bandwidth, low attenuation, and small size.

System line	4-kHz voice channels	Bit rate
T1	24	1.544 Mb/s
T2	96	6.312 Mb/s
T3	672	44.736 Mb/s
T4	4032	274.176 Mb/s

Broadband networks such as television are using fiber waveguides in commercial cable television (CATV), wired city (community), and dedicated communication systems. The advantages are simpler transmission-line characteristics and noninterference (security).

Computers for general electronic data processing and computer-based process control and instrumentation systems represent large-scale usage of optical waveguides. At a given bit rate, the benefits of optical waveguides versus metallic conductors increase with distance. This, together with optical waveguides' immunity to electromagnetic interference, suggests the greatest motivation to apply waveguides in computer systems operating in a high-EMI environment. Computer applications for fiber optics are dispersed systems, intersystem wiring, computer-based industrial process control, and instrumentation. The advantage is immunity to interference.

Military applications include wiring of weapons systems such as aircraft, helicopter, ship, or submarine. Technically, many of them entail data transmission within computer-based systems. Headquarters, field, and base communications are of special importance to the Army. Surveillance systems operated from shore, ship, or submarine may make a substantial impact on the Navy's capabilities to detect, identify, and evaluate underwater threats. Secure communications like those for missiles are receiving attention. Having recognized the potential of optical waveguides very early, the military laboratories and their contractors have an impressive number of programs underway. These programs are yielding significant technical results and are expected to conclude with a line of military-qualified hardware, demonstrated system feasibility, and know-how to apply the technology of optical waveguides. Again, the advantages are immunity to electromagnetic interference (EMI) and electromagnetic pulses (EMP), security, and the size/weight factors.

LARGE FIBER OPTIC INSTALLATION

One of the major installations using fiber optics today is the Lake Placid Winter Olympics Lightwave System. The system was a joint effort of New York Telephone, AT&T, Western Electric, and Bell Labs (see Figure 8–14). The new installation was obviously a tremendous success considering the clarity and quality of the television broadcasts of Winter Olympics 1980. Problems were few and primarily mechanical. The new installation was in place and operational by October 1979. In the figure the system connection is shown. Its first purpose was to transform the Lake Placid telephone facility into an ultrasophisticated communications center capable of handling a wide range of telecommunications services necessary to support the 1980 Olympic events.

The lightwave system extends 2 1/2 miles and links the Lake Placid telephone switching office, the Olympic ice arena, and the broadcast center. The broadcast center serves 25 mass-media agencies. The fiber cable, made by Western Electric travels ap-

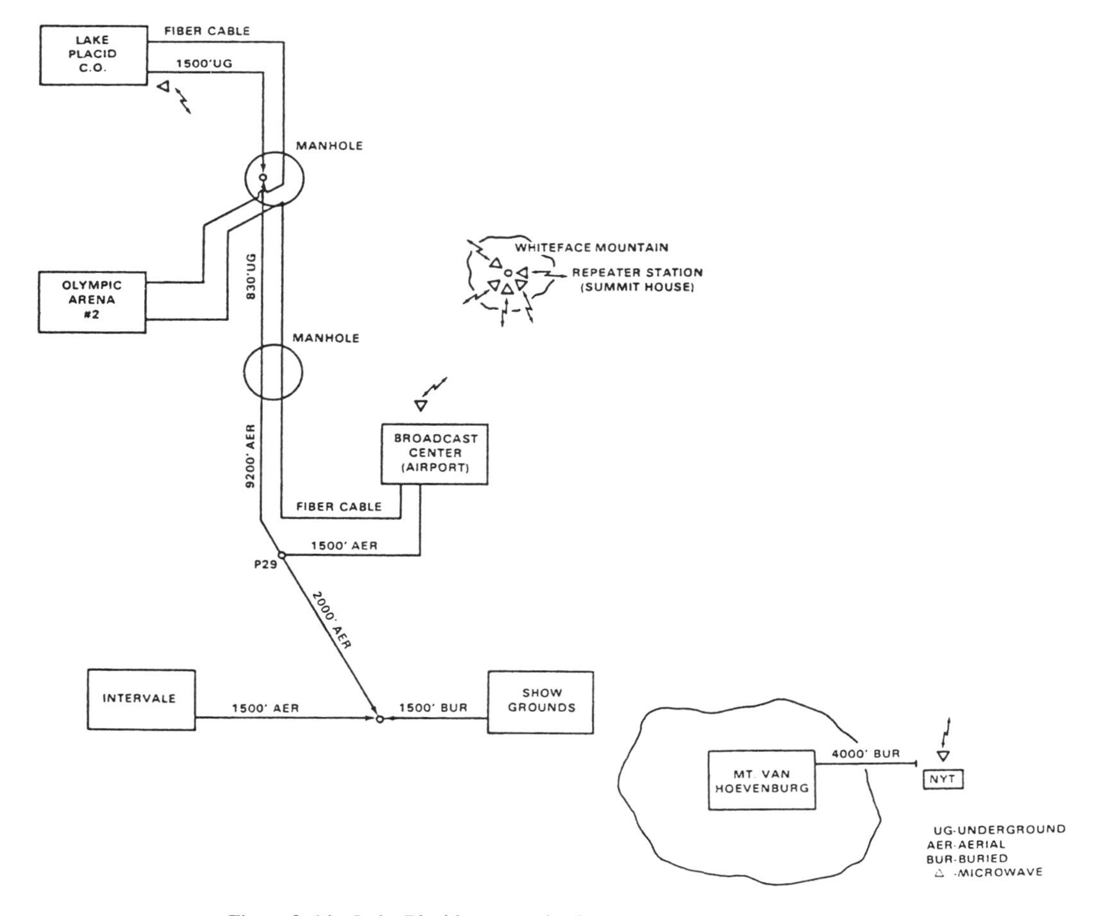

Figure 8–14 Lake Placid communication system (*courtesy of New York Telephone*).

proximately 2 miles on poles (aerial) and the other half-mile underground. Part of the underground cable is buried. The cable consists of 12 glass fibers carrying digital voice signals and television. Six of the fibers carry 288 two-way voice conversations, while two of the fibers transmit the video and associated voice signals. The remain-

Figure 8–15 New York City and Westchester fiber optic telephone link (*courtesy of New York Telephone*).

ing four fibers serve as backup in the event of failure. An added part of the Olympic system is a microwave installation which serves the entire facility.

A second major effort using fiber optics is the permanent installation linking telephone company central offices between White Plains and East 38th Street Manhattan, New York (see Figure 8–15). This is an extremely long system (30 miles). It was installed to meet with a growing volume of telephone calls between New York City and Westchester. The New York system will carry both voice and data communications between company central offices at White Plains, Scarsdale, Tuckahoe, and Mount Vernon. This part of the major link will cover 11 miles and have a potential for handling 14,000 calls simultaneously.

The second part of this major fiber optic system will enter New York City from Mount Vernon, pass through city central offices and terminate at the East 38th Street exchange. This second link will cover over 17 miles and is installed in existing cable ducts. The Mount Vernon–mid-Manhattan fiber optic link is probably the beginning of major usage of fiber optics in the telephone industry.

A BASIC T1 FIBER OPTIC SYSTEM

Early in 1979, Southern Bell Telephone and Telegraph and Florida Power and Light completed an installation of a T1 system. You may recall that a T1 system consists of 24 voice-frequency channels into a 1.544 megabit per second pulse code modulation (PCM) bipolar digital-bit stream. The system was manufactured by ITT Telecommunications of Raleigh, North Carolina. The unique feature of the system is the use of optical cable-to-metallic cable conversion in a repeater manhole environment (see Figure 8–16). The conversion repeaters convert infrared light pulses into electronic pulses for transmission over conventional digital span lines to the central telephone office.

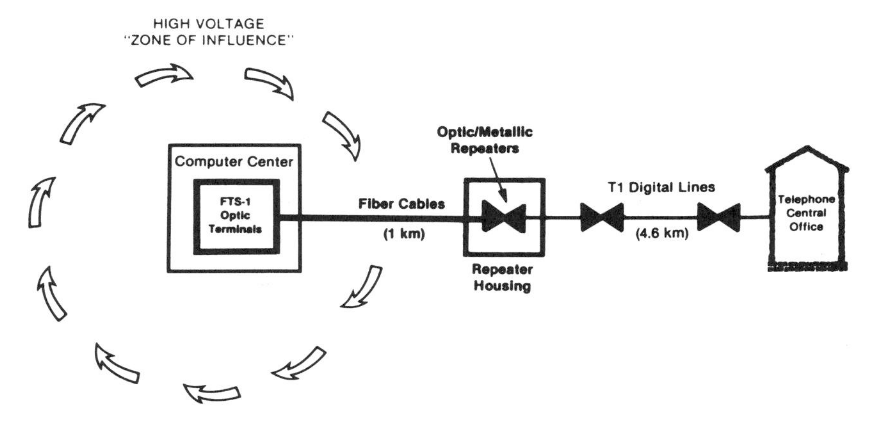

Figure 8–16 Basic T1 system (*courtesy of ITT Corporation*).

System Description

The computer center consists essentially of three pulse-code-modulation (PCM) channel banks, an automatic protection switch (APS) for three service lines with a spare, an optics shelf consisting of optical transmitter units (OTUs), office power converters (OPCs), and optical receiver units (ORUs) (see Figure 8–17).

Each PCM channel bank multiplexes 24 voice-frequency (VF) channels into a 1.544-megabit-per-second pulse-code-modulation (PCM) bipolar digital-bit stream for transmission. A special test-and-alignment unit plugs into the shelf of the PCM units to measure and adjust the receiver voice-frequency (VF) levels, provide a 1020-hertz test tone for setting transmit voice-frequency levels, and to measure idle channel noise, linearity, distortion, crosstalk, hybrid balance, and office equipment losses.

The automatic protection switch (APS) accepts the 1.544-megabit-persecond signals from the three PCMs. If any of the three optic service lines fails, the APS automatically transfers PCM signals to the spare line. After the fault is cleared, reset to normal is automatic. The optic line shelf consists of optical transmitter units (OTUs), office power converters (OPCs), and optical receiver units (OPUs). The optical transmitter units (OTUs) convert the 1.544-megabit-per-second-rate bipolar electrical signal to a 3.088-megabit-per-second optical signal suitable for launching over the fiber optic line. Wavelength of the launched signal is 850 nanometers. The optical receiver unit (ORU) detects the 3.088-megabit-per-second signal and decodes it back into a 1.544-megabit bipolar signal. The office power converter (OPC) provides the power for operation of the OTU and the ORU.

Three optical cables with five fibers each are routed through underground conduit from the optical line shelf to repeater housings in the manhole. The cables are

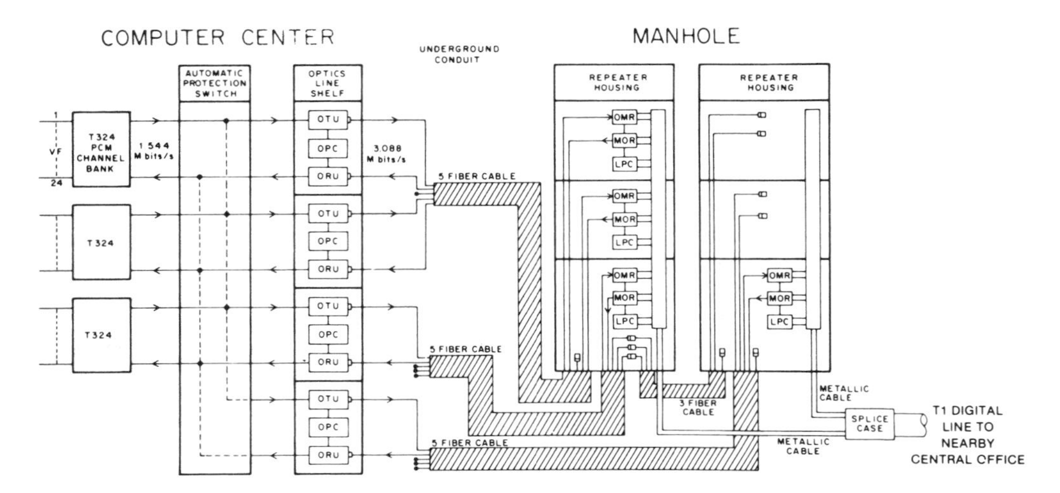

Figure 8–17 T1 system description (*courtesy of ITT Corporation*).

7 millimeters in diameter. The fibers are graded-index and have an attenuation of 6 decibels per kilometer.

An optical-to-metallic conversion repeater interfaces the opposite end of the fiber cables to the T1 span line. Each converter has capability of handling three single system repeaters. In this system the two repeaters are also connected together with three fiber cables allowing for expansion. The converter housing is stainless steel and is pressurized to 15 pounds per square inch to purge moisture. Each conversion repeater consists of three sets of conversion units. Each set has an optical-to-metallic regenerator (OMR) that consists of a 3.088-megabit-per-second optical receiver, a bit-rate converter, and a 1.544-megabit-per-second metallic span-line driver. The metallic-to-optic regenerator (MOR) contains a 1.544-megabit-per-second metallic regenerator, a bit-rate converter, and a 3.088-megabit-per-second optical transmitter. The line power converter (LPC) is a dc to dc converter which provides power for the OMR and MOR units. All the units in the conversion repeater are similar in function to the units on the optical line shelf. Metallic cables connect from the conversion repeaters to the T1 digital line in the central office.

A BASIC T3 SYSTEM

A typical T3 communications system spanning a distance of 22 kilometers was scheduled for completion by the Pennsylvania Commonwealth Telephone Company in mid 1979. The system, manufactured by ITT Corporation utilizes four intermediate repeaters between two north-central Pennsylvania communities of Wellsboro and Mansfield. The system carries toll, intertoll, operator, and special service traffic. Optical cables include 10 kilometers of direct buried (plowed) cable, 1.0 kilometer of underground (duct) cable, 3.0 kilometers of aerial cable lashed to existing cable, and 8.0 kilometers of aerial cable lashed to a new messenger strand.

System Description

The modified T3 communications system (see Figure 8–18) accepts a 44.736-megabit-per-second electrical signal from a digital radio or a multiplexer switch and converts it to an optical signal. The optical signal is launched into fiber optic cables and transmitted to a receiver at far-end optical terminal. Sufficient optical repeaters are placed into the link between the two terminals. The receiver detects the optical signal, converts it to an electronic level, and forwards it to the receive side of the far-end terminal where it is converted to T3-level signals and routed to radio or multiplexer.

In the lower part of Figure 8–18, the metallic equivalent of a T1 communication system is compared to the optical system. Electrical signals from digital radio or a digital switch are fed to repeater and transmitted over metallic lines to repeaters at the far-end central office and routed to radio channels. A typical T3 system conversion then consists of a pair of office terminal repeaters and their housings and the fiber optic cables with connections.

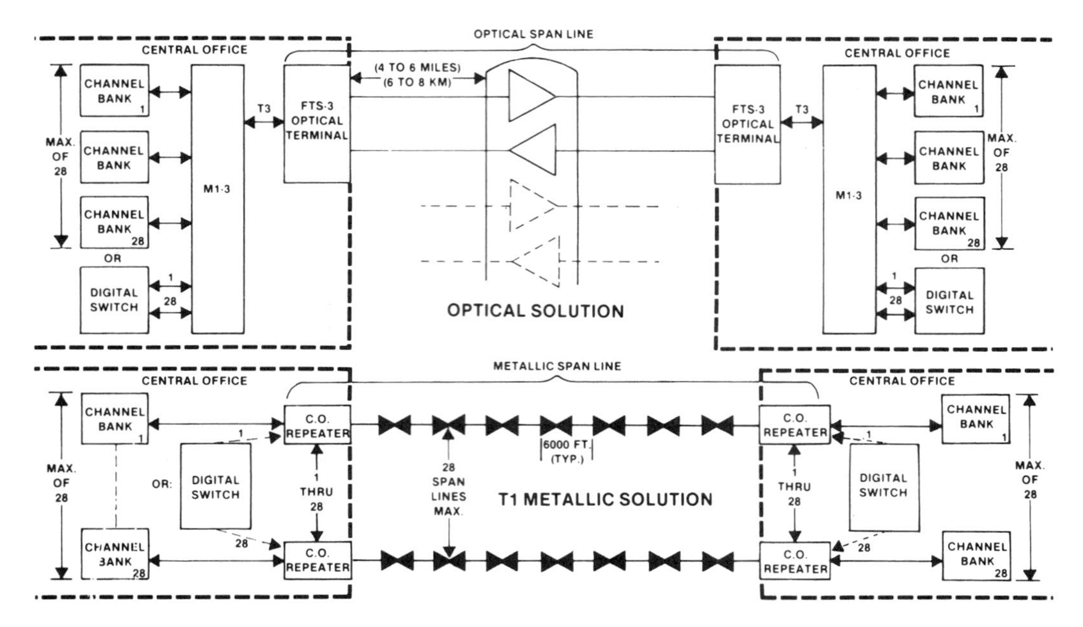

Figure 8–18 Basic T3 communication system (*courtesy of ITT Corporation*).

A T3 office terminal is pictured in Figure 8–19. Internally, the terminal consists of an optical transmitter, an optical receiver, a looping unit, scrambler and descrambler, and an alarm system. See Figure 8–20 for a line drawing functional block diagram of the T3 office terminal.

A T3-level signal is routed to and from the T3 terminal. The signal is scrambled to randomize the input, which improves repeater and receiver clock extraction and removes jitters. Parity bits are added to allow bit-error-rate (BER) monitoring and the transmission of system control functions. The 44.736-megabit-per-second signal (now a 47.367-megabit-per-second signal) is routed through the looping unit and to the optical transmitter for transmission. The looping unit allows out-of-service diagnostic testing from either end of the transmission link. One operator at one end of the system can test all units from the input to the output except for the optical receiver and transmitter.

The optical transmitter modulates an injection laser diode (ILD) at a near-infrared wavelength of 850 nanometers. Modulated light is launched into fiber cables and transmitted toward the far-end receiver. The modulated light moves through the fiber cable (through repeaters) and to the optical receiver in the far-end office terminal. The optical receiver employs an avalanche photodiode (APD) as a detector. The signal is amplified. Clock-timing information is extracted and the signal data stream is regenerated. Parity bits added at the input of the opposite-end office terminal are removed and signals are descrambled. T3-level signals are then routed to radio or multiplexer. All signals along within the office terminal are monitored in the alarm unit and line-fault information is collected between the optical transmitter and receiver.

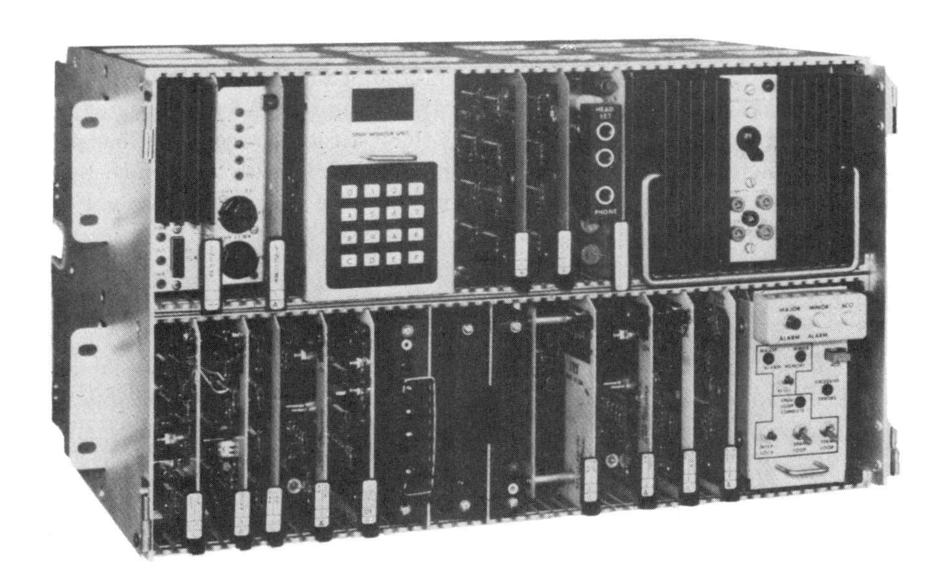

Figure 8-19 T3 office terminal (*Courtesy of ITT Corporation*).

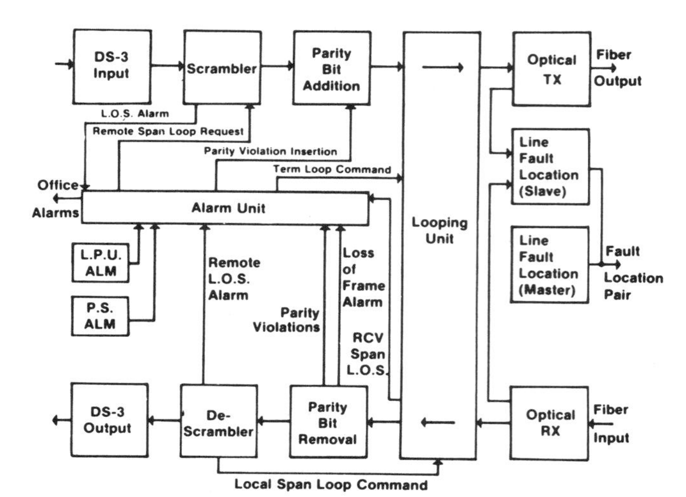

Figure 8–20 Functional diagram of T3 office terminal (*courtesy of ITT Corporation*).

Cable Routing Layout

The illustration in Figure 8–21 is a line drawing of the fiber optic cable and its metallic counterpart. Note that the expanse between metallic splices is typically 2 kilometers while the distance between optical cable splices is 1 kilometer. The practical loss allocations, along with a reasonable safety margin, result in a 5-kilometer expanse between the optics terminal shelf in the T3 office terminal and the repeater housing. Fiber cables are spliced in the field using a flame-fusion process. A fiber position fixture and a microscope are utilized to control alignment of fiber ends. Splice losses are around 0.5 to 1.0 decibel. Considering that the splicing takes place in the field, in varying kinds of weather, the results are more than satisfactory.

Optical Repeater Housing

Repeater housings like the one in Figure 8–22 are designed to be installed on poles or in manholes. When pole-mounted, a sun shield is used to eliminate heat rise that would reduce laser life. The housing contains the same optical receivers and transmitters as those in the office terminal. The case is pressurized. It has an internal electronic card cage designed to hold two bidirectional optical repeaters, power supplies, and repeater performance monitoring and reporting systems. The housing is equipped with metallic stubs for power leads, order wire, and system-monitoring wire pairs. Load coils and lightning protection for the copper pairs are also provided. Optical cables are installed in the field and with an epoxy dam. Power for the repeaters is ±130 volts direct current to provide a 140-milliampere loop current. Current is routed to the repeater using a 22- or 24-gauge copper pair. At the present time, optical span line electronics cost about the same per voice-channel mile as the metallic versions.

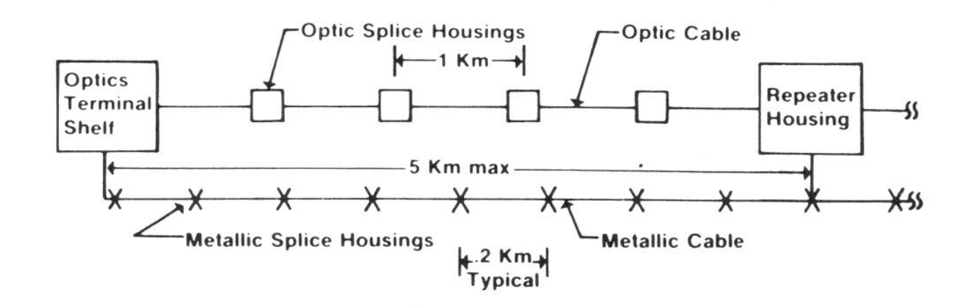

Figure 8-21 T3 system cable routing (*courtesy of ITT Corporation*).

Figure 8-22 T3 system repeater (*courtesy of ITT Corporation*).

Longer length and systems with large numbers of channels tend to make the optical solution a better purchase because of the cable cost savings.

SYSTEM COUPLING USING MULTIPLE-ACCESS COUPLERS

The purpose of an optical coupler is to distribute the optical signal into multiple branches of a data transmission system. Couplers (data buses) are usually star type, directional, or are specially made for a specific system. Input signals from numerous branches are mixed together and distributed. Multiple data are outputted by the receiver in a serial fashion. The concepts used are a mirrored surface, which reflects all of the inputs to each of the output channels and a ratio-splitting method using fiber bundles.

Star Couplers

Figure 8–23 shows the two basic star couplers, the transmission and the reflection star. These couplers are linear mixing devices, that is, input light signals are mixed together and then divided equally among the output parts. This being the case, the couplers can combine various signals together by a multiplexer or split one signal into various parts. The coupler may also insert light power into or out of a fiber optic link. Both types are composed of a set of input fibers, a set of output fibers, and a mixing region. The difference between the two star couplers is obvious.

In general, the reflection star is more versatile than the transmitting star coupler because the relative number of input and output parts may be selected or varied after the device has been constructed, while the number of transmitting star inputs and

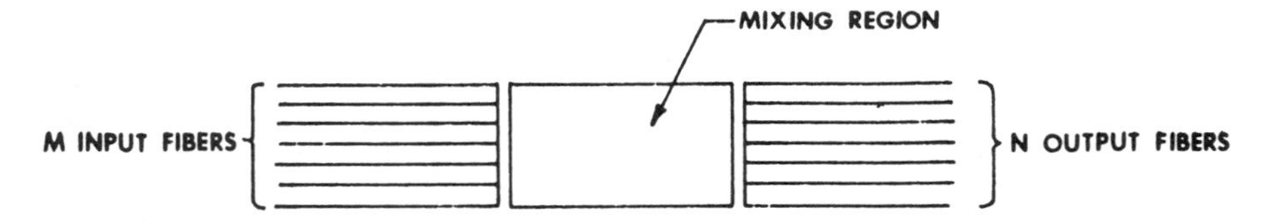

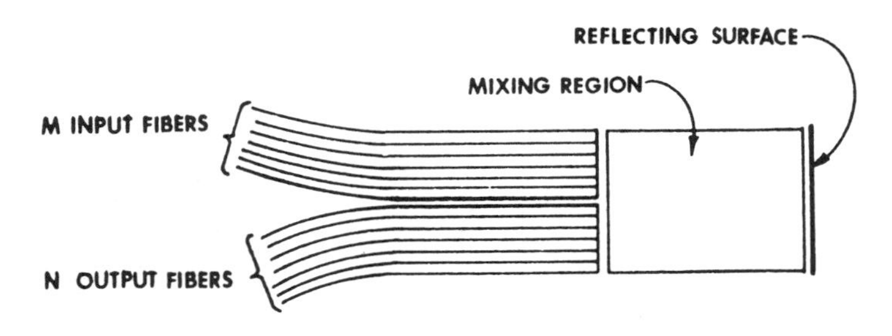

Figure 8–23 Star coupler types (*courtesy of Alcatel Cable Systems Group*).

outputs are fixed by initial design and fabrication. However, the reflecting star coupler is less efficient than the transmission star, since a portion of the light fed back into the coupler is injected back into the input fibers. Reflection and transmission star couplers are used where efficiency must be considered.

Directional Couplers

Directional couplers, also called tee couplers, can be used for signal insertion, monitoring branching, wavelength duplexing, or mixing in multiple-access, bidirectional, point-to-point, or data-bus systems. Optoelectronic directional couplers have a PIN diode within the coupler at a tap-off port (see Figure 8–24).

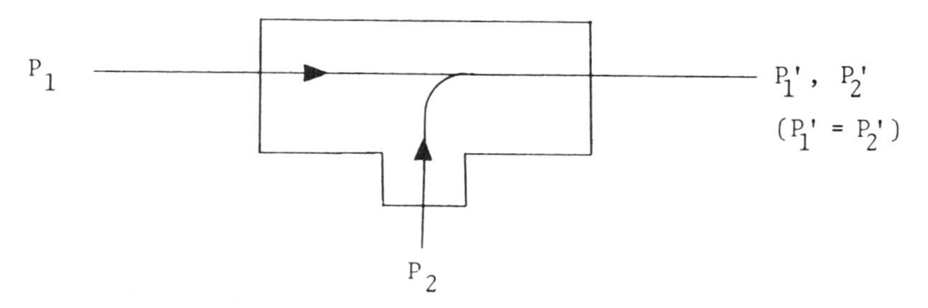

Figure 8–24 Directional coupler (*courtesy of Alcatel Cable Systems Group*).

Systems Using Couplers

With the coming of age of low-loss single-fiber tap-off, mixing, branching, and wavelength multiplexing, users of couplers can intercommunicate, link to a central terminal, or sample data streams along a single-optical fiber data bus (see Figure 8–25). Multiple-access couplers, when configured with appropriate light sources, fibers, and detectors allow the user to take advantage of the desirable features of optical fibers. These features, as outlined in Chapter 7, include wide bandwidth, low loss, immunity from radio frequency (RF) and electromagnetic pulse interference (EMI) along with small cable size, high transmission security, negligible crosstalk, ground isolation, and short-circuit protection.

In Figure 8–25a, an application using optical and optoelectronic couplers is illustrated. A in the illustration is a directional coupler being used as a laser-stabilizer feedback-loop element. This coupler has a PIN diode in its tap-off port. B_1 in the illustration is an optical directional coupler being used as a single-fiber information-drop element to a receiver. B_2 is also a directional coupler used as a single-fiber insertion coupler from a transmitter. Directional couplers may be compatible to plastic-clad silica fibers or glass-on-glass graded- or step-index fibers.

In Figure 8–25b, bidirectional couplers (C) are being used to place a duplex communication system on a single optical fiber. Two-color multiplexing provides required channel isolation. The bidirectional coupler is compatible to glass-on-glass fibers.

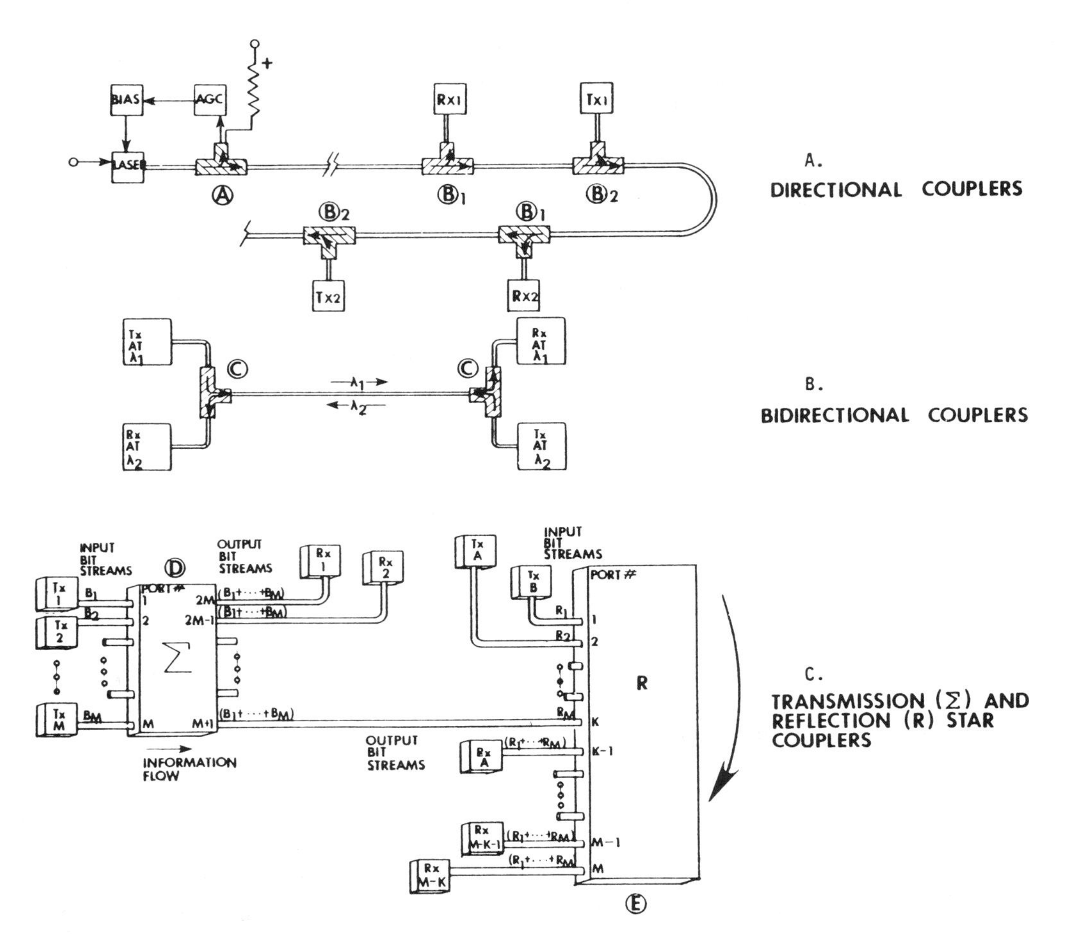

Figure 8–25 System using couplers (*courtesy of Alcatel Cable Systems Group*).

In Figure 8–25c, both transmission and reflection star couplers are used in a complex mixing system. The transmission star coupler's input parts are optically isolated from each other. The reflection star ports are all coupled. Star couplers are compatible to plastic-clad silica fibers.

While all couplers can be used with any fiber, power loss due to the mismatch of fibers may not be tolerable. Couplers can add that extra freedom necessary in unique system design.

Data Buses

The major use for fiber optic multiple-access couplers is in multiuser, multiservice fiber optic buses. Within a typical bus are terminals, nodes, multiplexers, and links that comprise the network. In addition, the bus is separated into access and center user networks. Local users are grouped and interact in access networks. Center users are interconnecting. Services that can be handled on a fiber optic bus include digital data, voice, video, and informational retrieval. These services can be segregated by frequency-division multiplexing. High-bandwidth fiber enables many channels to be multiplexed on a single-fiber bus.

Loop-Bus and Star-Bus Configurations

Both star and directional couplers can be utilized to develop a data-bus configuration. Figure 8–26 represents typical star-bus configurations, while Figure 8–27 represents a loop-bus configuraiton. The star bus requires a centralized selection process implemented by a branching element. The loop networks pass all information

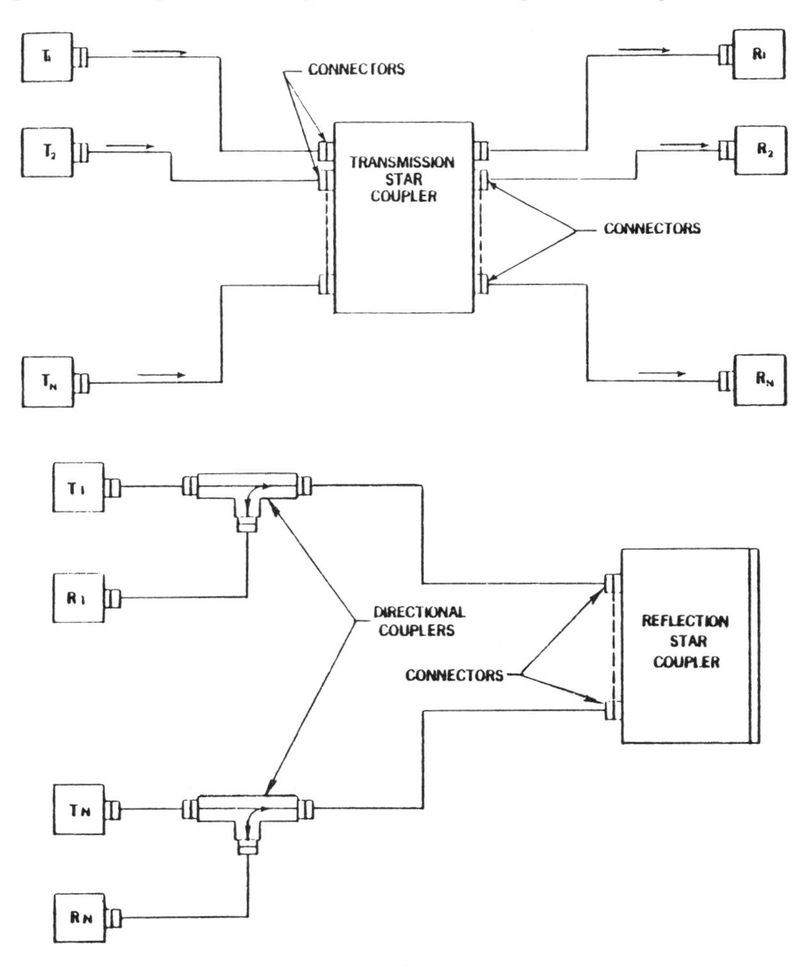

Figure 8–26 Star bus configurations (*courtesy of Alcatel Cable Systems Group*).

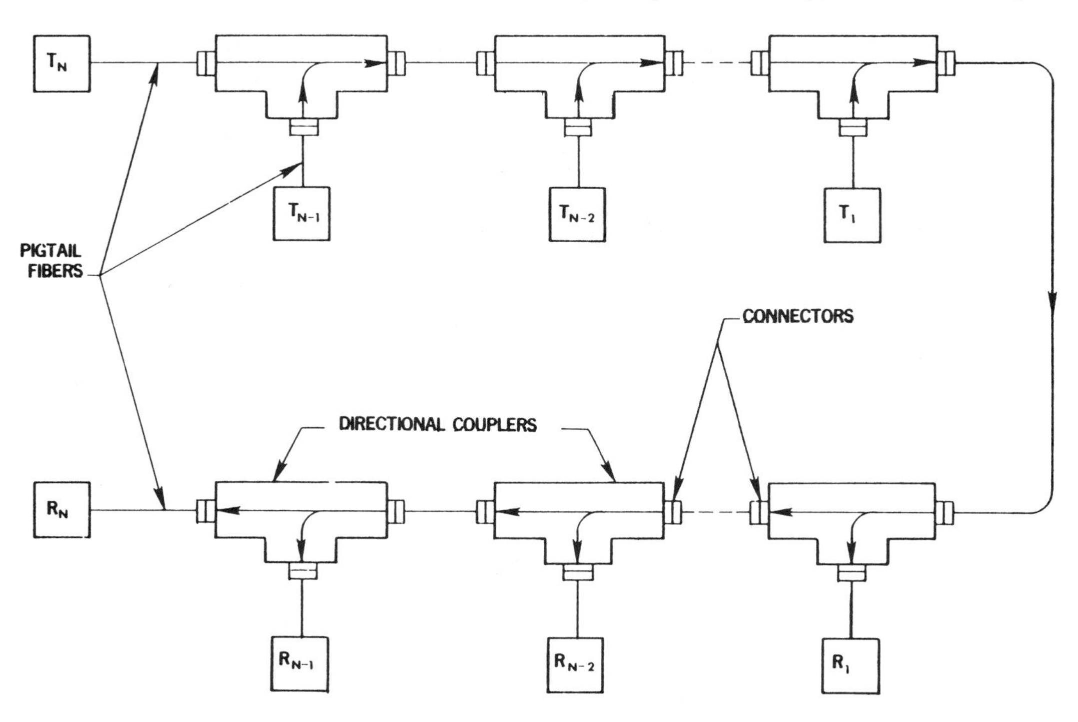

Figure 8–27 Loop bus configuration (*courtesy of Alcatel Cable Systems Group*).

through each node and require drop-and-insert elements to provide access to receiver and transmitter terminals. In Figure 8–26, both transmission and reflection star couplers are illustrated. The reflection-star configuration requires only half the cable used by the transmission-star configuration. However, power is sacrificed and an additional directional coupler is required for each channel. An alternative reflection-star coupler may have a single part for each transmitter and receiver. Figure 8–27 illustrates a loop-bus configuration consisting of three transmitter and three receiver directional couplers. Fibers are pigtailed. The choice between the star and loop configurations depends entirely on system requirements. Advantages of the loop configuration are expandability, ease of synchronization, and low cable cost. However, the loop configuration has accumulated insertion loss as a data stream passes drop points. Advantages of the star bus are the availability of a multiple of repeaterless terminals and a great savings in cable cost, therefore lower prices to consumers. It can also be easily expanded. The star bus has disadvantages in that synchronization is difficult because of the varying propagation delays from the user terminals.

Data-Bus Multiplexing

After the multiple-access couplers have been selected, a method for multiplexing several messages onto a single channel and a method to control and route data traffic are required. Multiplexing techniques in conventional wideband radio-frequency

(RF) networks are characterized by modulating carrier waveforms in the time domain. Two common methods of carrier modulation are time-division multiplexing (TDM) and frequency-division multiplexing (FDM).

INTERFACING DATA PROCESSING EQUIPMENT

One of the more typical and usable fiber-optic links in today's market is the Canoga Data Systems CRS–100 data link (see Figure 8–28). The data link operates in a data range from dc to 56 kilobits per second (Kbps) over a distance of 3000 feet with an operational bit error rate of 10^{-9} or better. This unit is powered by a standard ac wall outlet plug (the stand-alone unit) or may be mounted in a powered rack such as that in Figure 8–28. The link's transmitter and receiver operate at an input/output wavelength of 820 ± 15 nanometers (nm). The light source for the transmitter is an infrared LED, while the detector for the receiver is a PIN diode. The link has all the advantages found with fiber optic systems. These are elimination of electromagnetic interference (EMI), equipment electrical isolation, and data transmission security. Some of its applications follow.

Simplex Application

Probably the simplest fiber optic data link is a one-directional link between a CPU and a single card reader, plotter, printer, and paper-tape reader. The simplex version of the CRS–100 data link uses the CRS–101 transmitter and the CRS–102 receiver.

Figure 8–28 CRS–100 fiber optic digital data link (*courtesy of Canoga Perkins*).

The CRS–100 operates in transmit and receive modes simultaneously or can operate in half-duplex mode (receive or transmit) (see Figure 8–29).

Duplex Application

When two-way communication is required, the duplex CRS–100 link is utilized. A single terminal setup is shown on the top of Figure 8–30. The same communication link can serve several terminals if they are located within 50 feet of a common point and messages are short and/or infrequent enough that satisfactory operation is possible with only one terminal transmitting at one time. This can be accomplished in two different ways. The first is a straight polling configuration. See the center setup in Figure 8–30. Terminals are connected in a series/parallel arrangement and one terminal is connected to the data link. The second method is the use of a modern sharing unit. See the lower setup in Figure 8–30. Modem (link) sharing units typically have four or eight terminals and may be cascaded as shown in the lower illustration.

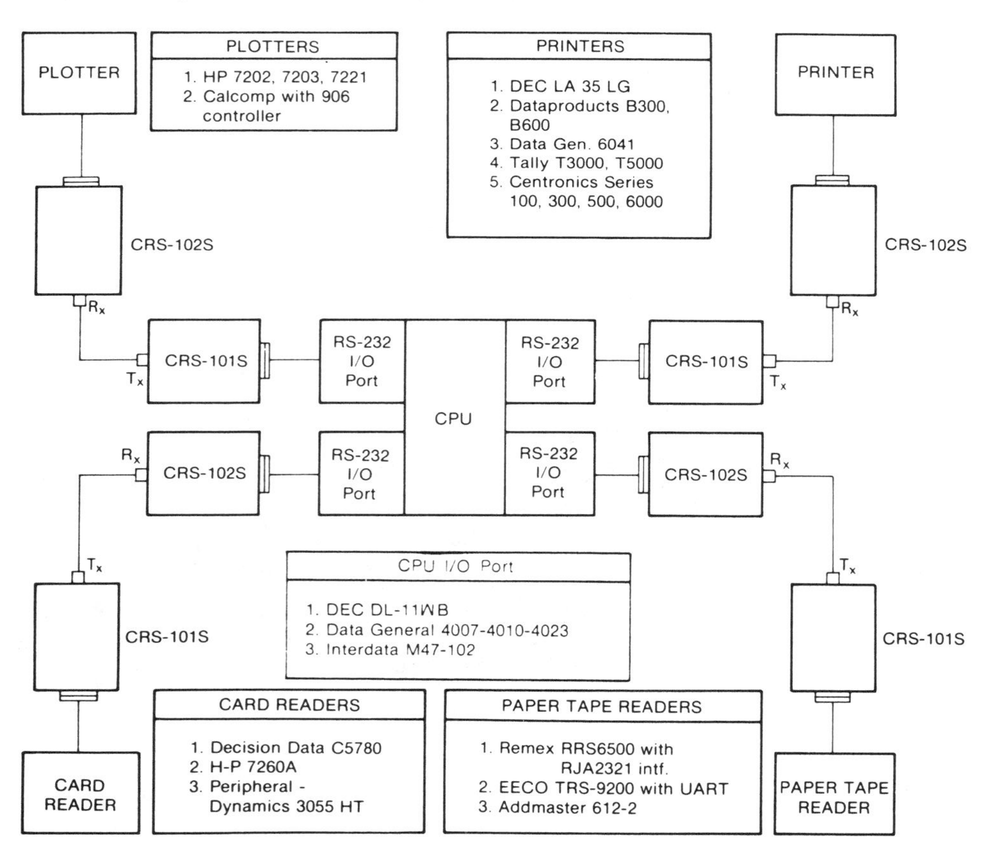

Figure 8–29 Digital simplex application (*courtesy of Canoga Perkins*).

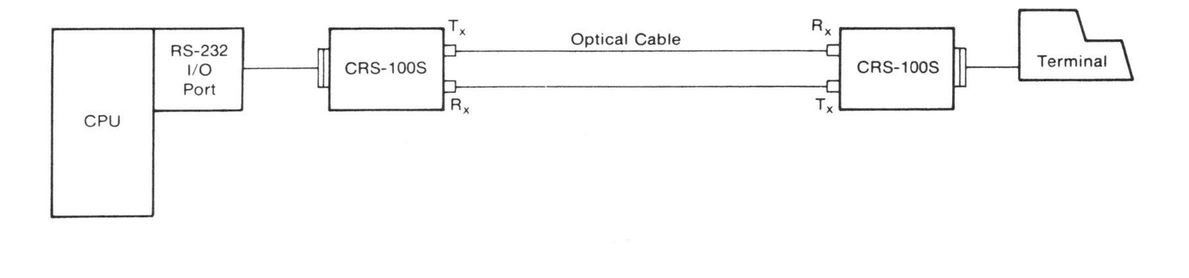

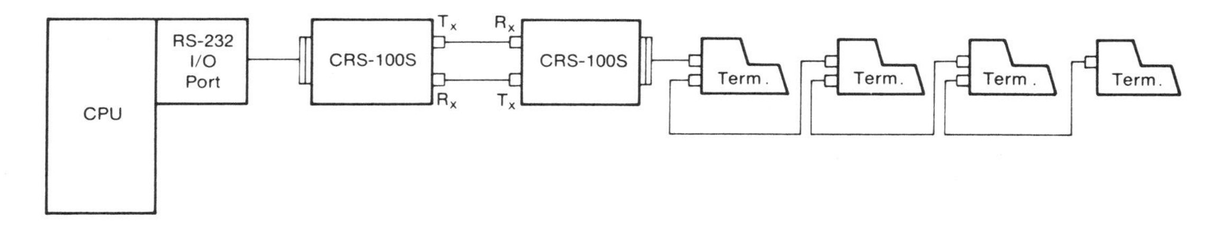

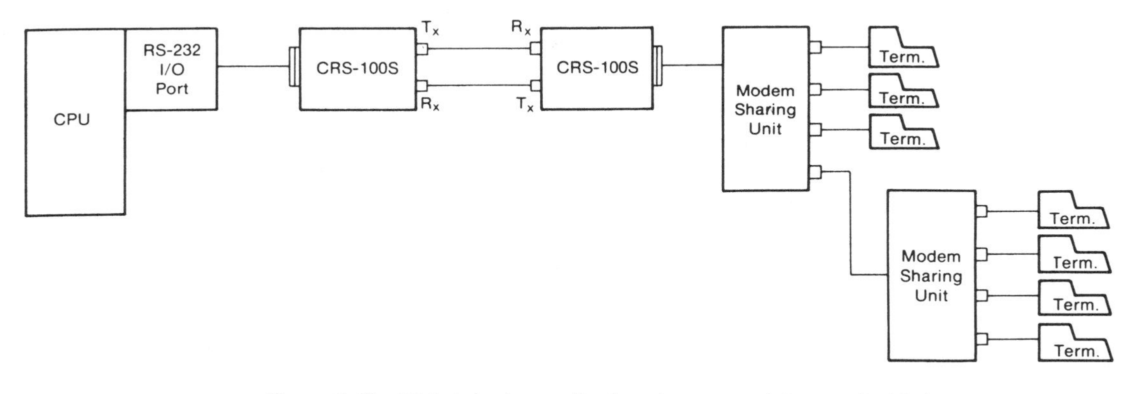

Figure 8–30 Digital duplex applications (*courtesy of Canoga Perkins*).

All the setups shown in these illustrations have one thing in common, the transmission lines between major units are optical fibers.

ON-SITE (FIELD) TESTING

The requirement for testing does not stop with the laboratory or the manufacturer's test line. Testing must also be performed during and after cable installation in the field. Testing of optical fiber systems is critical to ensuring overall system performance and integrity. Faults may be located and repaired. Parts must be replaced. Sometimes this involves piecing or replacing a section of cable.

Attenuation

One of the key parameters to consider in "On Site" field testing is the attenuation or optical power loss of the fiber itself.

Attenuation is caused by three primary factors: absorption, scattering, and externally induced losses.

Absorption and scattering are inherent to the fiber based on the structure and impurities of the glass. This type of attenuation is a function of the distance traveled and the wavelength of the transmitted light. The longer the length of fiber, the greater the total attenuation.

As a result, fiber quality is specified in terms of dB (decibels) per unit length, commonly dB/km, with a lower value reflecting a higher quality fiber.

Due to the characteristics of the glass, attenuation depends on the wavelength of the light being measured. Wavelength can be thought of in terms of the color of light, although the common operational wavelengths are outside what the human eye can see.

Externally induced losses are typically caused by bending of the fiber beyond a critical level, or outside forces placing stress or pressure on the fiber. Additional attenuation will also occur at splice points and connector interfaces.

The two methods of measuring attenuation are the power-through and the back-scatter techniques.

The power-through method involves a simple end-to-end test of the fiber. (See Figure 8–31). Using a stabilized light source (LED or laser) and an optical power meter, the relative amount of light lost as it travels down the fiber can be measured in dB. Variations of this test can also be adapted for maintenance and troubleshooting. This optical power loss will indicate what the end equipment will actually experience during system operation. The power-through method is accomplished in two basic steps.

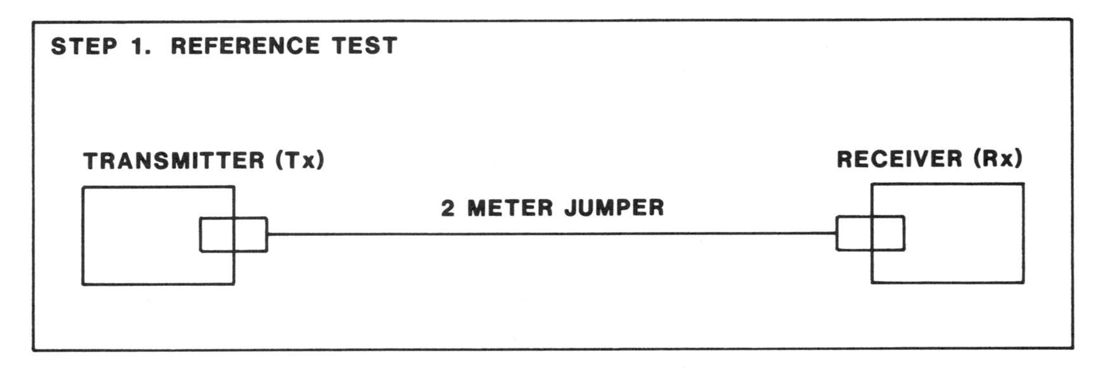

Figure 8–31A The power-through method for measuring attenuation (*courtesy of Siecor Corporation.*)

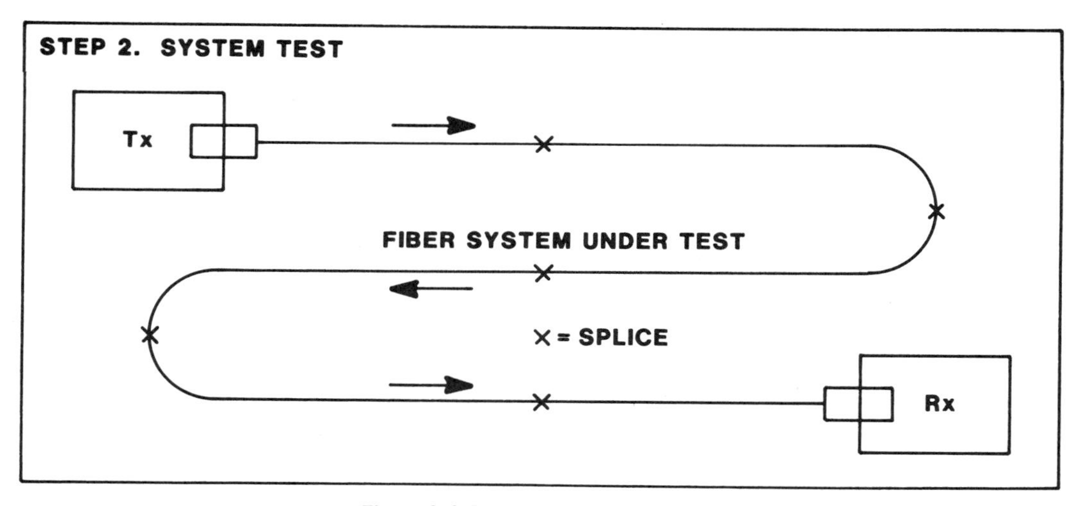

Figure 8–31B (*courtesy of Siecor Corporation.*)

1. A reading of the transmitter output level is taken as a reference.
2. The system is placed between the transmitter (Tx) and receiver (Rx). The level of power is measured. The difference between this power level and the reference level equals system loss (decibels).

Figure 8–32 illustrates the backscatter method for measuring attenuation. The backscatter method of testing utilizes an optical time domain reflectometer (OTDR). The OTDR injects a light pulse into the fiber. Backscattered signals

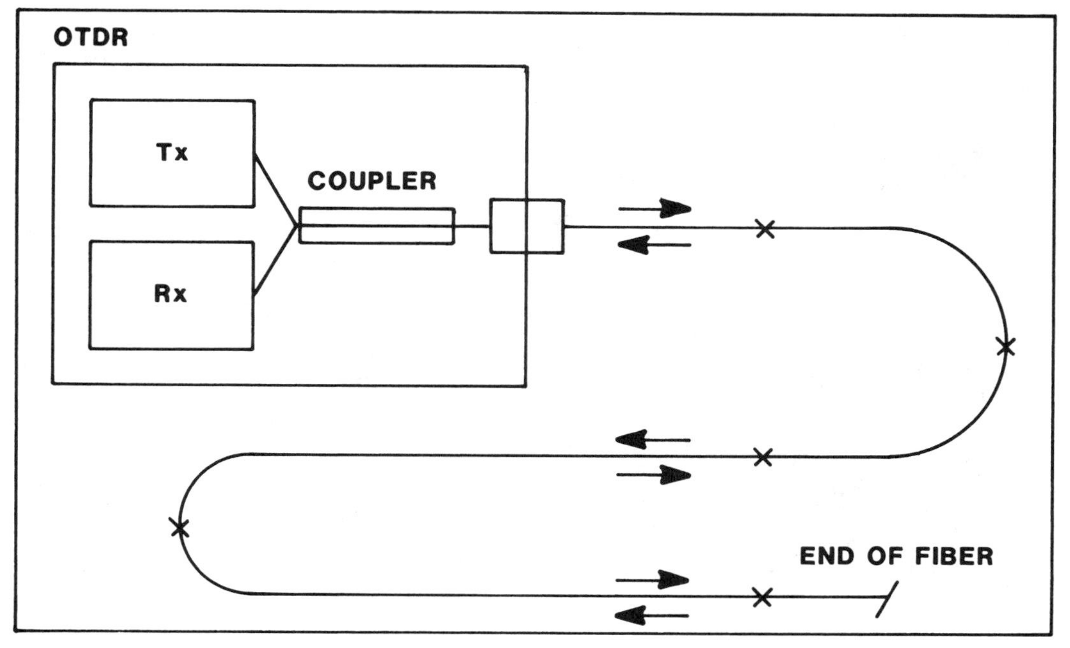

Figure 8–32 The backscatter method for measuring attenuation (*courtesy of Siecor Corporation.*)

and reflections return to the OTDR. The OTDR interprets the backscattered light to graphically display attenuation versus distance. As a result, the OTDR can be used to measure fiber, splice, and connector losses while providing the distances to locate system components, faults, and breaks. This is an indirect means of measuring the light transmitted down the fiber from a common end as opposed to an end-to-end measurement through the fiber.

The Optical Time Domain Reflectometer

The most important piece of field testing equipment for medium to long distance installations is an optical time domain reflectometer (OTDR). A typical OTDR available today is the Siecor model M51, as shown in Figure 8-33. Its measurement capabilities include length, determination of fibers or cables, location of fiber fractures and anomalies, evaluation of splice or connector loss, measurement of cable attenuation, and assessment of fiber homogeneity. It can be used to evaluate cable properties before, during, and after installation and

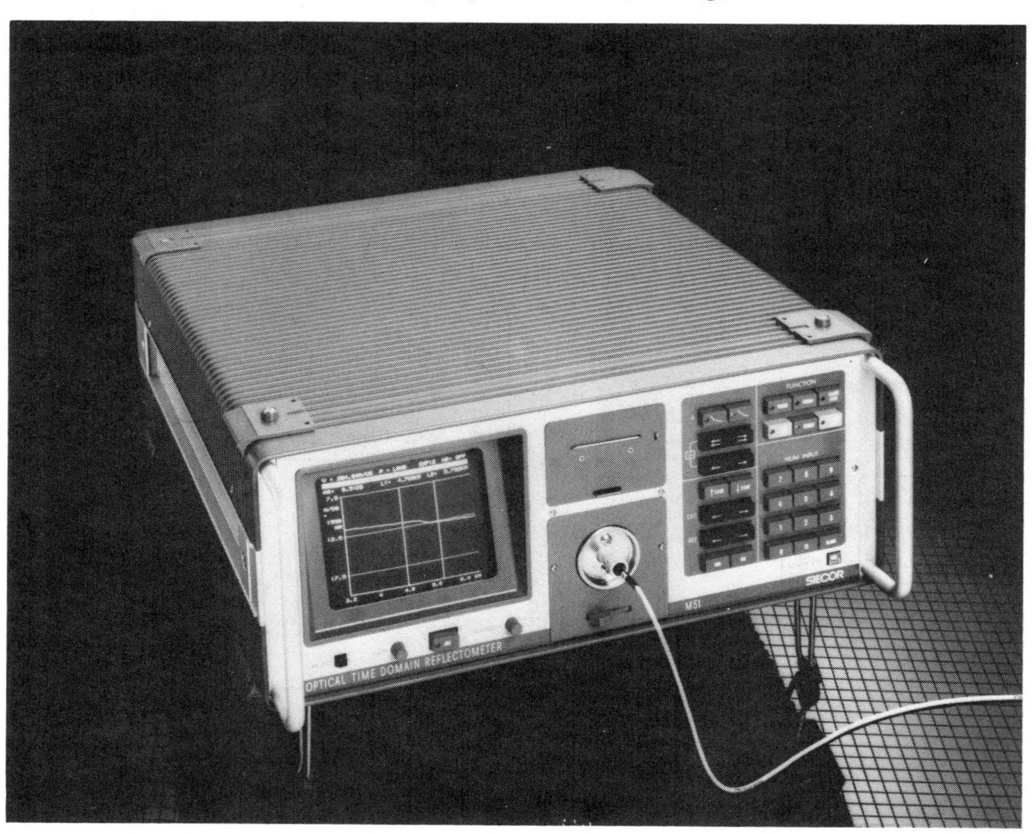

Figure 8-33 The Siecor M51 optical time domain reflectometer (*courtesy Siecor Corporation.*)

is one method of monitoring field splicing and connector interfaces. The OTDR is the primary fault locating and trouble shooting instrument for fiber optic cable systems.

The M51 OTDR is an example of the flexibility available in today's OTDRs. It utilizes plug-in optics module architecture which allows a variety of fiber types and wavelengths to be tested from one mainframe. High accuracy and distance resolution allow precise measurement capabilities in the field. Documentation capabilities range from built in printer to mass data storage and computer remote control. The selection of an OTDR can also depend upon the measurement application. Specialty OTDRs are now available for short LAN systems, aircraft and shipboard applications. These units feature cm resolution, and are optimized for analysis of very short systems.

Since today's optical systems range vastly in length and application, choosing the test equipment that meets your needs requires analysis of your particular application and testing requirements. Factors to consider include system length, fiber type, operating wavelength, and system loss tolerances. For example today's typical system lengths range from 500 meters (LAN) to 50 kilometers (long haul telco).*

* This article on ''On Site'' field testing was provided by Tim Yount and Todd Jennings of Siecor Corporation, Hickory, North Carolina

SELF-CHECK QUESTIONS

Self-check questions allow students to evaluate how well they have learned what they are studying

1. Of what does a most basic fiber optic system consist?
2. State the purpose of the optical transmitter (driver).
3. State the purpose of the light source.
4. Name the two light sources used in the fiber optic industry.
5. State the purpose of the light detector.
6. Name the three light detectors used in the fiber optic industry.
7. State the purpose of the optical receiver.
8. What specifications must the system engineer consider when changing a system to fiber optics or creating a new fiber optic system?
9. Review the procedures for design of a fiber optic system in the text. They are titled ''Design Procedures'' and are listed under ''Design Considerations.''
10. Describe the total attentuation in a fiber optic system.
11. What are transfer functions?
12. What are the several parts of a typical fiber optic cable?
13. List several tasks performed by the fiber optic connector.
14. What is the general construction of a heavy-duty cable such as the typical telephone communication cable?

15. The North American Digital Hierarchy provides standards for telephone systems within the United States. Of what do these standards consist?
16. The text illustrates three large fiber optic systems in the eastern United States. What are these? Can you recall other large U.S. fiber optic systems?
17. Review the several pages dedicated to the telephone system hierarchies T1 and T3.
18. What is the purpose of an optical coupler?
19. Name the two types of optical coupler.
20. What are the advantages/disadvantages of the star-type fiber optic coupling configurations?
21. What are the advantages/disadvantages of the loop-type fiber optic coupling configurations?
22. What is the difference between the simplex and the duplex digital data processing system?
23. Where are fiber optic systems tested? What is the most popular equipment used for testing, and what does it test?

Nine

Lasers

As described in Chapter 1, laser technology is the science that deals with the generation of coherent light in small but powerful beams. The word *laser* is an acronym originating from the initials of *l*ight *a*mplification by *s*timulated *e*mission of *r*adiation. The operating frequency range of the laser is in the optical frequency-spectrum from 0.005 to about 4000 microns. (A micron is equal to 10^{-6} meter.)

There are four basic laser types: crystal lasers, gas lasers, liquid lasers, and semiconductor lasers. Crystal lasers use a crystal such as the ruby for an active medium. Gas lasers use gases in combination, such as helium and neon, as an active medium. Liquid lasers use organic dyes such as rhodamine as an active medium. The semiconductor laser is made from semiconductor material such as gallium arsenide. The semiconductor laser was covered in detail in Chapter 4.

All lasers have some common characteristics. These are an active material to convert energy into laser light, a pumping source to provide power or energy, optics to direct the beam repeatedly through the active material to become amplified, and optics to direct the beam into a narrow cone of divergence. Other characteristics common to all lasers are a feedback mechanism to provide continuous operation and an output coupler to transmit power out of the laser.

Laser technology is extremely diverse and complex. This section of the book was added to give a general overview of lasers and is not intended to provide detailed physics and description of the laser field, which would take volumes.

OBJECTIVES

After studying this chapter and completing the self-check questions at the end of the chapter, the reader will be able to:

1. Explain several laser light properties.
   a. Monochromatic
   b. Coherent
   c. Directional
   d. Beam intensity
2. Describe the term *lasing action.*
3. State the various laser classifications.
4. Examine laser technical functions.
   a. Energy levels
   b. Stimulated absorption and emission
   c. Attenuation versus amplification
   d. Coherence
   e. Beam divergence
   f. The optical cavity
5. Describe the solid laser and its operation.
6. Describe the gas laser and its operation.
7. Describe the neutral-atom gas laser and its operation.
8. Describe the ion gas laser and its operation.
9. Describe the molecular gas laser and its operation.
10. Describe the liquid laser and its operation.
11. List the typical optics included in a laser set.
12. Explain the two purposes of laser alignment.
13. Name several uses of the laser.

LASER LIGHT PROPERTIES

Laser light has unique properties. A laser beam is one color or *monochromatic.* If the light is one color then it is also one wavelength, and is therefore *coherent.* An example of incoherent light is that of an ordinary light bulb, which emits white light. White light is incoherent because it has all visible colors and therefore all visible wavelengths. Figure 9–1 shows some laser light properties. Light such as the light bulb is panchromatic (has all visible wavelengths) and cannot be concentrated or controlled. Laser light can be controlled because it is monochromatic (one color), one frequency. The beams are in-step (in-phase) with each other.

The light in a laser beam is extremely *directional.* It is focused by a lens into a narrow cone and projected in one direction. The direction of a laser beam is the result of optics at the ends or sides of an active medium. The radiance of the beam can be compared with the intensity of the sun. The laser, when focused to a fine, hairlike beam, can concentrate all of its power into the beam. Hence the directionality of the beam has a bearing on its intensity. If the beam were allowed to spread out it would lose its power. Confinement of large amounts of power into a narrow

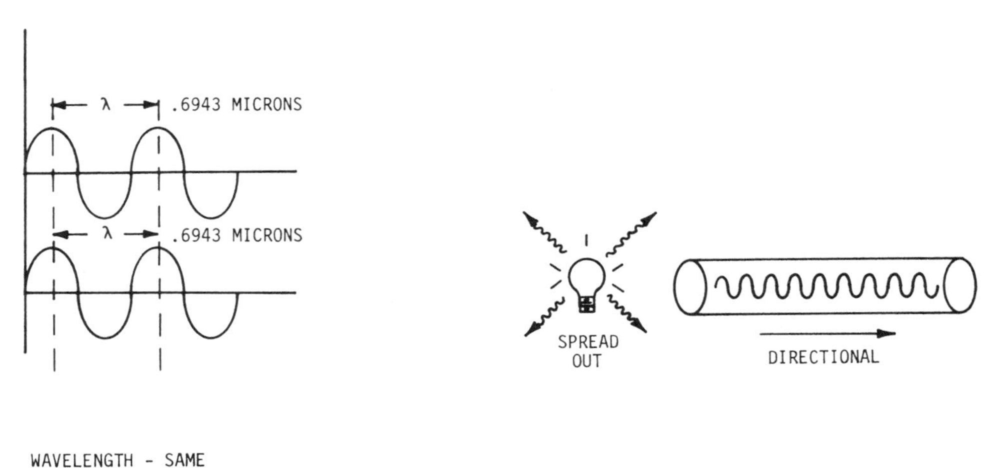

Figure 9-1 Laser light properties (*Mike Anaya*).

slit allows the laser to be used to drill holes in and cut metal, as well as welding applications.

Lasing Action

We have discussed the fact that incidence radiation could change the energy state of an atom from the ground state to an excited state. We also know that if left alone for a small time period in the excited state that the atom's energy will decay and fall to the ground state and produce a photon. For amplification by stimulated emission to happen, there must be a chance for the created photon to meet other atoms in the excited state and a *population inversion* must be created. That is, there must be more atoms in the excited state than in the ground state. This situation can happen by a process called *pumping*. In this process, energy is pumped into the active medium to raise atoms to an excited state. This must be done very quickly, because the atoms in the excited state may *spontaneously decay* to the ground state before amplification takes place. To get quick response action a high-intensity lamp is required with a very large power supply. With a low-intensity lamp only incoherent light will be produced. In a three-energy-level band such as a ruby laser, the pumping action raises the atoms to the *excited energy state* which absorbs a large range of wavelengths. These atoms in turn spontaneously drop to an intermediate energy state. Heat is dissipated. However, during the process, a population inversion has been created at the intermediate level. Now that the population inversion has been created, lasing action can take place. Soon atoms drop to the ground state and in doing so emit photons. These photons strike other atoms and two photons are created, and so on. Thus *stimulated emission* has started lasing action and soon the beam will build up to a large strength. When the beam strikes the mirror surfaces of the active medium, it

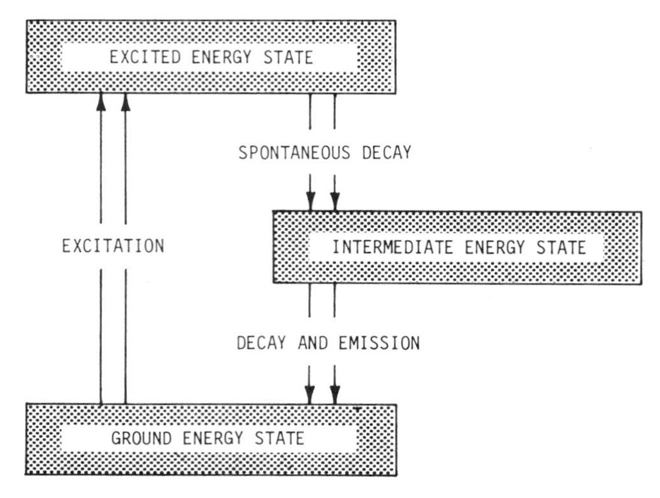

Figure 9–2 Lasing action (*Mike Anaya*).

is reflected back and continued lasing takes place. Because of this reflection and continual lasing action an intense beam is built up. Figure 9–2 shows a schematic of lasing action.

LASER CLASSIFICATIONS

Lasers are classified in three ways:

1. Their active mediums
   a. Solid
   b. Gas
   c. Liquid
   d. Semiconductor
2. Their wavelength
3. Their output time characteristics
   a. Mode-locked pulse
   b. Q-switched pulse
   c. Normal-mode pulse
   d. Continuous wave (CW)

Active Medium Classification

The solid-crystal ruby laser was the first one developed. As discussed in Chapter 1, Theodore H. Maiman, a scientist at Hughes Aircraft Company, created the first laser from a ruby crystal. The solid lasers in use today are characterized by their active medium (host material), the dopant they use for lasing ions, and their output. The major solid lasers are the ruby, Nd:YAG, and the Nd-doped glass. The medium for the ruby laser is aluminum oxide and its active ions are chromium. Chromium is also known as a transition metal. The Nd:YAG laser uses a rare-earth element called neodymium. The neodymium (Nd) is dissolved in the host crystal yttrium aluminum garnet (YAG). The third solid laser is the Nd-doped glass laser. Glass differs from other solid lasers in that it is not a crystalline material. It is, however, cheap to produce in any shape and can be more efficient than the ruby. The host glass is employed with an active ion usually neodymium (Nd). Another atom (uranium), part of a group called *actinides,* is used as an active ion.

Active mediums of gas are classified as neutral-atom, ion, and molecular. The neutral-atom gas lasers have active-medium atoms which have an equal number of charged protons and electrons. A type of neutral-atom laser is the helium-neon (HeNe) laser. An ion gas laser has an active medium of ionized atoms (have one or more electrons moved from their shells) and has a positive charge. A type of ion gas laser is the singly iodized argon Ar^+.

A molecular gas laser has an active medium which has two or more atoms either of the same or of different atomic species. A type of molecular gas laser is carbon dioxide (CO_2) laser. Liquid lasers typically use organic dye dissolved in alcohol or water and are usually pumped with an argon laser. A type of liquid laser is one using rhodamine 6G dye.

Semiconductor lasers are described in detail in Chapter 4.

Wavelength Classification

It is important that wavelength be considered when classifying lasers. The optical range of wavelengths, as stated many times in this book, have three major areas: ultraviolet, visible, and infrared. Wavelengths in the ultraviolet spectrum range from 0.005 to 0.3900 micron. Wavelengths in the visible spectrum range from 0.3900 to 0.7500 micron. Wavelengths in the infrared spectrum range from 0.7500 to 4000 microns. Table 9–1 lists typical lasers along with their wavelengths.

Output-Time Characteristics

Output-time characteristics fall into categories of time duration and power. The first of the characteristics (or *modes of operation* as they are called) is the mode-locked pulse. Mode locking is a process in which longitudinal modes circulating in a laser cavity have a fixed difference on separation. The effect is used in solid lasers by inserting a cell with a special organic dye. Mode-locked lasers operate in ultrashort pulses of about 10 picoseconds at 10^4 to 10^{12} watts of power.

Normal pulsed lasers, such as the first ruby laser, store energy in a capacitor bank that discharges into a flashlamp, which in turn radiates light into the ruby. This causes chromium atoms to move to their excited states. Single pulses are emitted with uneven spikes at a few milliseconds in duration at 10^3 to 10^5 watts of power. The pulses are not equal in amplitude.

TABLE 9-1 TYPICAL LASERS AND THEIR WAVELENGTHS

Laser type	Wavelength (microns)	
Hydrogen cyanide (HCN)	773.0	Infrared
Water vapor (H_2O)	118.6	
Carbon dioxide (CO_2)	10.6	
Hydrogen fluoride (HF)	3.1	
Yttrium aluminum garnet (Nd:YAG)	1.064	
Glass (Nd:glass)	1.060	
Gallium arsenide (GaAs)	0.9100	
Krypton (K)	0.7525	Visible
Ruby (Al_2O_3)	0.6943	
Helium-neon (HeNe)	0.6328	
Rhodamine (6G)	0.5900	
Argon (Ar)	0.5145	
Argon	0.4880	
Helium-cadmium (HeCd)	0.4416	
Molecular xenon (Xe_2)	0.1700	Ultraviolet
Hydrogen (H_2)	0.1600	

Q-switched lasers replace the 100% reflecting mirror on one end of the ruby rod with a high-speed rotating prism. The rotating prism develops a greater population inversion of atoms within the ruby. Pulses are cleaner than those of the normal pulse laser. Q-switched laser pulses are about 25 nanoseconds in duration at 10^6 to 10^9 watts of power. A continuously pumped repetitively Q-switched laser provides repetitive pulses of close to 50 nanoseconds in duration with up to 50 kilowatts of power.

SOME LASER TECHNICALITIES

Energy Levels

Each atom has its own light spectrum. That is, each material has a set of wavelengths called *spectral lines* that are representative of the material. Each atom also has its own set of energy levels, which is a result of transition between two of the energy levels. The energy of the levels is usually expressed in electron volts (eV), but may also be discussed in terms of joules or ergs. Some comparisons of these units are listed below.

$$1 \text{ joule} = 10^7 \text{ ergs}$$
$$1 \text{ volt} = 1 \text{ joule/coulomb}$$
$$1 \text{ electron volt} = 1.602 \times 10^{-19} \text{ joule}$$
$$1 \text{ electron volt} = 1.602 \times 10^{-12} \text{ ergs}$$

Measurement of these spectral lines is accomplished using a device called a *grating spectroscope*. The spectroscope has a narrow slit through which light enters. The light is partially collimated through an aperture and may be monitored on a wavelength scale. The wavelength scale is superimposed on a pattern of colored bright and dark lines representing the color spectrum of the light. The wavelength is measured in angstroms.

Stimulated Absorption and Emission

The ruby is usually a three-energy-state atom. The top shaded area in Figure 9-2 is the excited energy state. Dozens of energy levels are crowded at this energy state to form an energy band. The second energy state down is an intermediate energy state. The lowest energy state is ground. Pumping raises the chromium atoms from ground energy state to excited energy state. This is called *stimulated absorption*. The ions then decay and fall to a lower or intermediate energy state. When the population of ions in the intermediate state is greater than the ground state, population inversion is present. That is, there are now more excited atoms in intermediate than ground state. When the population inversion is present, laser action (lasing) can take place. In time, the excited chromium atoms will fall to the ground energy state. When they do so, photons are emitted. A photon you may recall is a packet of light caused by radiant energy. The frequency of the energy determines the strength of the photons. Higher frequencies cause greater-strength photons.

Now let us consider the term *stimulated emission*. Chromium atoms are pumped to an excited energy state, then go through a spontaneous decay to an intermediate energy state where an accumulation of excited atoms creates a population inversion. As time progresses, the excited atoms fall from the intermediate energy state to the ground state. When they do, photons are emitted. As the photons are emitted they strike excited atoms and two other identical photons will emerge. These two photons in turn strike other excited atoms and stimulated emission occurs. The more often this happens the greater the intensity of the light beam. The ruby laser just discussed produces laser light in pulses. It is possible for the ruby laser to produce continuous pulses. If this is the case, energy must be pumped continuously into the optical cavity to maintain population inversion. A problem of heating results but can be resolved using a four-energy-state laser.

Attenuation versus Amplification

In Figure 9-3, the difference between attenuation and amplification is explained simply with two separate mediums. When a light signal is directed or pumped into a medium of glass, part of the light signal is absorbed by the glass and part allowed to continue through the glass. This phenomena is called the *exponential law of absorption*. The law involves the intensity of the transmitted light signal, the thickness of the material, and the intensity of the incident light. In the ruby medium a light signal is directed or pumped into the ruby medium causing population inversion and stimulated emission rather than absorption.

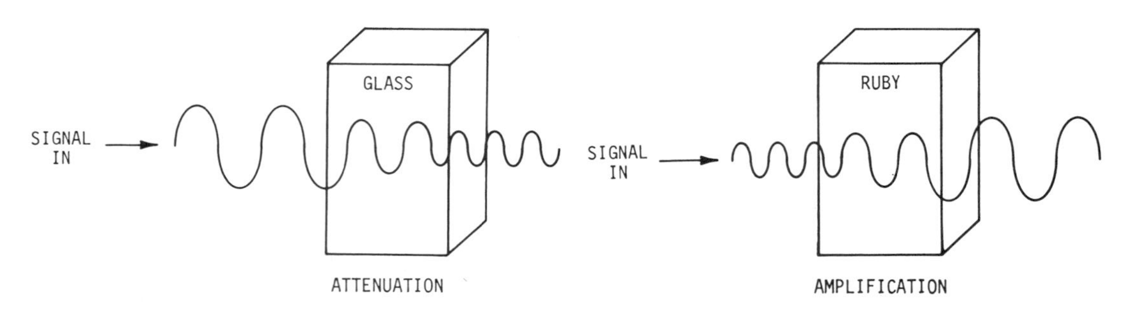

Figure 9–3 Attenuation versus amplification (*Mike Anaya*).

Coherence

Coherence of a laser beam must be explained by way of phase relationship. Spontaneous emission may move in any direction. This is not the case of stimulated emission. Stimulated emission has the same phase as the applied or incident radiation and indeed is the same shape. Furthermore, the radiation is moving in the same direction. When the incident and stimulated-emission wavelengths are in phase along the beam, this is called *temporal coherence.* When the same relationship occurs across the beam, it is called *spatial coherence.* The importance of coherence with lasers cannot be understated. A coherent beam means that maximum power can be focused in an extremely small area. With an incoherent beam this is not possible. The application of a monochromatic light source would seem imperative in laser applications.

Laser Beam Divergence

The light beam emitting from a laser has an extremely small diameter at its aperture. The beam does not continue at this same diameter but diverges (spreads out) into a larger diameter forming a *cone of divergence.* The divergence angle is usually very small. Gas lasers have the smallest divergence angle, somewhere near 1 milliradian. You may recall from mathematics that a radian is 57.3°. A milliradian is 0.0573°, which equates to 3.44 minutes of arc. Liquid lasers may have a 10- to 20-milliradian divergence angle. When specifying a divergence angle it is often given as a half-angle. Solid lasers have a divergence angle of 5 to 25 milliradians. Semiconductor lasers such as the injection laser diode often have a divergence angle of 500 milliradians. Laser beam divergence and lens focusing are important in laser alignment. Laser alignment is covered in greater detail later in this chapter.

The Optical Cavity

The amplifying medium of a laser is usually between a pair of reflecting mirrors. These mirrors serve as optics to direct the electromagnetic waves into and out of the active medium. Light waves are reflected back and forth through the cavity and bounce off the mirrors. At the surface of the mirror are incident waves (toward the mirror) and reflected waves (away from the mirror). You must remember that these waves are the same frequency. When they move in opposite directions, a pattern results if they interfere with each other. When they interfere a phenomena exists in that the waves appear to stand still. If two waves of the same frequency and amplitude interfere with each other by moving in opposite directions, the result is a standing-wave pattern (see Figure 9–4). The standing wave will only happen if the relationship between the wavelength and the length of the cavity are compatible. Lasing operation will only occur for typically two or three frequencies. Only those frequencies or modes of oscillation that satisfy the relationship are compatible. The formula below relates to this compatibility.

$$N = \frac{2n_0L}{\lambda}$$

where N = number of half-wavelengths or loops that will fit between the mirrors in the optical cavity
n_0 = refractive index of the active medium
L = length of the optical cavity
λ = wavelength

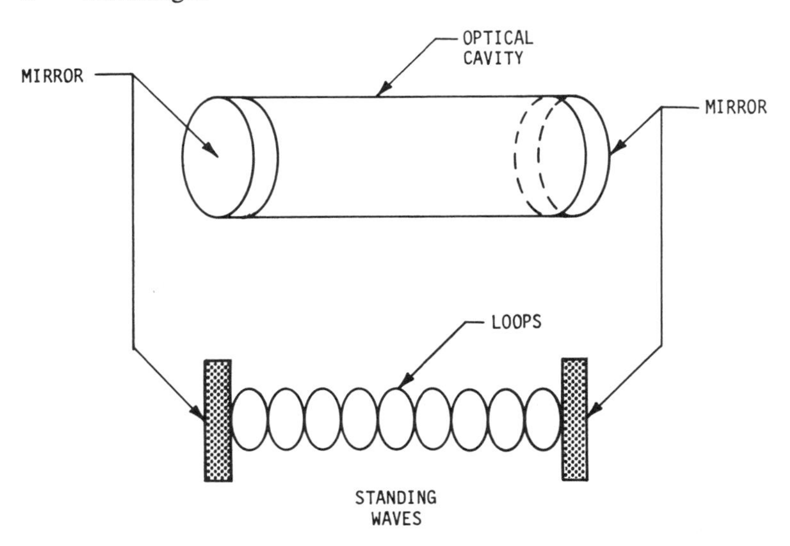

Figure 9–4 Standing waves in an optical cavity (*Mike Anaya*).

Four-Energy-State Lasers

The four-energy-state laser resolves the heating problem of the three-level ruby laser. Typical of the four-layer laser system is the Nd:YAG laser. The top shaded area in Figure 9–5 is the excited state. Dozens of energy levels are crowded at this energy state to form an energy band. The second state down is an intermediate state to where atoms spontaneously decay. A secondary energy state is a level of spontaneous decay and stimulated emission, while the ground state is at the lower energy state level.

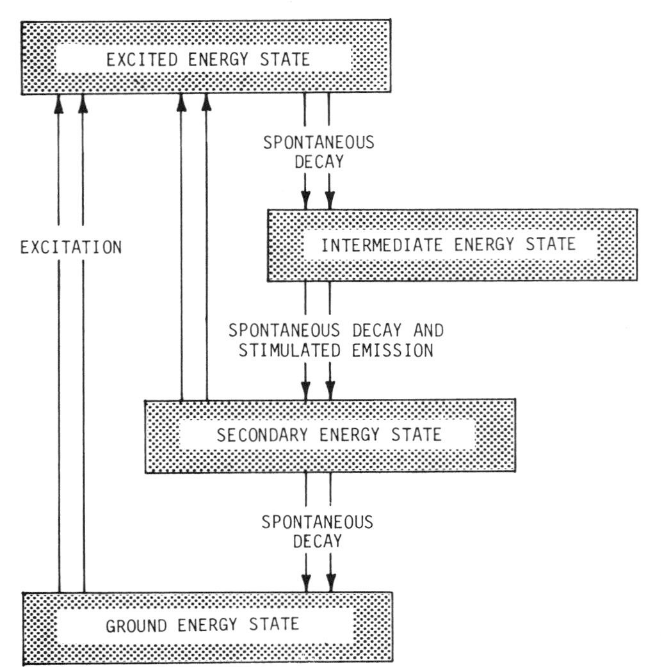

Figure 9-5 Four-energy-state laser (*Mike Anaya*).

The process of pumping raises the neodymium (Nd) from ground energy state to excited energy state. The ions decay and fall to a lower or intermediate energy state. No radiation takes place between the excited and intermediate states. Lasing action takes place between the intermediate and secondary energy states. These two center levels have a lower population than the ground state. The secondary has a lower population than the intermediate. The difference, however, between the secondary and intermediate states is still much lower than that between secondary and ground, therefore much less energy is required to maintain a population inversion between the two center levels. Much less energy has to be pumped into the laser cavity and the laser operates continuously instead of in pulses. It must be noted that all four-energy-state lasers do not operate continuously. Likewise, not all continuously operating lasers are four-layer.

SOLID LASERS

There are three basic solid lasers: the ruby laser, the Nd:YAG (yttrium aluminum garnet) lasers, and the Nd:glass laser. The ruby laser operates in the visible range of the electromagnetic frequency spectrum at 0.6943 micron wavelength. The Nd:YAG and Nd:glass lasers operate in the near-infrared range of electromagnetic frequency spectrum at 1.0640 and 1.0600 micron wavelength, respectively.

The ruby as an active medium consists of aluminum oxide with small amounts of chromium dissolved in it. Rubies used in lasers are formed in a high-temperature crucible by melting the aluminum and chromium and inserting a seed crystal of ruby. A drawing process is used with the seed crystal being pulled at a constant speed while rotating. The ruby acts as the host crystal while the chromium acts as the activator. The Nd:YAG laser uses a host crystal of yttrium aluminum garnet and an activator of neodymium. The Nd:glass laser uses a host material of glass and an activator of neodymium.

Solid Laser Operation

As an introduction to solid lasers let us take a cursory look at their operation. A typical solid laser consists of an active medium, an excitation source, a feedback mechanism, and an output coupler. The active medium in the solid laser illustrated in Figure 9–6 is a ruby. The typical active medium is a tubular-shaped crystal structure. The ends of the crystal are highly polished and are cut parallel to each other. These ends are coated with a reflective coating. One end of the ruby reflects 100% of the light internally. The other end acts as a transmitter, allowing only a small portion of the light to pass outward while reflecting most of the light back into the ruby for stimulated emission. The excitation source is usually a flashtube. The tube is fitted with a trigger which emits a burst of light to start the process.

Let's now go through the operation and consider how it all comes together. The flashtube is energized by the trigger. It produces a high-level burst of light similar to a camera flashbulb. The flash, being adjacent to the ruby cylinder, causes the chromium atoms within the crystalline structure to become excited. Population inversion, stimulated emission, and lasing action take place.

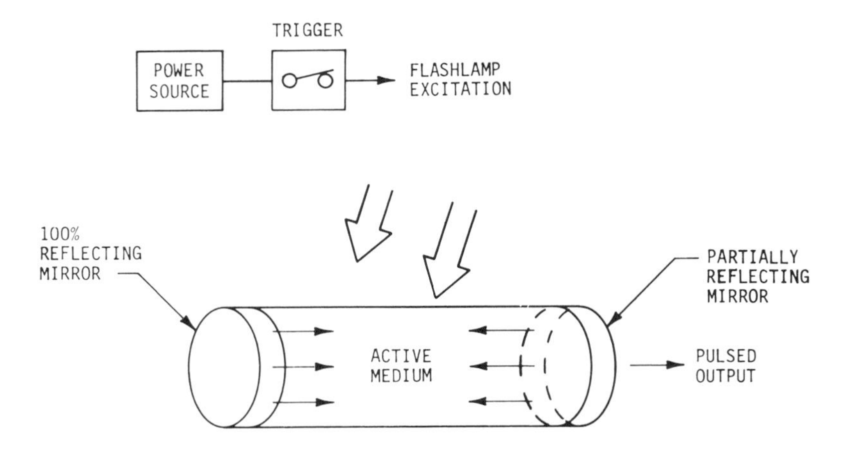

Figure 9–6 Solid laser operation (*Mike Anaya*).

When the beam of photons strikes either mirror ends of the ruby, the reflection of the light beam back into the ruby causes an even greater beam intensity because of photon action. The beam is red in color because the laser is amplifying at a wavelength of 6943 angstroms (0.6943 micron). The white light of the flashtube produces many wavelengths; however, only one wavelength can excite the atoms from ground state to excited energy state.

The amplified light is fed partially out of the laser by a transmitting mirror. The transmitting mirror reflects most of the light back into the ruby crystal to attain an ongoing process of stimulated emission and amplification. If too much energy is coupled out of the transmitting mirror the stimulated emission will decay and die. Therefore, the excitation mechanism must constantly pump energy into the laser.

Q-Switched Solid Laser Operation

The most powerful of all laser pulses are those developed by the Q-switched laser (see Figure 9–7). The phenomenon is called the giant pulse and operates on the premise that the atoms in the active region remain at the excited state much longer than in other lasers (perhaps 2 or 3 thousandths of a second). In the initial moments of operation, the light path between the 100%-reflecting mirror and the partially reflecting mirror are blocked so that lasing action cannot take place. You may recall that these mirrors may be external to the active region or may be coated on the regions ends. On the Q-switched solid laser, the 100%-reflecting mirror has been replaced by a prism.

The active material is pumped by a high-power light source to ensure that a large population of atoms is in the excited state. Population inversion is ensured and lasing action can take place. All the energy previously stored is released in one giant pulse. The pulse lasts between 10 to 100 millionths of a second. These are short pulses and dependent on the laser.

The prism is rotated at high speed. Lasing can only occur when a population inversion takes place. Stimulated emission amplification far exceeds losses. The

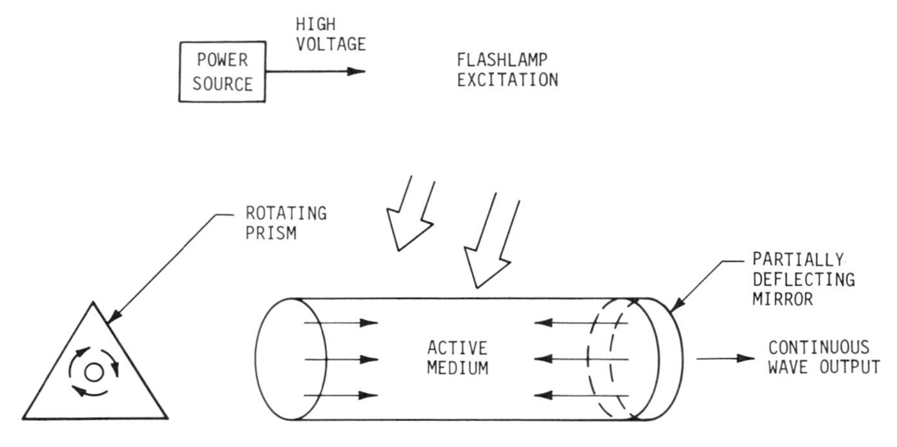

Figure 9–7 Q-switched solid laser operation (*Mike Anaya*).

rotating prism is called a Q-switch. Lasing action can only take place when this mechanical switch rotates to a point where it is aligned with the partially reflecting mirror on the opposite end of the cavity. The Q-switch can only be rotated at a mechanically controlled speed. Therefore there are limitations to mechanical hookups and switching times. The function of the switch is to allow time for the active medium to be pumped up before the next pulse. The active medium then becomes a storage place for energy between pulses.

Electro-optical Q-switches provide faster switching times by changing optical characteristics with the use of polarized voltage. These are the Pockel's cell and the Kerr cell. Chemical Q-switches are used for extremely short pulses of a few picoseconds.

STATE-OF-THE-ART SOLID-CRYSTAL LASERS

A typical solid-crystal laser is the Nd:YAG MY-series designed by William D. Fountain and manufactured by the Molectron Corporation, Sunnyvale, California. This laser is now manufactured by Laser Photonics, Orlando, Florida. We shall take a longer look at the Fountain design later on, but let us first consider the development of the Nd:YAG laser and its growth patterns.

Origins of the Nd:YAG Laser

You may recall that the first solid laser was a ruby laser developed in 1960 by T. H. Maiman, a scientist with Hughes Aircraft Company. It was through Maiman's research that other scientists were encouraged to look for solid materials that could be used as laser material. One of these solids was from a group of rare-earth materials. The rare-earth material neodymium was dissolved in yttrium aluminum garnet and was found to easily operate within a laser. In 1964, the Nd:YAG laser was created. The Nd:YAG laser operates in pulsed or continuous-wave (CW) modes. Early versions produced a pulse 100 microseconds in duration. These pulses were a group of much smaller pulses. Better control of the laser was made by addition of the Q-switch. The Q-switch resulted in shorter pulses (10 to 30 nanoseconds duration) and with extremely high peak power. Pulse lasers today easily produce megawatts of peak power output.

The first Nd:YAG lasers produced an output beam with a bell-shaped (Gaussian) profile and low divergence (see Figure 9–8a). The Gaussian TEM_{00} or fundamental mode beam is called the optimum beam profile. A great deal of effort is spent in laser design to produce the Gaussian beam. Unfortunately, the TEM_{00} cavity is limited in the amount of energy that can be extracted from a YAG rod. The only way to achieve high energy and low divergence with this cavity is to add amplifier stages.

A diffraction-coupled positive unstable resonator was introduced in 1976 as an alternative to the multiple amplifier TEM_{00} system (see Figure 9–8b). This cavity is known as diffraction-coupled because the beam exits and diffracts around a small

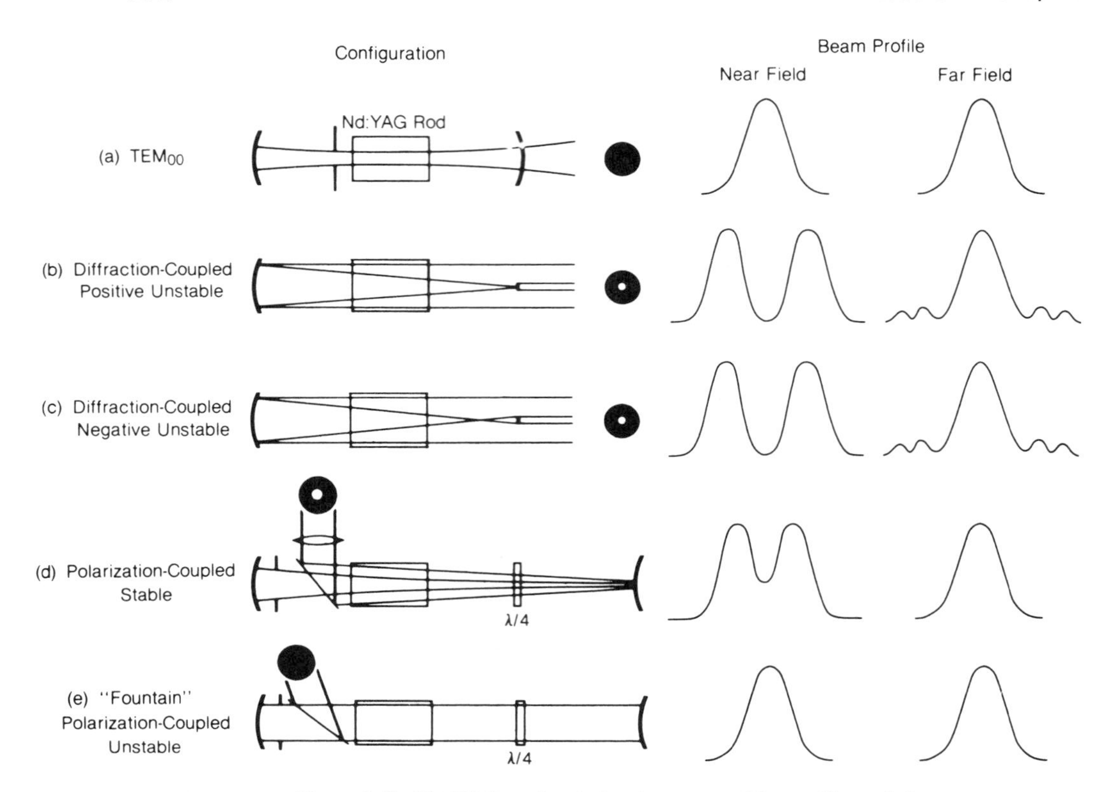

Figure 9–8 Nd:YAG cavity design (*courtesy of Laser Photonics*).

total reflecting output mirror. The term positive unstable refers to the positive curvature of the output mirror. This cavity produces high pulse energy and low divergence in a beam that has a planar or uniphase wavefront. The pulse duration depends on both cavity length and pump energy. The beam profile is that of a doughnut in the near field. Fresnel rings also appear in the intermediate field. These rings are caused by diffraction from the edges of the Nd:YAG rod and fade away in the far field. One of the characteristics of this design is that the doughnut does not persist. Because the beam is diffracted sharply around the output mirror the beam profile constantly changes as a function of distance from the laser. In the intermediate field, a small central spot appears which has extremely high-power density, up to four times that of the doughnut. This spot is potentially dangerous to optical components. The spot causes a discontinuity in the refractive index of transmitting materials and leads to self-focusing and material damage. In the far field (10 meters from the laser) the beam profile has only 60 to 80% of the energy of the central lobe.

Figure 9–8c shows the optical schematic of the negative unstable cavity. It is termed negative because of the concave output mirror. The optical and mechanical tolerances of this design are less stringent than those of the positive versions. As a result, it is well suited for lasers on military equipment. For mirror curvatures identical to its positive counterpart, the cavity is twice as long and produces a longer pulse. Otherwise, the output is identical to the positive version.

Late in 1977, a stable resonator using a polarization output coupling was introduced (see Figure 9–8d). This design employs two factors different from previous design. First, polarization techniques couple the beam from the cavity. Second, it uses the oscillator rod to increase the output energy. The basic resonator is a TEM_{00} configuration formed by a concave rear mirror and partially reflective front surface of a quarter-wave plate. The aperture restricts the oscillator to the center of the rod and the resulting beam polarized horizontally is low in energy. This beam exits the cavity through the quarter-wave plate $\lambda/4$, which converts it to circular polarization. When reflected off the convex mirror and passed back through the $\lambda/4$ plate, the expanded and diverging beam is polarized vertically, then amplified in the YAG rod and coupled out by the polarizer. A lens then recollimates the beam. Although the oscillator output is the same as the TEM_{00}, the beam exiting the polarizer is not. The oscillator has depleted the gain at the rod center so that the near-field beam has a prominent dip in the center. Fresnel rings are also present. All of the horizontally polarized oscillator beam is converted to vertical polarization for coupling out of the cavity. At high pump energies, unwanted feedback upsets cavity oscillations and limits the energy and quality of the output beam.

All these cavity configurations proved to have good and poor points. In 1978 a beam configuration was designed by William D. Fountain which has no modulation on the beam and its intensity decreases away from the beam center. Beam profiles were close to Gaussian (bell) shaped. The beam maintains high pulse energy and low divergence (see Figure 9–8e).

The Fountain Design

Mirrors M_1 and M_2 in Figure 9–9 form the optical cavity. In the region to the left of the intracavity polarizer, the beam is polarized horizontally. In passing the quarter-wave plate ($\lambda/4$), the beam becomes elliptically polarized. After reflecting off mirror M_2 and passing back through the wave plate, the beam is polarized 75% in the vertical plane and 25% in the horizontal plane. The vertical component is coupled out of the cavity by the polarizer, while the horizontal component remains in the cavity as feedback. The horizontal component of the beam that remains in the cavity for feedback must pass back through an aperture before reaching mirror M_1. The aperture acts as a loss mechanism, which allows energy to build up. Energy lost at the aperture is small.

The Fountain oscillator magnifies the concentric rings produced in previous designs so that the rings are eclipsed by the aperture leaving only a uniform central spot. After the beam exits the cavity, it is incident on a second polarizer which is used to improve the polarization ratio before amplification. Two turning mirrors direct the beam along the rear beam line of the laser.

In an oscillator-alone configuration a beam-reducing telescope is used to increase the power density on the harmonic crystals. A complementary beam-expanding

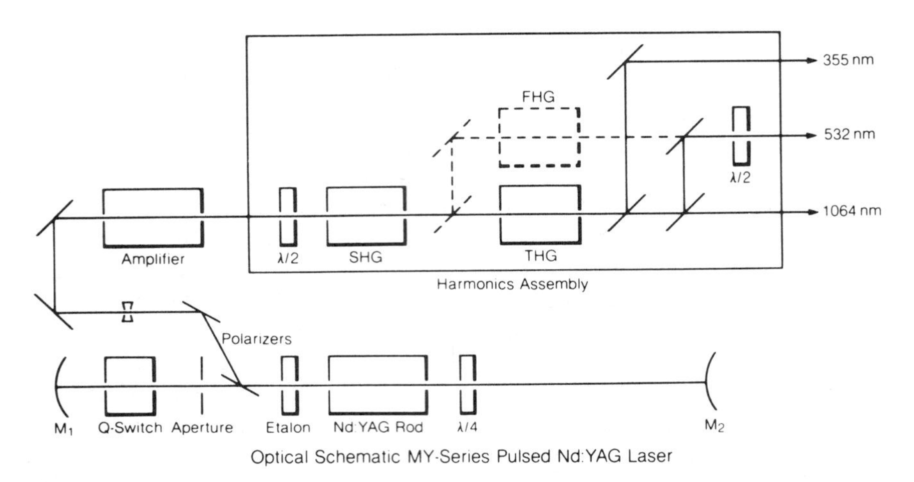

Figure 9–9 Nd:YAG laser operation (*courtesy of Laser Photonics*).

telescope on the output brings the beam back to its original size and also permits adjustment of the output divergence.

In an oscillator/amplifier configuration the amplifier stage replaces the beam-reducing telescope, which is not needed. A negative lens is also added before the amplifier to expand the oscillator beam and to compensate for the positive thermal lensing of the amplifier rod.

In both configurations the beam enters a sealed harmonics box where second, third, and fourth harmonics are generated. These are SHG, THG, and FHG respectively. Each harmonic crystal is hermetically sealed in a cell and kept at a stable temperature. The crystals are angle-tuned with thumbwheels accessible from outside the laser. Before the crystals, a half-wave plate ($\lambda/2$) rotates the polarization of the beam to an axis appropriate to the crystal used. Following the crystals, beamsplitter pairs separate sequentially the shortest wavelength from the combined beam. Following the beamsplitters, half-wave plates ($\lambda/2$) may be inserted to produce any desired polarization axis. The beamsplitters and wave plates serve to permit any combination of these beams to be directed out of the three exit parts of the laser. The center beam line pumps a dye laser directly. The rear beam is used to mix with the dye laser outputs. The front beam is reserved for the residual fundamental beam.

The MY Series Laser

The MY series laser provides high efficiency and high reliability with an output beam that is near Gaussian (bell-shaped). The output pulse duration is 15 to 30 nanoseconds with excellent pulse-to-pulse energy and peak-power stability. The laser has a programmed Q-switched power supply. The oscillator and the amplifier both use high-efficiency power supplies which have a switching rate of 12 kilohertz. Both supplies are current and voltage regulated. The oscillator power supply has a maximum power delivery of 600 watts. The amplifier power supply has a maximum power delivery of 1200 watts. A closed-cycle cooling system of deionized water is used to cool the laser heads and the beam pump. The system is filled with an alcohol–water mixture for shipment. Tap water is required for operation at 2 liters a minute.

A control panel controls power and indicates that the beam is on. A service panel contains voltage controls for the oscillator and amplifier, repetition rate controls, and Q-switch controls. The pump cavity contains the Nd:YAG rod and flashlamps. The Q-switch is a Pockel's cell (an electro-optical device). The Q-switch power supply applies high voltage to the switch crystal only during a flashlamp pulse. A laser safety and interlock system mounts on the outside of a laboratory door to indicate the status of the laser inside. A green light on the interlock indicates that the laser is off. Yellow, of course, means caution and entry by permission only. A red light means that the laser is on. A door interlock will close the beam shutters on the laser if the laboratory door is accidently opened.

The beam is 8 millimeters in diameter. Pulses are selective at 1064, 532, 355, and 266 nanometers (visible and infrared). Pulse durations are 15 nanoseconds or 15 to 30 nanoseconds variable. Maximum average output power is 20 watts. The laser is a class IV laser product and can cause damage to the skin and/or the eyes.

Design Characteristics of Pulsed Nd:YAG Lasers

The optical cavity must have mirrors that are perfectly aligned. Initial alignments must be within 25 microradians and must be stable for long-term operation. Focal length of the YAG rod varies inversely with the average power delivered to the rod from the flashlamps. If the focal length deviates from design parameters, performance deteriorates in divergence then pulse energy (see Figure 9–10).

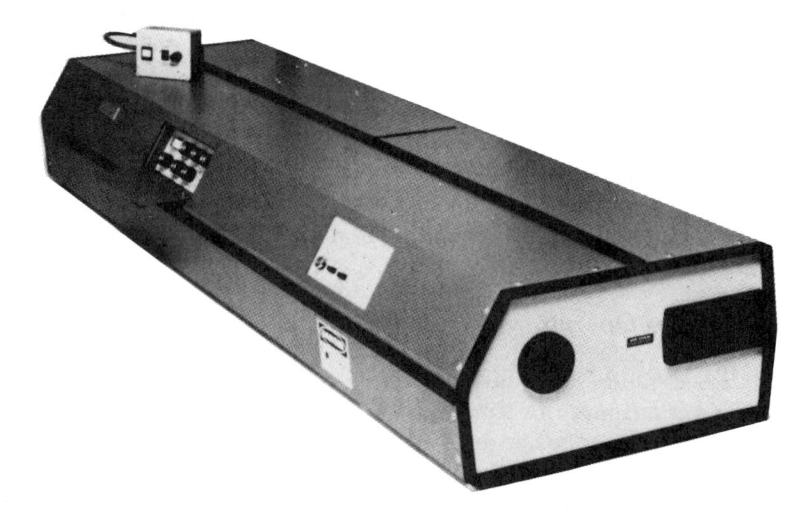

Figure 9–10 Nd:YAG laser (*courtesy of Laser Photonics*).

Q-switches have crystals with a limited lifetime at high voltage due to ion migration from electrode into crystal, which reduces transmission at 1064 nanometers. Programmed Q-switch power supplies can turn off the power after each pulse.

Optics must be kept extremely clean. High-energy pulses place a severe strain on optical components. The smallest amount of dirt on a mirror surface can cause coating damage. Most lasers use a sealed beam to protect the optics from dirt.

If one mode of the laser is active, the pulse is smooth. If more than one mode is active, beating between modes produces a modulation on the temporal profile (phase relationship at two points in time). Short cavities produce shorter pulses than long cavities, but the transit time is less so the beats are closer together. Nd:YAG rod size is dependent on whether the rod is used as an oscillator or an amplifier. Oscillator rods should be fine quality. Amplifier rods need not be as high quality, but the size of the two should be matched. A typical size is 8.5 millimeters in diameter.

GAS LASERS

Gas lasers were supposed to have been the first lasers built. In the theoretical paper written in 1958 by Arthur Schawlow and Charles Townes, they described an active material of potassium vapor enclosed in a tube. Maiman's ruby laser was created in mid-1960. Shortly after (2 or 3 months), Ali Javan and associates of the Bell Telephone Laboratories made the first gas laser. Gas lasers include three types: the neutral atom, the ion, and the molecular. Neutral atom lasers are electrically balanced and have an equal number of protons and electrons. An example of the neutral-atom laser is the helium–neon (HeNe) laser. This laser operates in the visible range of the electromagnetic frequency spectrum at a wavelength of 0.6328 micron. An ion gas laser has an active region whose atoms have missing electrons from their valence rings. An example of the ion gas laser is the singly ionized argon Ar + . This laser operates in the visible range of the electromagnetic frequency spectrum at wavelengths of 0.4880 and 0.5145 micron. A molecular laser has an active medium whose molecules have two or more atoms either of the same or different atomic species. An example of the molecular laser is the carbon dioxide laser. This laser operates in the infrared range of the electromagnetic frequency spectrum at a wavelength of 10.6 microns. Other prominent gas lasers are krypton, an ion laser (0.725 micron); helium–cadmium, an ion laser (0.4416 micron); and nitrogen, a molecular laser (0.3371 micron).

Gas Laser Operation

As a beginning, let us take a cursory look at the gas laser operation. Each gas laser has several things in common. A typical gas laser consists of an active gas medium within a glass tube, an excitation source, a feedback mechanism, and an output coupler. The active medium in the gas laser illustrated in Figure 9–11 is helium–neon (HeNe). The gas is enclosed in a sealed glass tube called a *plasma tube.* The glass windows on each end of the plasma tube are made at an angle to transfer maximum

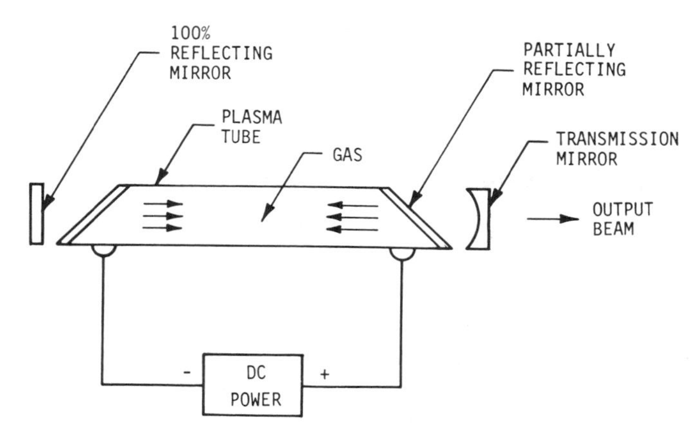

Figure 9–11 Gas laser operation (*Mike Anaya*).

light from the reflecting mirrors. The glass ends are called Brewster windows. One end of the plasma tube has a 100% reflective mirror while the other end is partially reflective and transmits most of the light energy back into the tube for more lasing action.

Now let's go through the operation and consider how it all comes together. Helium is added to neon (90% to 10%, respectively) in a plasma tube for maximum pumping efficiency. Direct current is supplied to a negative and positive end (anode and cathode) of the plasma tube. Population inversion, stimulated emission, and lasing action take place. The beam of photons go through the Brewster windows on each side of the plasma tube. The beam strikes the 100%-reflecting mirror and is reflected back into the active medium. The beam strikes the partially reflecting (transmission) mirror and most of the beam is reflected back into the active medium. A small amount of the beam is released and coupled through an aperture for external use. If too much energy is coupled out of the transmitting mirror, the stimulated emission will decay and die.

The output of most gas lasers is extremely coherent, and continuous wave (CW) in operation. Most gas lasers have very low beam divergence and can easily be tuned. They also have low comparative prices and are of fairly rugged construction.

NEUTRAL-ATOM GAS LASERS

Typical of a neutral-atom gas laser is the helium–neon (HeNe) laser. As previously discussed, the gas laser developed in late 1960 at Bell Laboratories was an HeNe laser. This first gas laser had a gas mixture of 90% helium and 10% neon. Neon gas, however, is the active medium in the HeNe laser. When helium is added to the neon, it aids in the pumping action. It seems that there is some difficulty in pumping neon atoms to their excited state. Helium atoms are very light in comparison to neon atoms and are easily excited. It so happens that the excited state of the helium atom is the same energy level as the neon atom's excited state. Therefore, when a helium atom in an excited state collides with a neon atom in the ground state it can transfer all its energy to the neon atom. These collisions happen frequently in the gas mixture at high pressure; therefore, population inversion can take place and be maintained by pumping action.

Energy-Level Diagram for Helium–Neon Laser

The energy-level diagram shown in Figure 9–12 is extremely simplified. It must be noted once again that the excited energy states of the helium and the neon atoms are almost the same. There are several of these excited states, but they have purposely been left off the energy-level diagram for better understanding. As the electrons are released from the dc power supply, they are drawn through the active region by the power supply's positive pole. The electrons collide with the gaseous helium atoms. The helium atoms are raised to an excited energy state. Those excited helium atoms in turn collide with neon atoms, and transfer all their energy to the neon atoms. The collisions occur more often at high pressure. This process is called *resonant excitation.* When this transfer happens often enough and at a fast enough rate to be larger than the decay rate to ground energy state, then population inversion can take place. In the figure, stimulated emission occurs between the excited energy state and an intermediate state. This stimulated emission is occurring at a wavelength of 0.6328 micron. Stimulated emission can occur in neon at several wavelengths. Others are 1.152 and 3.390 microns. Some lasers have these in the output simultaneously.

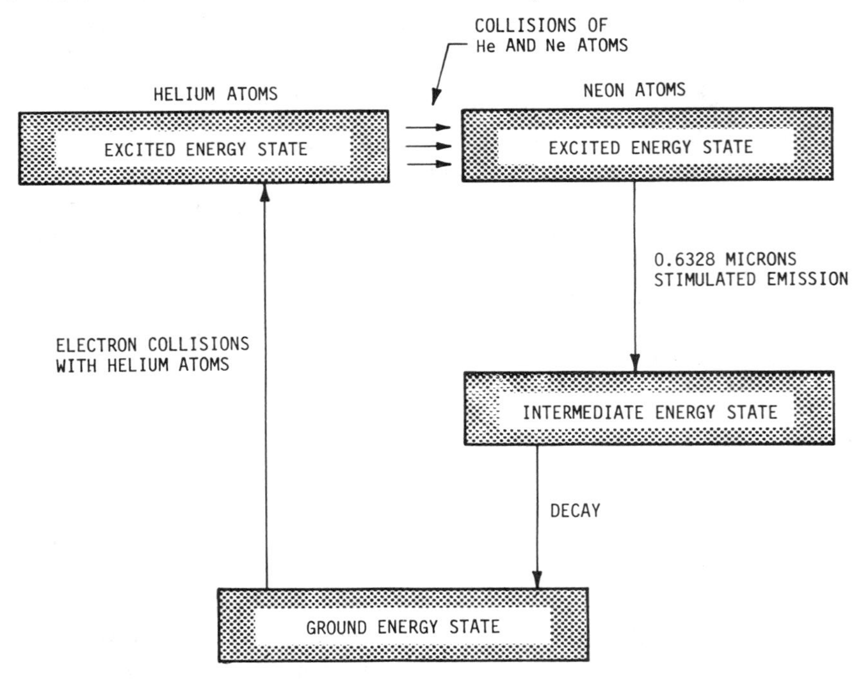

Figure 9–12 Energy levels for helium–neon laser (*Mike Anaya*).

The Plasma Tube

Plasma tubes are made in many different configurations. The line drawing in Figure 9–13 depicts a simplified version of the HeNe plasma tube. A gas reservoir contains a mixture of helium and neon gases (HeNe). Just below the reservoir and connected to it is a capillary tube. The tube is where lasing takes place. The reservoir ensures that the supply of gas in the capillary tube does not fall low. For maximum power output, a constant pressure should be realized. The aluminum cathode tube is a hollow cylinder. On one end is a path for gas to the capillary tube. An element attached to the cathode lead prevents contaminants from entering the gas by collecting them electrically. The anode connected to the capillary tube is made of nickel. As in other lasers there are 100%-reflecting mirrors on one end of the tube and the partially reflecting output mirrors on its opposite end.

In operation a power supply supplies direct current to the anode (+) and cathode (−) for continuous-wave (CW) operation. The high-voltage pulse initiates the glow discharge (breaking down of the neon gas). Ionization of the gas involves removal of electrons from gas atoms. The ionization is synonomous to weak ionization. Stimulated emission can take place. The process of stimulated emission has been previously described in several passages of this section.

Helium-Neon (HeNe) Laser Applications

The HeNe laser envelope can be made of aluminum or glass. Aluminum is, of course, a stronger material. Problems in application are centered around the prevention of gas contamination. The HeNe laser is by far the most widely used. There are literally tens of thousands installed annually around the world. Use of HeNe lasers can be placed in three main categories. Low-power laser beams are routinely directed hundreds of feet away for alignment purposes. The construction industry accounts for

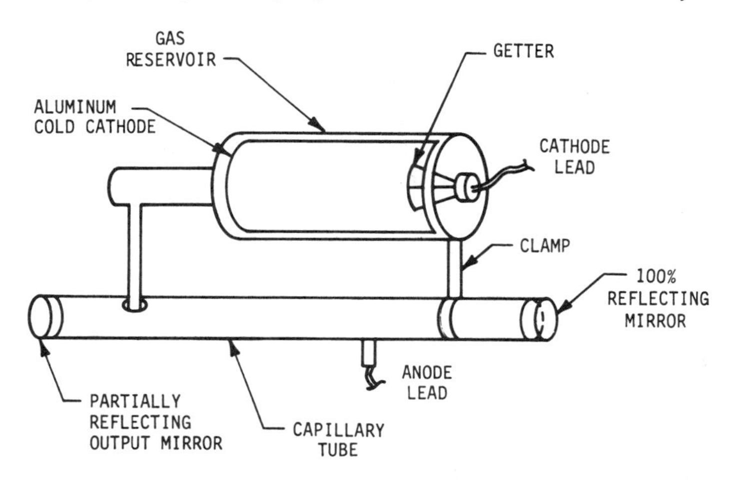

Figure 9–13 Plasma tube (*Mike Anaya*).

the greatest numbers of all lasers in use today. It is used in trenches where it facilitates the setting of grades for sewer pipes. With a rotating beacon optical accessory the laser line can be rotated in a plane around a central point. It then provides a very convenient reference plane which can even be tied to the hydraulic controls of massive leveling equipment. With a simple low-cost optical accessory which fans the beam into a sector, lines can be projected against walls for leveling drop ceilings inside buildings, or for aligning lumber and planks in sawmills. On factory floors, the alignment of many parallel cylinders or rolls, like the ones found in production machinery in paper and steel rolling mills; the alignment of giant rotary kilns used in the cement industry; the alignment of complex assemblies such as turbine generators or linear accelerators; the alignment of another (invisible) laser beam (CO_2 or YAG laser) is finding widespread acceptance in machining materials such as metals, plastics, and others. In hospitals the HeNe laser is now used to precisely align patients for diagnosis or treatment by sophisticated x-ray equipment.

The HeNe laser can measure. Its light can be analyzed after it has been reflected off a variety of surfaces, after it has been shadowed by a number of objects, by the interference patterns due to its monochromaticity, or by the time it takes to travel over a given distance. Accurate dimensions can be recorded. Parts can be inspected for defects without part contact. This means no part damage, no gauge wear, no pressure contact error, and faster remote inspection. Lasers often perform better than their pneumatic and electronic counterparts in accuracy, resolution, and measurement range. They are not affected by radio-frequency interference, and their output is compatible with photodiodes, TV cameras, and other high-level output devices.

The HeNe laser efficiently detects surface defects (scratches, pits, breaks, burns, and missing threads on bolts). It can sort parts and measure thicknesses, diameters, profiles, flatness, and orientations. The HeNe laser also contributes to such analytical measurements as blood-cell sorting and small-particle measuring as required for pollution monitoring.

The major laser recording applications in use today are in facsimile transmission, telecopiers, computer line printers, oscillographic chart recorders, newspaper plate-making, industrial and supermarket label readers, optical character readers, and computer output microfilm. They "write" on a variety of media ranging from film to dry-silver paper, electrostatic film, and plain paper via electrostatic transfer techniques. Research is now developing new media with optimum response to the reliable and inexpensive red light of helium–neon lasers. The laser light offers significant advantages for recording: high speed and excellent resolution. More important, it can be directly controlled by a computer when coupled to a device such as an acousto-optic (A/O) modulator.

Typical Neutral-Atom Gas Laser

Typical of a neutral-atom gas laser is the Spectra-Physics Model 155 (HeNe) laser shown in Figure 9–14. This low power (also, low cost) laser has output power ranging between 0.4 and 0.76 milliwatts. The Model 155 laser operates at a 632.8-nanometer wavelength, with a beam diameter of 0.9 millimeter and a beam

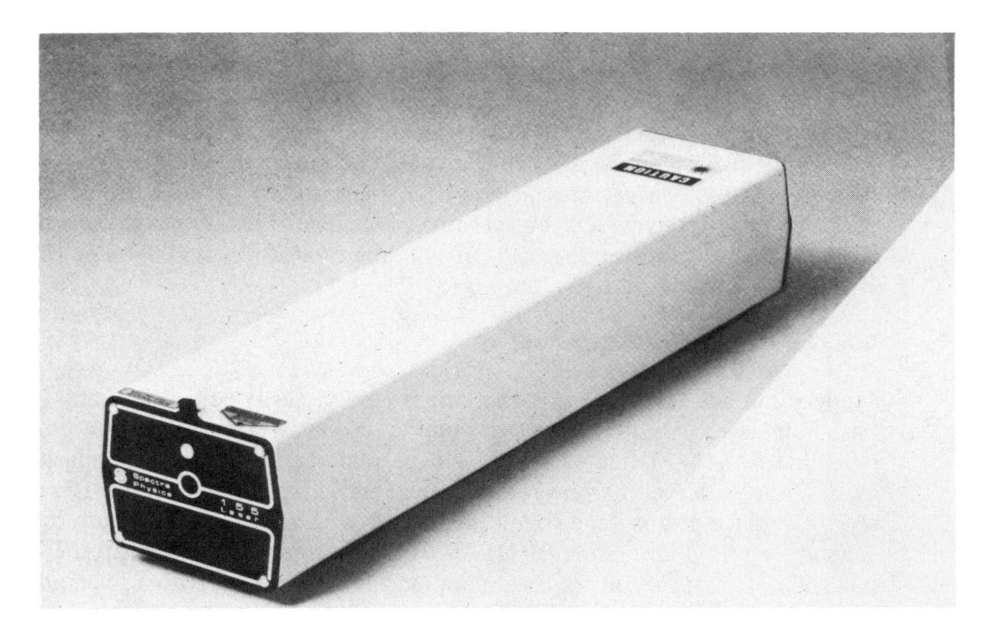

Figure 9–14 Spectra-Physics Model 155 helium–neon (HeNe) neutral-atom gas laser (*courtesy of Spectra-Physics, Inc.*).

divergence of 1.0 milliradian. The power supply is contained within the laser head. A special neutral density filter within the supply assumes that the output will never exceed 1.0 milliwatt as established by the Bureau of Radiological Health (BRH) for class II lasers. The Model 155 laser is used in laboratories and for educational purposes where low-power lasers are required. For user safety a key-actuated switch prevents unauthorized use. An emission-indicator light provides a signal to show that the laser is operating. A beam attenuator enables the user of the laser to block the beam at the output aperture without shutting down the laser.

ION GAS LASERS

Ion gas lasers are usually operated with an active medium of argon or krypton. Argon operates at 0.5145- and 0.4880-micron wavelengths. Krypton operates at a wavelength of 0.7525 micron. Several years after the helium–neon (HeNe) laser was developed (probably 1963) the ion gas laser came into being. An American scientist, W. E. Bell, developed a laser using mercury ions. He found that it was fairly simple and easy to remove electrons from the outer shell of the mercury atom. Once the electrons were removed, the atom was left with a net positive charge. This atom then became a positive ion.

Ion gas lasers in their early stages were pumped by electron beams used also to ionize the gas. Heating of a metal cathode in turn boiled electrons from the cathode surface. The electrons were accelerated to high speeds while being focused into a dense, tiny beam. When the beam struck the gas atoms it knocked electrons from them and drove the new ions to an excited state where population inversion could take place. Problems with this system occurred because of poor concentration of atoms and high temperatures. The plasma tubes physically pumped gas from one end to another. A new tube fitted a return pipe which tended to equalize the end pressures. A magnetic field was used to concentrate the laser beam. These methods were short-lived, because the large current developed caused unbearable heat. Furthermore, the constant battering of positive ions on the walls of the tube caused physical damage. Units failed and new methods had to be developed.

Ion Gas Laser Operation

Modern ion gas lasers are extremely high power, easily controlled, and stable. The basic ion laser consists of a plasma tube, a resonator structure, and a power supply (see Figure 9–15). The most important part of the ion laser is the plasma tube. It alone controls the output power and general performance of the laser. It is also expensive to replace and critical to the longevity of the laser. Modern ion lasers utilize graphite or beryllium oxide (BeO) for use in plasma-tube design. In the illustration you will note that the ion gas laser has a baffled plasma tube with a solenoid surrounding the tube. A reservoir is installed to the plasma tube and an ac power supply is tied to the anode and cathode. The ac power supply excites the anode and cathode, which in turn releases electrons. The electrons are accelerated through the gas in the plasma tube. The electrons bombard the gas atoms, which release electrons and become positively charged ions. The new ions raise to an excited energy state. A solenoid is placed around the plasma tube to concentrate the ion beam so that higher power can be developed. Baffles in the plasma tube allow for cooling. The concentrated beam produces extreme heat, while baffles act as a heat sink. Modern plasma tubes utilize cooling water jackets. At the end of the plasma tube are Brewster win-

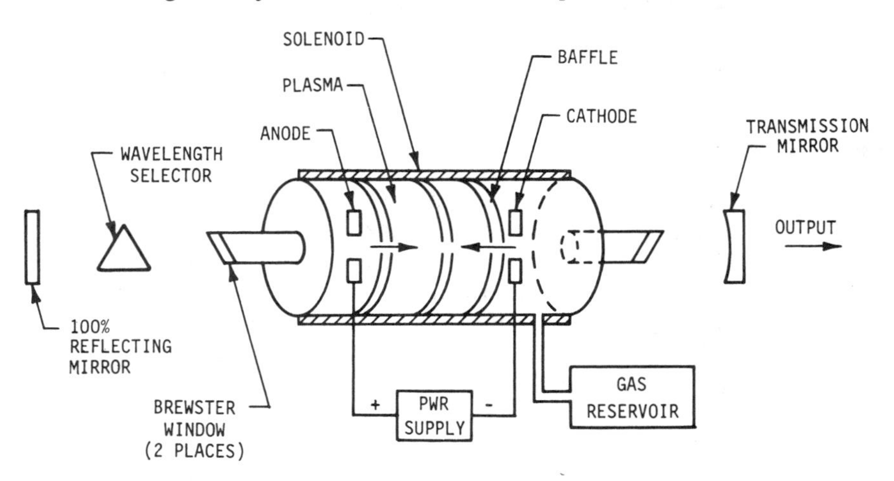

Figure 9–15 Ion gas laser operation (*Mike Anaya*).

dows that allow maximum light transfer to optics. A prism external to the plasma tube acts as a wavelength selector. The selector is tilted in respect to the plasma tube axis. This causes high cavity losses except at one specific wavelength. This allows the laser to be tuned to only one wavelength of those available from the gaseous active medium. Highly reflective and partially reflective mirrors at either end of the tube ensure the normal feedback operation. A gas reservoir attached to the plasma tube maintains the tube with a constant supply of gas at a predetermined pressure.

Spectra-Physics STABILITE® Resonator Structure

The two most important considerations in the design of a laser resonator are the angular and length stability of the optical cavity. These determine the ultimate amplitude and frequency stability of the laser. The best technique for achieving optimal performance is to select a material that has a very low thermal coefficient of expansion and to thermally isolate this material from ambient temperature changes and heat-generating components in the laser. Another technique is to select a material that adjusts uniformly to temperature changes so that optical alignment is always precisely maintained. The Spectra-Physics resonator in Figure 9–16 uses both techniques in its design. The aluminum extrusion structure utilizes the property of excellent thermal conductivity to produce amplitude stability. The structure is kinematically mounted so that thermal and mechanical stresses cannot be coupled into the resonator to disturb the mirror or the plasma-tube alignment. Three spherical bearings kinematically isolate the entire optical cavity from mechanical stress applied by the outer case, end plates, and the feet of the laser. These bearings also relieve thermally generated mechanical stress that might originate in the resonator structure.

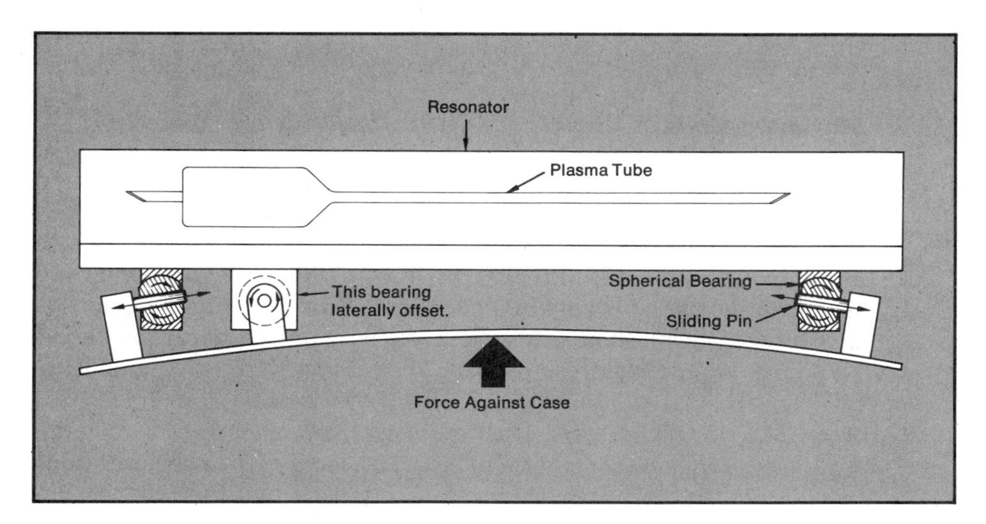

Figure 9–16 Spectra-Physics STABILITE® resonator structure (*courtesy of Spectra-Physics, Inc.*).

Ion Laser Applications

The ion laser has a multitude of applications which are dependent on the power required. Ion lasers are used in the fields of holography, spectroscopy, data recording and retrieval, eye surgery, dye laser pumping, biological and cell-sorting experiments, and printing plate exposure systems. High-power ion lasers are ideally utilized for isotope separation. The ion laser is again one of the most useful tools for general laboratory work that requires a strong, versatile light source.

MOLECULAR GAS LASERS

The molecular gas laser was developed in 1965 by C. K. N. Patel, a scientist at Bell Telephone Laboratories. The molecular laser is the most powerful of the gas lasers and most efficient. As early as 1967, North American Aviation had produced a CO_2 laser with an output of 4000 watts. The molecular CO_2 gas laser is so powerful that it can be used for drilling holes in metal, welding, and even rock crushing. The molecular nitrogen (N_2) gas laser is used to pump tunable dye lasers. The molecular gas laser utilizes the molecule for basic operation. You may recall from high school physics that a molecule is made up of atoms which may or may not be the same. An example of a molecule you may be familiar with is water (H_2O).

There are two primary molecular gas lasers. These are carbon dioxide (CO_2) and nitrogen (N_2). The CO_2 laser operates around 10.6 microns in the infrared range of the spectrum. The nitrogen laser operates at 0.3371 microns wavelength in the near ultraviolet range. CO_2 lasers can be continuous wave (CW) with peak power to 1000 watts, repetitively pulsed at with peak power to 3000 watts and with Q-switching may reach peak power of 100,000 watts. Nitrogen molecule gas lasers (also called *superradiant*) operate at peak power of 100,000 watts.

Molecular Carbon Dioxide (CO_2) Gas Laser Operation

Molecular gas lasers operate in much the same manner as the neutral-atom gas lasers. Molecular gas laser systems are excited by electrical current flow in the direction of the light source, the same as in the HeNe laser. In the CO_2 laser, a mixture of carbon dioxide (CO_2) and nitrogen (N_2) is used. The nitrogen serves as the exciter, similar to the use of helium in a neon laser. Nitrogen molecules are easily pumped to an excited energy state by electrical discharge. In turn, carbon dioxide molecules are raised to an excited energy state by collision with nitrogen molecules.

The original molecular gas lasers had the active gas such as CO_2 flowing between reflection mirror ends. The flow would encounter nitrogen molecules that were previously excited by some form of electrical energy. The two gases would clash and energy stored in the nitrogen atoms would transfer to the CO_2 molecules. Lasing action could take place. The flowing gas CO_2 laser develops greater power with higher rates of gas flow. Carbon dioxide (CO_2) was later found to be able to function standing still and the gas flow was no longer required.

Another carbon dioxide laser is the TEA CO_2 laser. The TEA CO_2 laser is excited at several points perpendicular to the optical flow between reflective mirrors. TEA is an acronym for *t*ransversely *e*xcited at *a*tmospheric pressure. The TEA CO_2 laser has a high-voltage direct-current power supply that charges a large capacitor. The capacitor is discharged into the gas in the plasma tube at predetermined times. Switching action controls the capacitor charge and discharge times and therefore the laser pulse.

Molecular Nitrogen (N_2) Gas Laser Operation

The molecular nitrogen (N_2) operates in much the same manner as the TEA CO_2 laser. Nitrogen molecules are brought to an excited state by a high-speed electrical discharge from capacitors. The discharge channel can be so fast that mirrors are not required for reflections. In most cases the mirrors are used and a superradiant output results. The N_2 laser does have limitations in that its pulse duration is short at 10 nanoseconds. Furthermore, nitrogen gas has a low energy-storage capacity. Capacitors may be charged in series and discharged in parallel or charged in parallel and discharged in series. The latter, called a Blumlein circuit, provides higher peak currents and pulses of 3 to 5 nanoseconds duration. This is compared to the 10 nanosecond pulse of the parallel discharge method.

N_2 lasers use either channelized or open discharge channels. In the channelized design, a double-discharge beam is projected along the outer walls of the discharge channel. In the open-discharge channel, the channel is made large and the beam is projected through the center of the channel. The double-discharge beam is not suited for dye laser pumping because of limited output energy.

As a dye laser pumping source the nitrogen laser has no equal. The most important parameter that governs dye laser efficiency is the pulse duration of the nitrogen laser. For all cavity lengths and pumping peak powers, dye laser efficiency increases with pumping pulse duration. A long pulse is always more efficient than a short pulse.

Molecular Gas Laser Applications

Because of the high power of the molecular gas laser it has become an important tool in machining operations. CO_2 lasers are used for a great number of cutting, drilling, and welding tasks in materials which include metal, rubber, and plastics. Beams of the CO_2 laser can be focused to as small as 0.005 inch for drilling and cutting. Most materials are instantly vaporized and may be immediately removed. There is no debris from laser drilling and cutting. There is no tool contact, meaning no tool wear or replacement. Metal welding includes materials of such hardness as monel metal and titanium. Nitrogen lasers are primarily used as pumping devices for dye lasers. More on these applications will follow later in this chapter.

LIQUID LASERS

Liquid lasers, also called dye lasers, utilize organic dye as an active medium. The advantage of using a dye laser over others is that the dye laser can be tuned over a wide variety of wavelengths for each dye used. The variety of the dyes literally extends over the entire visible range of the spectrum with some dyes in the near-infrared ranges. This latter attribute makes the dye laser a highly respected tool in biomedical research.

Some of the dyes utilized in liquid laser operation and their lasing range are as follows:

Carbostyril	419–485 nanometers
Coumarin	435–565 nanometers
Sodium fluorescein	538–573 nanometers
Rhodamine	540–690 nanometers
Cresyle violet/rhodamine	675–708 nanometers
Nile blue	710–790 nanometers
Oxazine	695–801 nanometers

The center of the range provides the user maximum dye absorption. Optimum dye laser performance is achieved by spectrally matching the dye laser pump with the absorption band (range) of the dye, then tuning the concentration of the dye for maximum absorption. The dye is generally mixed in an alcohol or water solution or some special solution which provides a better mix. Liquid detergent is often added to aid in lasing performance.

Dye lasers can be continuous-wave (CW) or pulsed. In operation dye lasers are pumped by gas lasers or flashlamps. A few are pumped by ruby or Nd:YAG solid lasers.

TRANSVERSE ELECTROMAGNETIC MODES (TEM) LASER BEAM PROFILES

There are dozens of beam profiles with which lasers operate. Several of these patterns are illustrated in Figure 9–17. These patterns are perpendicular to the beam in its direction of operation. Longitudinal modes are in the direction of wave propagation in a spatial *Z*-axis orientation. The field which represents the light beam consists of all three spatial coordinates *X, Y,* and *Z*. The *Z*-axis is perpendicular to the *X-Y* coordinates, which makes the *X-Y* fields perpendicular to the direction of laser beam travel.

The basic mode of lasers is called the *fundamental mode* or the *transverse mode.* In Figure 9–17a, this mode is illustrated. The pattern is labeled TEM_{00}. It would seem that all laser beam profiles should take this pattern and rightly so. Once the bore characteristics of a laser have been decided upon, the beam profile is determined

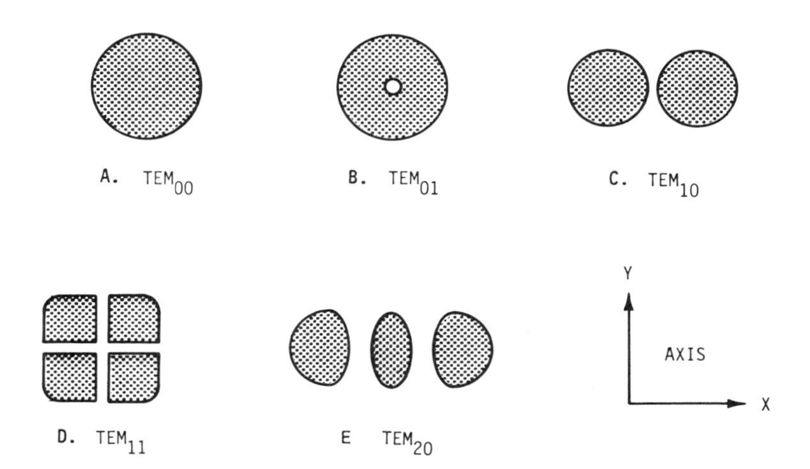

Figure 9–17 Transverse electromagnetic modes (TEM) laser beam profiles (*Mike Anaya*).

by spacing and curvature of reflective resonator mirrors. The mirrors are chosen to ensure a TEM_{00} pattern at the longest wavelength. The mode diameter varies with the square root of the wavelength. Variations in the cavity mirror diameters can cause wavelength lines to be TEM_{00} or fundamental mode.

Subscripts represent the number of times the magnetic field crosses the *X, Y,* and *Z* axes of the laser beam. They also represent the number of modes in the standing-wave pattern along each axis.

Figure 9-17b is called a *doughnut mode.* It is the superposition of one mode on another, with the small circle as the second mode. This mode, TEM_{01} can also be caused by contamination present on a mirror face. Figure 9–17c, d and e are representative of multimodes. A cavity that operates in multimode operation is said to be "lossy," because modes other than the fundamental mode TEM_{00} do not have maximum power density. The reason for multimodes is the inability of the laser to focus due to phase changes across the beam and dispersion or spreading out of beam power. The figure simply illustrates several of the beam profiles. In actual practice these beam profiles may make up a multitude of patterns.

LASER OPTICS ALIGNMENT

With the understanding that alignment is a fundamental requirement to all lasers we can make two general statements. First, to establish maximum power output the laser mirrors must be perfectly aligned. Second, alignment is said to be at its best when an imaginary line can be drawn perpendicular to the center of each mirror. When these lines are in coincidence with each other and with the axis of the active region the cavity is said to be aligned (see Figure 9–18). This fundamental alignment is dependent upon the laser in need of alignment. Some essential requirements fit all lasers. The cavity of the laser along with its optics must be placed on an optical bench that

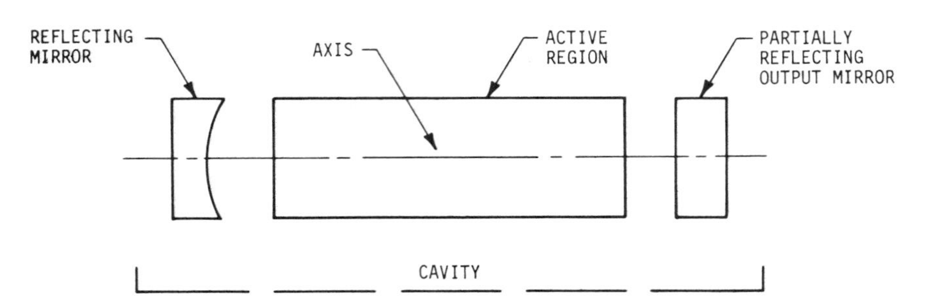

Figure 9–18 Laser optics alignment (*Mike Anaya*).

has adjustable supports. In the case of some lasers, alignment can take place in the laser mount because adjustment devices are built in. Power is applied to the cavity and adjustment screws are used to mechanically move the laser optics in the horizontal, vertical, and directional (X, Y, and Z) axes. Again this is dependent on the laser under alignment. An index card may be used for visual output lasers to align for a sight profile of the laser. The card is placed in the output beam line while adjustment takes place. Adjustment continues until the specific beam profile is achieved.

Special equipment such as a laser power meter may be placed in the laser beam path to measure power output as alignment adjustments are made to monitor output power. Some laser cavities may be removed from the laser head. This cavity [such as in the helium–neon (HeNe) laser] may be aligned with another laser. The HeNe plasma tube is mounted on an optical bench on supports.

An operational laser is mounted on a three-way adjustable support on the optical bench so as to direct a beam through the center of the plasma tube. As the beam is projected through the cavity, a black index card is placed on the opposite end of the plasma tube under alignment. The card will show the beam profile if it is partially aligned. Alignment of the operational laser with the plasma tube is necessary. If the index card shows a profile, the operational laser beam is not hitting the walls of the plasma tube. The 100%-reflectance mirror of the laser is placed in a mount on the optical bench. This mirror is adjusted in the same manner as the operational laser with mount adjustment screws. An index card is used to show adjustment.

SPECIAL LASER APPLICATIONS

Holography

Holography was invented by Dennis Gabor, a professor at London University in 1947. Holography is the production of three-dimensional displays with the use of laser beams. Three-dimensional pictures have long been the desire of the motion picture industry, photographic technicians, inventors, and entrepreneurs of that art form. In the early 1950s, the 3D motion picture subjected the public to the wearing of special glasses in order to view pictures in the third dimension. It did not take hold. Even prior to that a special set of glasses was used to look at a double picture which gave a three-dimensional appearance. These were not acceptable after a time. So the search went on to find a 3D device that would allow you to look behind an image without the special help of spectacles or other devices and with no effort on the viewer's part except for looking. We shall briefly describe holography and then discuss its applications.

When light shines on an object the object either absorbs the light or reflects it. The reflection of the light from the object takes the shape of the object in three dimensions. A single eye sees only the intensity and the color of this light, as does a photographic film. Distance does not enter into the picture that one eye or a photograph sees. Two eyes, however, view the object at two angles. The human brain provides the intelligence to compute the various terms of distance. We see 3D because of our brain.

It is possible, with holography, to use a third component of light, its phase, to produce a three-dimensional record of incident light. The process involves two basic steps. These are the development of a hologram of an image and the reconstruction of the image. A hologram is made by reflecting half of a laser beam as a reference with a plane mirror on to a photographic plate. The second half of the laser beam is reflected off the image onto a photographic plate. The photographic plate records the interference pattern (phase relationship) of the two half-beams on the plate. The result is a photograph. This photograph contains intensity and phase variations. The record is called a *hologram* and is unlike any photograph you have ever seen. Several holograms can be recorded on one plate if the light beams are reflected at different angles. The hologram in itself is a complex-looking mass of patterns that represent the image that was reflected. To retrieve the image, a laser beam is driven through the hologram in the same direction as the original reference beam. If you look at the hologram from its opposite side a three-dimensional form will be obvious. If you move your head, you can observe the image from another angle just as you would view any three dimensional real image.

As you may have concluded, holography, though a fairly simple concept, requires a great deal of technical expertise. The development of an image to a hologram and then the reconstruction of the image is a new technical field and is worthy of a life's career. Moving objects are being recorded and displayed by holography. Holographic television was patented as early as 1940 by Fritz Fischer. In industrial manufacturing, holography is used for measuring vibrations and small deformations in metal. In early 1980s, making a hologram will be just like shooting a roll of film, at least so say the experts.

Welding, Drilling, and Cutting

Lasers are not common in every small plant in the country to do the jobs of welding, drilling, and cutting because of the initial expense. The time is coming when the precise nature of the laser beam coupled with its intense power will be more than an expensive dream. The laser can confine high intensity. This and the characteristic of non-contact is extremely important where delicate parts have to be joined; contamination must be eliminated and thermal distortion cannot be tolerated.

Industrial welding has utilized two laser types, the CO_2 gas laser and the Nd:YAG solid-state laser. The CO_2 laser is used where high power and high efficiency are required. The Nd:YAG laser is used where high precision and minimum thermal damage are demanded. The Nd:YAG is also utilized where difficult welds are needed. Difficult welds are those which involve high-melting points, high conductivities, or metal mismatches. There are two basic welding modes of operation utilizing the laser. These are conduction and penetration. In conduction welding, the laser beam is impinged on the material at a power density that allows fusion but does not cause vaporization. Energy of the beam is conducted to the base metal. The fusion zone provides an aspect ratio of unity (1:1 ratio of depth to width). Penetration welding takes place with laser power necessary to drill a hole in the material and the hole is propagated through the material to form a fusion zone. Vaporization does take place. Aspect ratios are larger because the energy is not limited to the surface weld as in conduction welding.

The pulsed laser can take the place of hand soldering. Soldering is slow, tedious, and messy. Furthermore, repeatability is almost impossible to achieve and quality control is difficult. The pulsed laser can allow welding in close proximity to delicate components. Seam welding and spot welding are both used without changing laser energy levels. Production lines in laser welding now weld such things as pacemakers, instrumentation, gyroscope bearings, and sensor probes.

Laser driling can offer tremendous savings for the manufacturer because of its speed, but there are limitations to its use. Precision holes are not possible in metals more than a few thousandths of an inch thick. The diameter of the hole is uncontrollable and the face of the hole tends to become funnel-shaped. Thin layers of recast metal cling to the side of the holes. This recast metal is structurally different than the original metal. It may be harder and susceptible to cracking. Despite the problems, lasers are used to efficiently drill holes. Diamonds are drilled in minutes with a laser, where hand drilling required hours. Mass production and speed are the attributes of the laser drill. Holes that need a better finish are sized or polished after the laser operation.

Cutting operations have the same general characteristics as drilling operations. Cutting, however, may be the wrong term, for the laser does that job by heat rather than physical force. Temperatures for cutting may be up to 150,000 °C. Vaporization does take place. Lasers have been used in the clothing industry to cut material.

Wood and Ceramics

The CO_2 laser is used in cutting, drilling, and etching of ceramics and wood. Energy from the laser is focused as a beam through a lens onto the ceramic or wood. This energy is high intensity in the megawatts-per-centimeter range. The ceramic and wood absorbs the high energy in localized areas. The material quickly melts and vaporizes.

Resistor Trimming

The Nd:YAG laser is utilized for resistor trimming along with several others such as the xenon ion laser and the CO_2 laser. Resistor trimming is categorized as thick and thin film technologies. The Nd:YAG laser is used throughout thick-film trimming because of its high repetition rate and high pulse power. The xenon ion laser is utilized with thin-film trimming because it is highly accurate and can be used in small areas. Resistor trimming with a laser is accomplished by focusing the laser beam on the film to be removed, and heating the film in the specific area. During heating film droplets are formed. The droplets move from the center of high intensity and evaporate. The resulting clean area is called *kerf*. The kerf is film-free.

Medicine

In medicine there is no technology that has advanced to the extent of the laser. Ruby lasers for years have been used for eye surgery. Argon ion lasers are useful as scalpels. The body skin fat and muscles are transparent to selective laser wavelengths, therefore cancer cells can be destroyed without harm to other parts of the body. Decayed areas of the teeth can be destroyed or removed without harm to healthy teeth. Laser beams have been made narrow enough to perform to microsurgery on red and white blood cells.

Alignment

As an alignment mechanism the laser has no equal. Placing a beam only microns in diameter several hundred feet to align an aircraft wing is significant. While all alignment is not this precise, the laser is indeed capable. Alignment of construction ditches is a case in point. Alignment in assembly of precision equipment is another. Lasers are used to align other lasers. An ultimate use in laser alignment is in military fire control and stability of precision direction-locating equipment.

STATE-OF-THE-ART ION LASER

General

(See Figure 9–19). The INNOVA series of ion lasers represent state of the art in ion laser technology. This is supported by the development of the INNOVA® ion plasma tube.

Ion lasers have three key elements: the plasma tube, resonator, and power supply. The ion laser plasma tube is the most critical component requiring the highest level of technology. Low operating pressures, high current densities, and high input power require the correct choice of materials and design to insure high output powers and reliability.

The INNOVA tube uses tungsten/copper disks brazed to a one-piece alumina

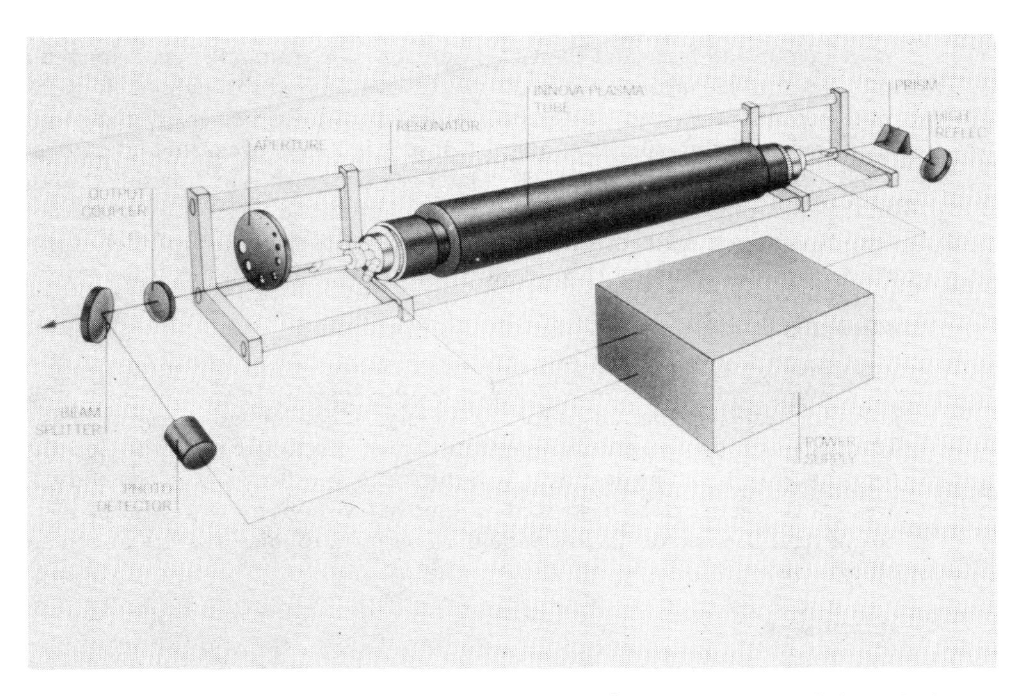

Figure 9–19 Functional block diagram of the INNOVA® ion laster (*courtesy Coherent Inc.*).

envelope that resists the high current densities and at the same time efficiently conducts waste heat to the cooling water. All metal and ceramic parts are used in the construction of the tube bore to insure a rugged design.

The resonator holds the two laser mirrors in a precise orientation such that an optical cavity is formed. An extremely rugged and stable resonator is imperative for fast warm-up and stable output power.

The power supply delivers the electrical energy required to sustain the plasma discharge. Additionally, it must not induce noise on the laser beam. INNOVA lasers use linear passbank power supplies for proven reliability and low noise.

Operating Wavelengths

Ion lasers use either argon or krypton gas as the lasing medium. Argon systems are used primarily for their high output power in the blue/green region. The principal wavelengths are 514.5 nm and 488.0 nm. Additionally, there are several other wavelengths in the blue/green region as well as output in the ultraviolet and near infrared.

Krypton tubes are used principally for their red output at 647.1 nm. Krypton lasers have an advantage over argon lasers in that they have lasing wavelengths over the entire visible spectrum as well as operation in the ultraviolet and near infrared.

Modes of Operation

There are three modes of operation for which an ion laser cavity can be constructed: multiline, single-line, and single-frequency.

Multiline operation in an ion laser utilizes the simplest cavity: a curved output mirror and a flat high reflector. In this configuration, many of the laser lines will oscillate simultaneously. Generally, in an argon laser the lines from 514.5 nm to 454.5 nm will lase, and in a krypton laser the most common configuration has output from 568.2 nm to 676.4 nm.

Single-line operation uses a prism intracavity to allow only one wavelength to oscillate at a time. In single-line operation, the bandwidth of the laser output is 4–12 GHz. By tilting the prism wavelength selector, different single-lines may be selected; this tuning is easily achieved with the vertical rear resonator control.

Single-frequency operation uses an intracavity etalon to force the laser to operate with a single longitudinal mode. An ion laser in single-line operation oscillates on 10–20 longitudinal modes with a frequency stability determined by the stability of the cavity length. The etalon allows only a single longitudinal mode to lase, thus narrowing the laser output to an effective linewidth or jitter of several megahertz.

Transverse Modes

The lowest order transverse mode, TEM_{00}, has a light path on axis with the laser tube. The intensity distribution of the cross section of the beam is Gaussian. A laser operating in the TEM_{00} mode has the smallest beam diameter, the least beam divergence and can be focussed to the smallest spot size. It is conventional to rate the output powers for an ion laser in TEM_{00} operation.

Operation in the fundamental mode is achieved by limiting the bore size of the plasma tube, correctly selecting the radii and spacing of optics, or using an intracavity aperture.

As the configuration required for TEM_{00} operation is wavelength dependent and changes with plasma tube age, a well-designed ion laser will employ all of the above techniques and allow multi-mode as well as fundamental mode operation.

Tube Construction (See Figure 9–20 and 9–21)

The INNOVA tube uses tungsten/copper disks brazed to a one-piece tube envelope to confine the plasma discharge. Nickel headers with metal/ceramic electrical feedthroughs are brazed to the ends of the tube envelope. Hard-sealed crystalline quartz Brewster windows are mounted to the headers using a flexible metal vacuum seal.

Tungsten was chosen as the bore material because of its exceptional resistance to sputtering which for many alternative tube designs is the primary cause of the tube failure. Even though the tube structure will withstand very high operating currents, the INNOVA system is run at conservative current densities in order to maximize tube life.

To efficiently conduct the plasma heat to the cooling water, the tungsten disks,

Figure 9-20 INNOVA® plasma tubes (*courtesy Coherent Inc.*).

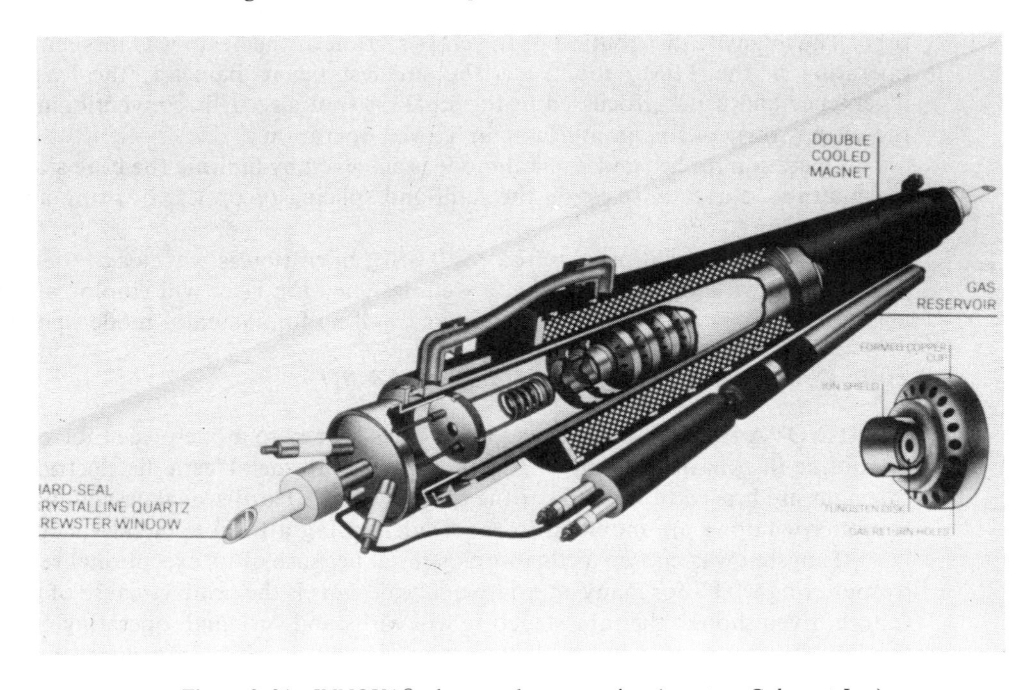

Figure 9-21 INNOVA® plasma tube crossection (*courtesy Coherent Inc.*).

which are in contact with the plasma discharge, are brazed to copper cups. The copper cups, directly brazed to the ceramic tube envelope, efficiently conduct the heat into the cooling water. The low thermal mass and operating temperature of the CoolDisk construction enable the INNOVA tube to achieve operating conditions rapidly after turn-on. If a water failure were to occur during system operation, an interlock shuts down system power, and the quantity of water in the tube water jacket would be sufficient to absorb the heat in the tube without any damage being caused to either the tube or magnet.

The anode is conductively cooled and completely isolated from the cooling water. This design allows for cool operation of the anode and eliminates the problem of corrosive electrolysis which can lead to tube failure. The large surface area of the tube envelope enables efficient transfer of heat to the cooling water even when carbonate minerals (found in most water systems) have been deposited on the tube. The conductively cooled anode and one-piece ceramic envelope allow the INNOVA to operate with ordinary tap water for cooling at most locations throughout the world.

Automatic Gas Fill (See Figure 9–22)

Some of the INNOVA series lasers use Autofill, an automatic gas fill system, to maintain constant tube pressure. Ion lasers consume gas when the discharge drives ions into the tube wall. If an ion laser is to have long life, the gas must be replaced with gas from a fill system. If optimum output performance is to be achieved, the gas must be replaced as required. The Autofill system maintains the tube at the correct pressure.

The dual valve solenoid system is sequenced so as to fill the tube in small increments. When the system senses that gas is required, the "ready" valve opens, allowing enough gas for a 10 millitorr pressure increase in the tube to fill the metered volume. The "ready" valve then closes and the system waits three minutes to con-

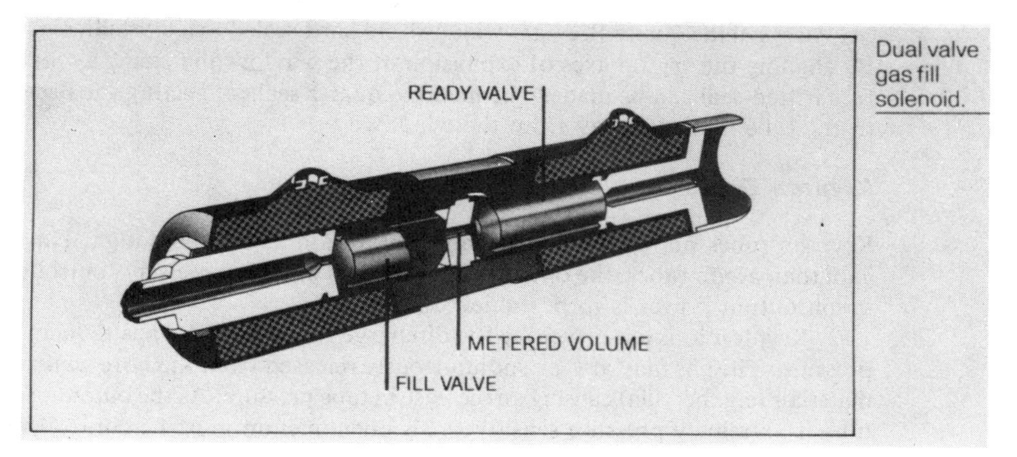

Figure 9–22 Automatic gas fill (*courtesy Coherent Inc.*).

firm that a low pressure condition exists. If confirmed, the "fill" valve will open allowing the metered volume into the tube. When closed, the dual valve insures that gas from the reservoir does not leak into the tube.

An all-metal fill system using vacuum tight welds eliminates the need to use fragile glass and insures the ruggedness and reliability necessary for long tube life.

Internal Gas Return

The ability to maintain uniform gas pressure along the length of the tube is critical in achieving high output powers. The discharge pumps ions to the cathode end of the tube causing a pressure gradient along the length of the tube. Gas must be returned to the low pressure area to equalize the pressure.

The INNOVA uses an internal gas return system to equalize the pressure along the entire length of the tube. Return holes around the outer edge of the disks return the gas to low pressure areas.

Hard Seal Crystalline Quartz Brewster Windows

Hard-sealed crystalline quartz Brewster windows are mandatory if a tube is to have long life and high output powers. High energy vacuum UV generated in the plasma discharge can cause color center formation in Brewster windows, which will ultimately degrade the output of the laser, especially in UV operation. An advanced window material selection method eliminates the problem of color center formation.

To insure long tube life, crystalline quartz Brewster windows must be hard-sealed, using a glass frit material, to the Brewster window stem. If the Brewster window stem is made from a material which has different thermal expansion properties than the window itself (as is the case with many other lasers), thermal expansion will cause strain on the window. This in turn creates birefringence, increasing the window losses and reducing the output power of the laser.

Crystalline quartz Brewster windows are hard-sealed to a crystalline quartz stem. By aligning the crystal axes of expansion of the window and stem, a reliable stress-free fritted-seal can be made. The metal to quartz seal connecting the Brewster stem to the tube is made away from the window.

Krypton Tubes

Krypton tubes put additional demands on an ion laser tube design. Having lower gain than argon tubes, the operating parameters must be precisely controlled if maximum output power is to be achieved.

Krypton ions are driven by the discharge into the bore walls reducing the tube pressure. The gas may also be spontaneously released from the bore walls with some materials (e.g. beryllia) causing an increase in tube pressure. As the output of a krypton tube is extremely pressure sensitive, this phenomenon must be controlled.

To effectively control these pressure changes, the total volume of gas within the INNOVA tube is large compared to the volume of the bore. When ions are lost into the bore walls, the tube pressure changes slowly due to the relatively large volume of ballast gas.

The INNOVA® 70 Series Ion Laser (See Figure 9–23)

Typical of the INNOVA® Ion Lasers is the Coherent model 70 series. It is designed as a low cost, high output power, fast warm-up, stable, and reliable laser systems.

The INNOVA 70 uses CoolDisk tube construction. Tungsten/copper disks, brazed to a one-piece ceramic tube envelope, confine the plasma discharge and efficiently conduct the plasma heat to the cooling water. Metal/ceramic electrical feedthroughs and brazed joints eliminate fragile glass work, making CoolDisk tubes exceptionally rugged.

To insure that the INNOVA 70 lasers operate at the proper gas pressure and have sufficient gas for the life of the tube, a ballast tank with extended gas supply

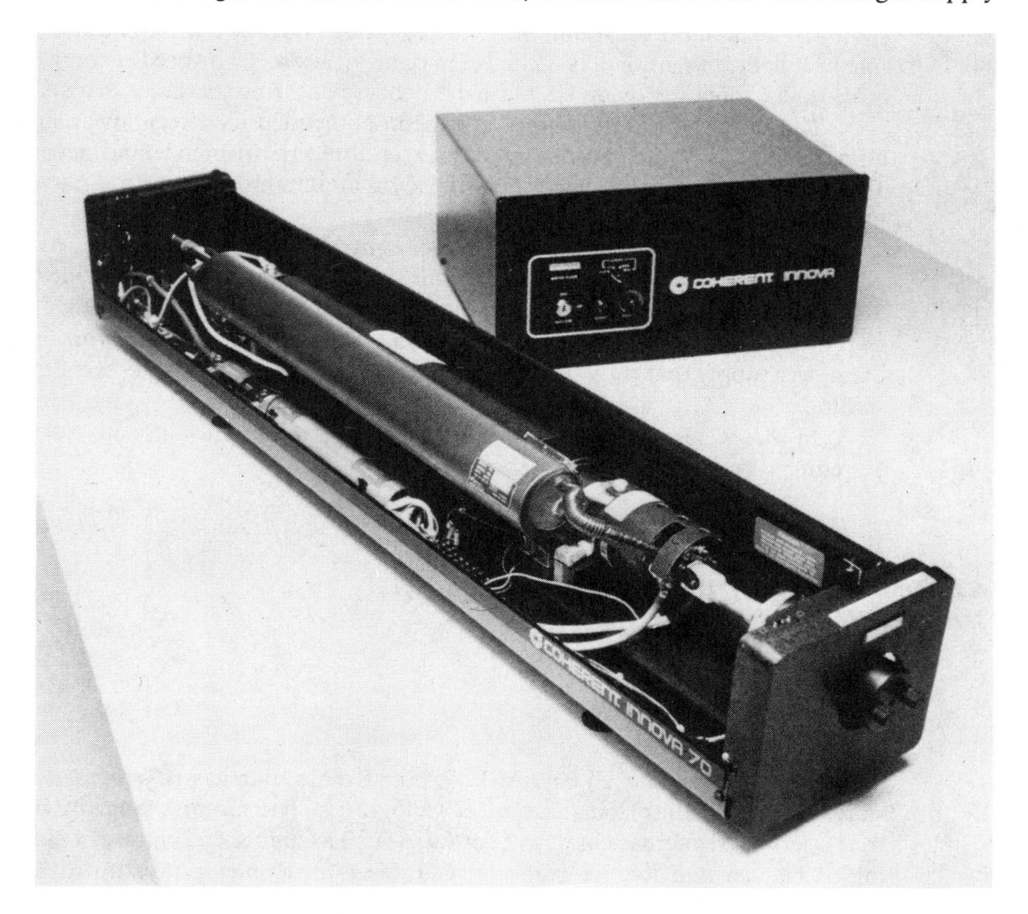

Figure 9–23 The INNOVA® 70 series ion laser (*courtesy Coherent, Inc.*).

is used. When the gas pressure in the laser tube reaches the low side of the operating window, a low pressure warning light will flash on the front panel of the power supply. By breaking open one of four gas cartridges on the ballast tank, a new supply of gas restores the tube pressure to the original value.

An aluminum resonator is used for its excellent warmup and pointing stability. A resonator that is intended for single-line/multiline operation does not need the length stability required for single-frequency lasers. The high thermal conductivity of aluminum insures that heat from the tube/magnet is distributed evenly over the resonator, minimizing temperature induced distortion of the resonator. In addition, the resonator is kinematically isolated from the feet and head covers to eliminate mechanically induced stress from being transmitted to the resonator.

The intracavity space is completely sealed to keep the optical elements free from dust and other contaminants, dramatically reducing the need to clean optical surfaces.

An adjustable aperture ensures TEM_{00} operation at any wavelength or power setting. The high reflector mirror holder functions both as a multiline and single-line mirror holder, requiring only a change in optic location. Two-speed rear mirror mount adjustment screws provide backlash-free coarse and fine tuning with a single knob.

The INNOVA 70 uses an SCR (silicon-controlled rectifier) power supply with a linear transistor passbank to achieve exceptional performance and dependability. The linear passbank reduces optical noise by maintaining a constant current to the plasma tube.

Local control systems have the laser controls located on the front panel of the power supply. Precise adjustment of the system's output power is achieved with a 3-turn current control. The laser output power can be read by connecting a voltmeter to the BNC on the back panel. In addition, there is a terminal strip on the back of the power supply that provides detailed diagnostic information in the event of a system failure.

In the event of water failure, a waterflow switch automatically turns off the system, protecting it from overheating.

STATE-OF-THE-ART MOLECULAR GAS (CO_2) INDUSTRIAL LASERS

General

(See Figure 9–24) The EVERLASE series of molecular gas (CO_2) lasers represent state-of-the-art in molecular gas laser technology. The major change in CO_2 industrial lasers is the increase in power for tasks such as high speed cutting, welding, drilling of heavier and thicker material. CO_2 lasers now have power up to 2500 watts and may cut steel of 0.50 thickness while penetrating steel with welds of 0.10 inch.

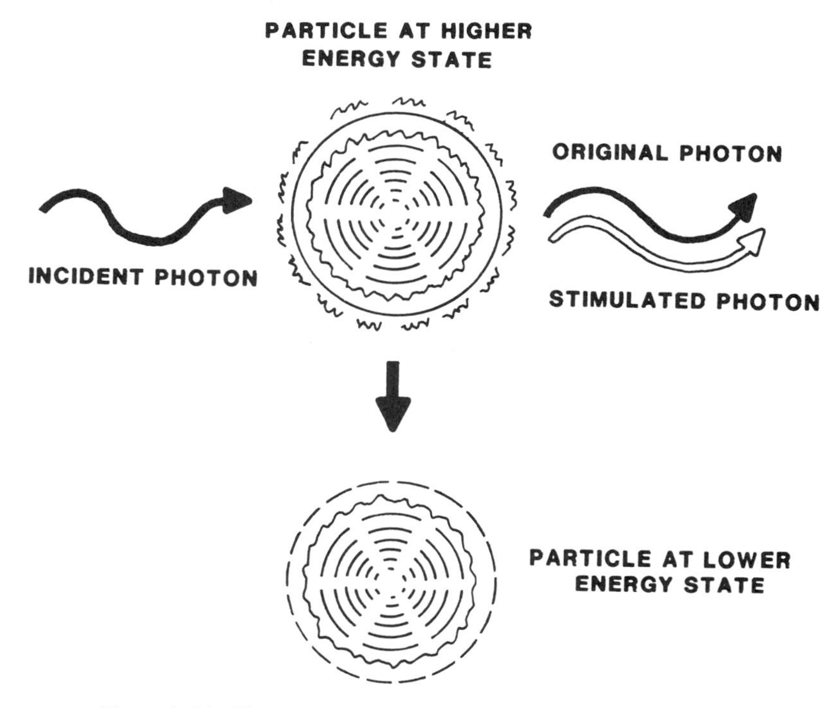

Figure 9–24 The CO_2 lazer lazing process (*courtesy Coherent Inc.*).

The CO_2 Laser Lazing Process (See Figure 9–24)

Lasing is a process which converts energy from one form to another, more organized, more controllable, and more useful form. The fundamentals of a jack hammer operated by compressed air are very similar—a naturally resonant mechanical system is supplied with the raw energy of compressed air, and the resulting oscillations provide powerful, yet controllable and useful results.

The lasing process differs from this example in that the energy conversion is provided by molecular excitation. Energy from the electrical discharge excites nitrogen molecules which in turn cause the single carbon atom in the CO_2 molecules to vibrate back and forth between the two oxygen atoms on either side of it. This vibrating CO_2 molecule can give up a fixed amount of energy in changing to another lower energy vibrational pattern. This transition from the higher energy vibration state to the lower state results in energy being given up as a fundamental particle of light called a photon. Each of the photons thus emitted can move along to the next excited molecule and stimulate it to give up a photon in changing to its lower energy vibrational state.

This process of photons causing "stimulated emission" of yet more photons is directional in nature. Each new photon travels in the same direction as the photon that triggered it. If mirrors are placed on both ends of this assembly of CO_2 molecules, the photons will emerge from a partially-transmissive mirror on one end of the tube in a single, well-collimated beam. A partially-transmissive mirror ensures that there will always be a supply of photons reflected back and forth in the cavity to keep the process going.

Lasing, the process by which the EVERLASE CO_2 laser emits its powerful and easily controllable beam, makes use of the naturally resonant qualities of the CO_2 molecule to "organize" the raw electrical energy that powers the laser.

Typical CO_2 Gas Laser Configuraton (See Figure 9-25)

A mixture of CO_2, He and N_2 in a glass tube at low pressure provides the lasing medium.

A high voltage power supply is connected to electrodes in the discharge tube to provide the excitation source.

Mirrors reflect light back and forth along the optical axis forming an optical resonator. One mirror is partially transmissive and allows some of the light out in the form of a laser beam.

CO_2 Industrial Laser Operational Modes

Laser machine tools provide advantages over their mechanical counterparts in several ways:

— CONTINUOUS OPERATION—Except for routine maintenance, laser machine tools need not stop production for recalibration, parts replacement, or debris clean-up. Because lasers are non-contact tools, the wear and breakage common with mechanical tools are eliminated.

— AUTOMATED OPERATION—Lasers are readily integrated with parts handling subsystems and are well-suited to computer numerical control (CNC) operation.

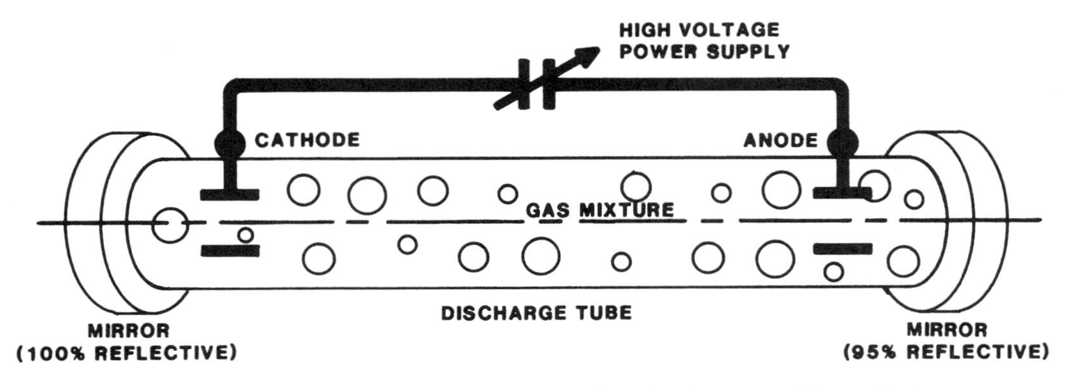

Figure 9-25 Typical CO_2 gas laser configuration (*courtesy Coherent Inc.*).

- ENERGY EFFICIENT—Using equivalent energy comparisons, industrial lasers use energy more efficiently than most conventional industrial processes.
- NO SPECIAL ENVIRONMENT—Lasers do not require special environments, such as vacuum chambers, to operate.

The advantages of laser machine tools are most apparent in terms of specific applications. The primary applications areas for laser tools are cutting, drilling, welding, scribing, and heat treating.

Metal Cutting

Metal cutting with CO_2 lasers produces an extremely narrow kerf, providing unmatched precision for thermal cutting of small holes, narrow slots, and closely spaced patterns. Kerf widths of 0.004″ to 0.010″ (0.10 to 0.25 mm) for thin metals up to 0.375″ (9.52 mm) thick are typical. By reducing the kerf, cutting speeds are enhanced, heat affected zones are reduced, and less energy is required from the laser. Applicable functions are:

- No tool wear or slippage in comparison with contact tool processes;
- Narrow kerf width, from 0.005″ (0.127 mm) to 0.012″ (0.305 mm);
- Minimum Heat Affected Zone (HAZ);
- Minimum dross produced when cutting metal. Limited dross produced when cutting aluminum and brass;
- Cuts steel up to 0.375″ (9.47 mm) thick with highest quality;
- Cuts steel up to 0.50″ (12.70 mm) thick with lower quality;
- Cuts aluminum and brass up to 0.125″ (3.18 mm) thick;
- Cutting quality and rate not dependent on material hardness.

Welding

Welding operations can be performed in either the cw or Enhanced Pulse modes. Using the Enhanced Pulse mode results in exceptional bead control, minimal HAZ, and negligible part distortion. CW operation results in faster welding speeds, with a trade-off in weld bead width and total heat input.

Power densities of 10^5 to 10^9 watts/in^2 (1.6×10^8 to 1.6 to 10^8 watts/cm^2) are required for laser metal welding operations. At these power densities, most metals become molten and can be welded. As with metal cutting, some materials, such as gold and silver, do not absorb beam energy well and may not be suited to laser welding. Applicable functions are:

- Because laser machining is a non-contact process, no pressure is applied to the part;
- No filler material required;

— Reduced HAZ and distortion due to minimal heat input to the part;
— Minimum HAZ often results in stronger welds than with conventional methods;
— Weld bead placement is very accurate;
— Penetration to 0.1″ (2.54 mm) in steels;
— Penetration to 0.05″ (1.27 mm) in some aluminums

Drilling

Holes ranging from 0.003″ (0.08 mm) to 0.042″ (1.0 mm) in diameter are drilled rapidly with excellent repeatability using the Enhanced Pulse mode to limit heat input. Larger holes up to 1.0″ (25.4 mm) in diameter are produced by trepanning the beam.

Continuous strips of natural or synthetic material are drilled by rapid pulsing of the beam as the material passes under the laser. Hole elongation, a potential problem at very high transportation speeds, can be avoided by using a high-speed beam deflector, which causes the beam to track the material during each pulse. Applicable functions are:

— High drilling rates (up to 50 million holes per hour in some materials);
— Hole diameters down to 0.003″ (0.076 mm) with CO_2 lasers;
— Hole diameters down to 0.001″ (0.025 mm) with Nd:Glass lasers;
— Clean holes with minimal recast material when drilling alumina.

Typical of the EVERLASE CO_2 lasers is the Printed Circuit Board Laser Drilling System shown in Figure 9-26.

The EVERLASE® CO_2 Lasers (See Figure 9-26)

Typical of state of the art CO_2 lasers are the Coherent EVERLASE® models. These models perform total industrial applications such as cutting, scribing, drilling, welding, heat testing, engraving and perforating. They perform these tasks on a variety of materials including metal, ceramic, glass, paper, and wood. Unique to the EVERLASE® CO_2 lasers are:

— An exclusive Enhanced Pulse feature provides optimized material processing capability;
— A unique, thermally-compensated resonator structure assures unmatched stability of beam power and mode quality, and eliminates heavy, unstable granite bases used in older generation CO_2 lasers;
— An exclusive modular resonator design offers multibeam economies not available from any other system;
— Up to four output beams are available from a single laser enclosure;
— A gas recycling system cuts gas consumption by 90 percent.

Figure 9-26 The EVERLASE® CO_2 laser (*courtesy Coherent Inc.*).

The resonator structure incorporates a temperature-stabilized circulating oil system that maintains the resonator temperature to within $\pm 0.1\,°C$ ($\pm 0.18\,°F$). The coolant circulates through the resonator, discharge tubes, mirror mounts, and water-cooled heat exchanger.

A key concept of the EVERLASE CO_2 laser family is modular resonator design. This is a major advance for multibeam applications, permitting construction of single resonators to emit multiple beams. The multiple beams share a common electrical, gas, and water system in a single cabinet and can be pulsed simultaneously.

Multiple-beam laser configurations offer significant advantages over the two conventional approaches to multibeam applications—beam splitting and ganging of multiple lasers.

Multibeam lasers do not require complicated "beam splitters" or suffer the power loss characteristic of beam splitting, and the total power delivered is greater than that model's single beam power rating. Multibeam lasers offer economic benefits for many applications.

All laser folding mirrors are mated to the resonator structure by means of matched mirror mounts, allowing them to be removed, checked, and replaced without realignment. The mirror blocks at the ends of the discharge tube do not require adjustment in the field. In fact, only periodic adjustment of the output mirror is necessary to maximize the power output.

This unique Enhanced Pulse feature enables EVERLASE CO_2 lasers to perform many machining operations that would normally require lasers of much higher power ratings. The laser can enhance-pulse amplitudes by as much as eight times the continuous power level. An additional high-voltage power supply is used to achieve increases in pulse power.

The Enhanced Pulse feature is part of an extensive set of beam control functions available in EVERLASE laser systems. These functions include control of pulse repetition rates and burst patterns, as well as control of individual pulse configurations.

With a standard gas mix, lasers can be operated to 1000 pulses per second. Pulse rates higher than 2500 pulses per second are possible when a special, commerically-available gas mixture is used. This special mixture is needed to shorten the Enhanced Pulse's decay time and eliminate pulse overlap. Pulse width is adjustable from below 100 microseconds to continuous.

A burst pattern control allows the operator to specify a number of pulses of a selected duration to be repeated at a selected frequency. This technique provides, for instance, an efficient means for drilling holes on a production part that requires a pattern of holes to be drilled while moving, or in deep hole drilling to minimize heat input to the part and optimize hole shape.

The continuous mode is useful for heat treating metal, welding, and for many cutting applications where a fast cutting speed, clean vaporization, or smooth edge is desired. A single, short Enhanced Pulse assures clean drilling and cutting of plastics, rubber, and metals. Material is instantly vaporized, leaving no slag or melt products. Single long pulses with enhanced leading edge spikes are ideal for welding metals. The leading edge spike quickly melts the metal, which is necessary to overcome its very high reflectivity. This allows the pulse's subsequent continuous segment to complete the welding process at a much lower level.

Many cutting, drilling, and welding operations are accomplished with considerably less heat coupled into the workpiece. Lower heat input results in much less thermal distortion of the part, reduced HAZ (heat affected zone) and stronger welds. Because thermal distortion and HAZ are minimized, these processes are often performed to final dimensions with the traditional finishing steps eliminated.

The control panel can also be installed in a standard 19-inch electronics instrument rack, connected to the laser electronics by remote cable. This optional arrangement provides remote control of the laser and is useful for applications where centralization of various control panels is desired.

Gas Recycling

EVERLASE CO_2 lasers feature an exclusive gas recycling system which recycles 90% of the system gas. This reduces the major operating expense of gas consumption costs ten-fold.

Snap-In/Out Glass Connections

All gas, coolant, and electrical connections are made through metal manifold blocks with snap-in/out "O" ring-type connections. This simplifies the discharge tube construction and eliminates the strain-inducing connections found in conventional lasers.

The discharge tubes are fitted into snap-in/out mounts on the pre-aligned end plates. These can be easily removed and replaced in the field without realignment.

Correct matching of the focussing lens to the application is an essential part of the laser specification process. Coherent offers a wide range of standard anti-reflection-coated lenses in four focal lengths: 1.5″, 2.5″, 5.0″, and 10.00″ (38.10 mm, 63.50 mm, 127 mm, and 254 mm). Each focal length is available in three substrate materials: germanium, gallium arsenide, and zinc selenide. All three substrate materials exhibit excellent transmittance at 10.6 microns.

Lens focal length determines the spot size of the focussed beam and, hence, its power density and processing speed. The smaller the spot, the faster the speed. The shorter the focal length, the smaller the spot size and the higher the power density. Shorter focal lengths also mean less depth of focus and a greater change in power density with respect to the distance from the focal plane. This imposes limits on the thickness of material that can be processed and on acceptable variations in the part's working distance, i.e., the distance between the lens and the work surface.

STATE-OF-THE-ART LIQUID LASERS (DYE)

Liquid Laser Operation (See Figure 9–27)

An optical schematic of the Coherent Inc. Model 702 is shown in Figure 9–27. The model 702 is continuous-wave (CW) and operates in either mode, locked synchronous pumped operation or cavity dumped operation. The CW dye laser functions by rapidly moving dye molecules through a CW laser pump spatially coincident with the optical cavity of the dye laser. Each molecule in effect sees a pulse of light as it passes through

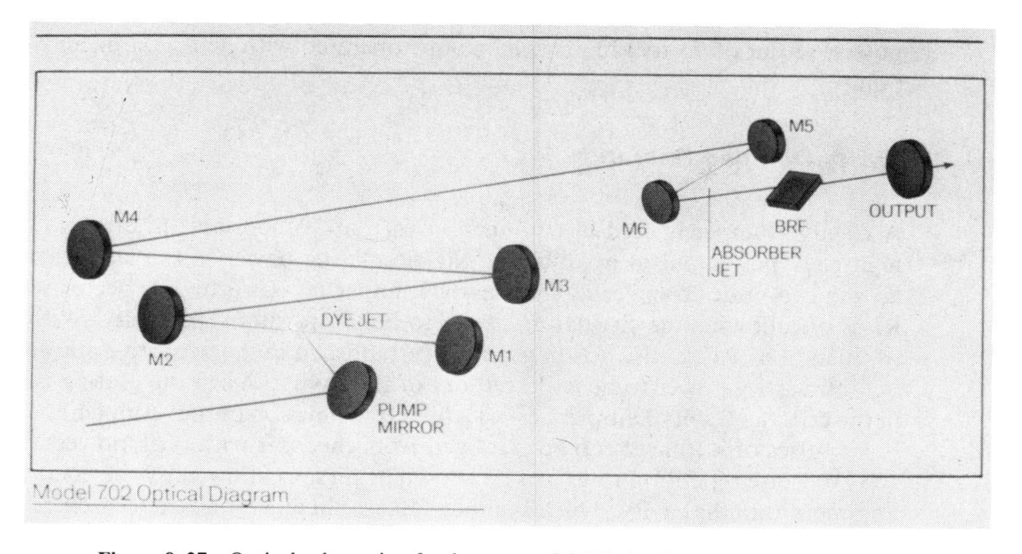

Figure 9–27 Optical schematic of coherent model 702 dye laser (*courtesy Coherent Inc.*).

the pump beam. Synch pumped dye lasers need to be cavity length matched to the mode-locked pump laser. Mechanical stability and ease of alignment are essential for repeatable and reliable performance in the picosecond and femtosecond regime.

The 700 Series Dye Laser has been developed specifically to offer optimum performance in the sub-picosecond region. A stable resonator is constructed on a 50 mm diameter Invar bar. Simple alignment is ensured by using a Z-shaped cavity with two independently adjustable flat mirrors. The different length requirements when using different pump lasers or a cavity dumper are achieved by sliding one of the flat mirrors to different, pre-marked positions on the Invar bar.

The radii of curvature of the high reflector and low fold mirror have been chosen to optimize power density in the dye stream for the shortest, cleanest pulses.

One, two, or three-plate birefringent filters allow flexibility in controlling pulse duration and tunability. Conventional synchronous pumping of a dye laser with a mode-locked ion or Nd:YAG laser allows pulse duration of less than 6 psec. It is generally not possible to reach the shortest pulses theoretically allowed because of the appearance of secondary, or satellite pulses.

To overcome this limitation, a saturable absorber can be added inside the cavity, whose function is to prevent the formation of secondary pulses. The Model 702 features a second jet assembly with a separate pump module, especially designed for the circulation of a saturable absorber. The second jet is placed in an auxiliary cavity waist generated by two curved mirrors. In this configuration, and with a one-plate birefringent filter, the system delivers pulses on the order of 500 femtoseconds.

Typical Tuning Data (See Figure 9–28)

The Coherent Model 702 Dye Laser may operate at ranges from 570 to 620 nm[9] with power output of 75 to 120 mW. It may be operated with an argon or an Nd:YAG pump.

Cavity Dumped Operation

A Cavity Dumper is used in conjunction with a CW ion laser in order to provide high peak power pulses by dumping the circulating power in the laser cavity with an acousto-optic Bragg cell. The Cavity Dumper acts as a high reflector when the RF is off allowing the circulating power to build up within the cavity. With the introduction of RF, a diffraction grating is established in the acousto-optic cell causing intracavity power to be deflected out of the cavity. When the grating is present in the cell, the Cavity Dumper acts as a high transmission output coupler and, therefore, pulses of $\leqslant$10 nsec can be extracted from the laser with peak powers of up to 200 W. Pulse repetition rates can be chosen as high as allowed by the energy buildup time within the cavity, which is about 150 nsec in an argon laser. Faster repetition rates can be chosen, but lower peak powers will be produced.

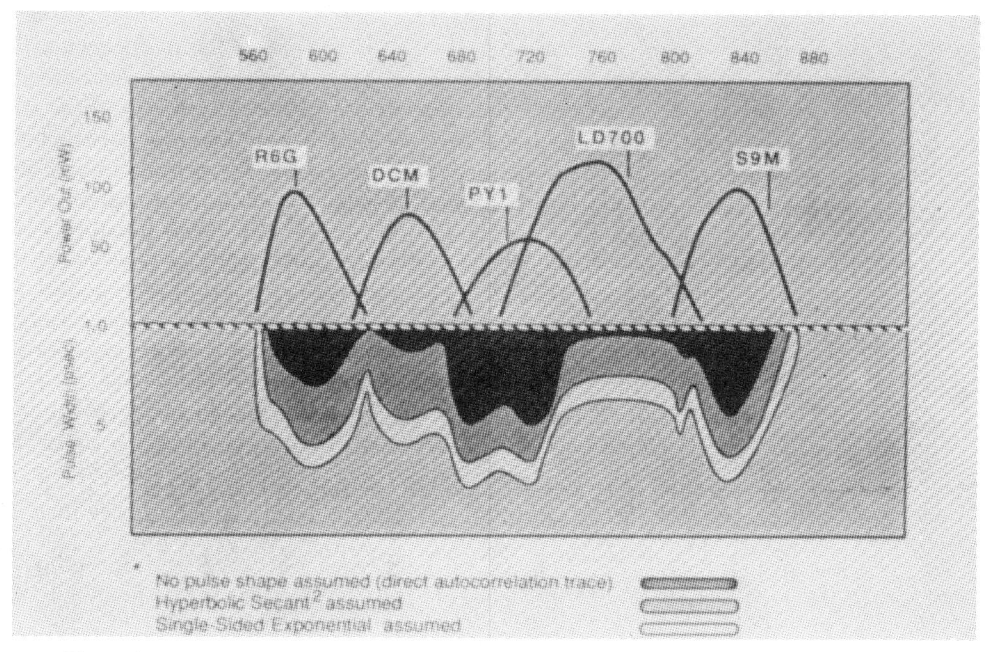

Figure 9–28 Tuning curves for the Coherent model 702 dye laser (*courtesy Coherent Inc.*).

The Coherent 700 Series ULTRAFAST Dye Laser (See Figure 9–29)

The 700 series dye laser was designed to allow scientists to concentrate on their experiments in ultrafast phenomena while the laser system operates in synchronous pumping operation. It has several prominent features.

— It may operate with any CW mode locked ion or Nd:YAG pump laser.
— It may operate with dual jet option for sub-picosecond performance.
— It may operate with a cavity dumper to provide high peak power pulses.

SELF-CHECK QUESTIONS

Self-check questions allow students to evaluate how well they have learned what they are studying.

1. What is a laser?
2. Define the following laser light properties.
   (a) Monochromatic
   (b) Coherent
   (c) Directional

Figure 9–29 The Coherent 700 series ultrafast dye laser (*Courtesy Coherent Inc.*).

   (d) Lasing action
   (e) Pumping
   (f) Excited energy state
   (g) Stimulated emission
   (h) Spontaneous decay
   (i) Population inversion

3. How are lasers classified?
4. What are the four active mediums used with lasers?
5. The optical range of wavelengths are stated as three major areas. What titles are given to these major areas?
6. *True or False:* Wavelengths in the visible spectrum range from 0.3900 to 0.7500 micron.
7. Explain the output of a pulsed laser.
8. Explain the output of a continuous wave laser.
9. What are spectral lines?
10. Define the term *stimulated absorption.*
11. Define the term *stimulated emission.*
12. Define the term *attenuation* as to do with lasers.
13. Define the term *amplification* as to do with lasers.

14. Define the term *coherence.*
15. What is the difference between temporal and spatial coherence?
16. Explain the laser beam as to its cone of divergence.
17. What is the purpose of the optical cavity?
18. Why would an engineer utilize a four-energy-state laser?
19. What are three types of solid lasers used in the laser industry?
20. Describe the operation of a solid laser.
21. What is a Q-switched solid laser?
22. Describe the operation of a gas laser.
23. What are the characteristics of a neutral atom gas laser?
24. What are the characteristics of an ion gas laser?
25. What are the characteristics of a molecular gas laser?
26. Describe the operation of a liquid laser.
27. What is an optic set?
28. Why are different laser beam profiles used?
29. What are two basic statements that explain laser alignment in general?
30. What are some modern practical uses for the laser in today's life?

Ten

Photometry/Radiometry

Photometry is the science that deals with the measurement of visible light, that is, wavelengths from 390 to 770 nanometers, depending on what reference you use. Radiometry is defined as the science that deals with the measurement of the entire optical spectrum including ultraviolet, visible, and infrared light. Wavelengths of light range from 0.005 to 4000 microns. (A micron is equal to 10^{-6} meter.) You may wish to review the section "Optoelectronic Operational Spectrum" in Chapter 1.

In the discussion of light measurement there are several important measurements of light necessary to understanding the science. These are flux, illumination, intensity, and luminance. Other terms are also important and will be defined in detail later in this chapter. In order to assess the major quantities involved in photometry and radiometry we must consider what it is we are going to monitor and how we can measure it accurately. In this chapter we deal with the relationship of light-measurement quantities and the techniques involved in the measuring of these quantities.

OBJECTIVES

After studying this chapter and after completing the self-check questions at the end of the chapter, the reader will be able to:

1. Define terms that relate to photometry.
2. Define terms that relate to radiometry.

3. Explain the several functions that are involved with measurement and the photometer/radiometer.
   a. Light suppression
   b. Light attenuation
   c. Photometric filtering
   d. Radiometric filtering
   e. Illuminance/irradiance and luminance/radiance measurement
   f. Light flash measurement
   g. Densitiometry measurement
4. Become familiar with the portable and the laboratory photometer/radiometer.
5. Become familiar with the laser power meter.
6. Become familiar with the LED display measurement.
7. Become familiar with photometer/radiometer accessory equipment.

PHOTOMETRIC AND RADIOMETRIC TERMS

The relationship of photometric and radiometric terms is that they both deal with light energy or flux. Visible light energy is called *luminous flux* and is measured in *lumens.* Radiant light energy is called *radiant flux* and is measured in *watts.* Wavelengths of visible flux fall in the range 390 to 770 nanometers in length. Wavelengths of radiant flux include all of the optical spectrum, that is, ultraviolet, visible, and infrared wavelengths. Luminous flux is related to radiant flux by a wavelength factor developed by the Commissioner Internationale de'Clairage (CIE) in France in 1924 (see Figure 10–1). The photoptic eye response curve, as it is called, was developed from a sample of the population under daylight conditions. The response curve represents an average eye. The curve relates wavelength (λ) to relative luminosity or visual response and is called the luminosity function ($V\lambda$). The average eye sees best (is at its maximum efficiency) at 555 nanometers. The conversion factor for maximum efficiency of the average eye is 680 lumens per watt. Wavelengths other

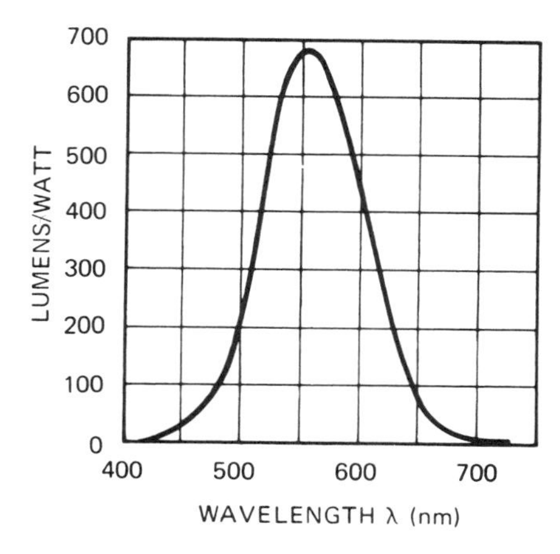

Figure 10–1 CIE standard photopic luminosity function (*courtesy of United Detector Technology, Inc.*).

than 555 nanometers within the visible range must be converted to the luminosity function in relation to the CIE spectral response curve. Test equipment utilized in photometry must use sensors that match the CIE curve.

Light Sources

Light energy is emitted in straight lines from one of two source types, the point source and the extended source (see Figure 10–2). The point source is idealistically a single point; however, even an extremely small point of origin would be finite. The extended source of light energy illuminates a point in space from several directions.

The Solid Angle (ω)

A sphere contains 4π steradians. The steradian is the unit measure of solid angles just as the radian is the unit measure of plane angles. A circle contains 2π radians. A sphere contains 4π steradians. A hemisphere contains 2π steradians. The solid angle ω is that angle with an apex in the center of the sphere and a radius r which subtends on the surface of the sphere (see Figure 10–3).

$$\text{Solid angle } (\omega) = \frac{S}{r^2}$$

$$\text{Solid angle } (\omega) = 1 \text{ steradian (sr)} \qquad \text{when } S = r^2$$

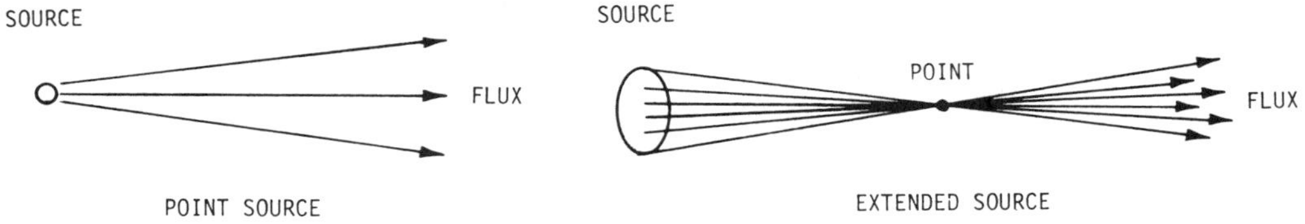

SOURCE
FLUX
POINT SOURCE
SOURCE
POINT
FLUX
EXTENDED SOURCE

Figure 10–2 Light sources (*Mike Anaya*).

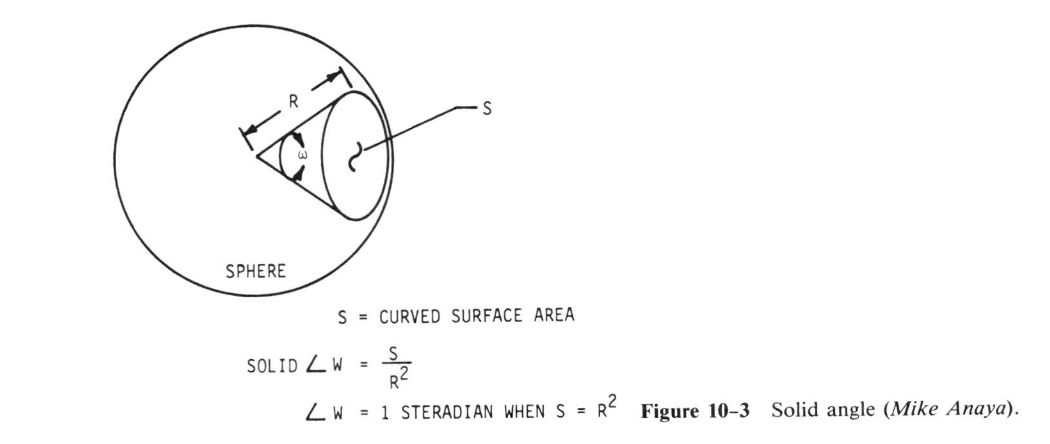

Figure 10–3 Solid angle (*Mike Anaya*).

Terms Relating to Photometry and Radiometry

The following terms are related to both optical measurement sciences (see figure 10–4).

Flux (ϕ): Continuously changing light energy to, from, or through a surface
Incidence (E): Flux per unit area incident (to) a surface
Exitance (M): Flux per unit area exiting (from) a finite source
Intensity (I): Flux per unit solid angle radiating from a finite source
Sterance (L): Flux intensity per unit area leaving a light source

Terms Relating to Photometry

The following terms are peculiar to measurement in the visible part of the optical spectrum, that is, the science of photometry. The reader will note that descriptive words in photometry such as prefixes will use the word luminous.

Candela (cd). The basic unit of luminous intensity or flux per unit solid angle. The luminous intensity, in the direction of the normal, of a blackbody surface of 1/600,000 square meter in area, at the temperature of solidification of platinum under a pressure of 101,325 newtons/meter^2. The term replaces the term candle, the obvious source. In 1948, the CIE redefined the term, as stated previously.

Lumen (lm). The basic unit of luminous flux. The lumen is equal to 1 candela (cd) of flux generated by a source point into a solid angle of 1 steradian (see Figure 10–3).

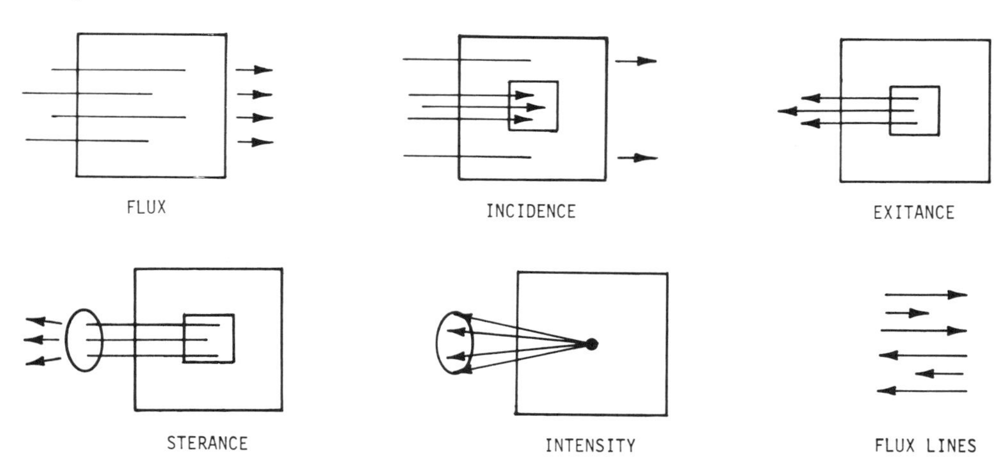

Figure 10–4 Terms relating to photometry and radiometry (*Mike Anaya*).

The lumen in relation to the solid angle. You may recall that a sphere contains 4π steradians. As was just discussed, one lumen of flux is generated from a point source into a solid angle of 1 steradian. A point source of 1 candela (cd) of luminous flux will generate 4π lumens when the point source is generating flux in all directions (over the sphere). A point source of 1 candela of luminous flux will generate 2π lumens when the point source is generating flux in all directions (over the hemisphere). The luminous flux (lumens) is dependent on the luminous intensity (candelas) and the projection area.

Luminous energy (Q_V). Luminous energy is the generic term for luminous flux.

Luminous flux. (ϕ_V). Luminous flux is the photometric measure for luminous energy. The SI unit of luminous flux is the lumen (lm).

Luminous intensity (I_V). The measure of luminous flux per unit solid angle from a point source. The unit of luminous intensity is lumens per steradian (lm/sr). The SI unit for luminous intensity is the candela (cd). The SI unit is recommended.

Luminance (B). A measure of photometric brightness. The unit of luminance is lumens per steradian per square meter ($lm/sr/m^2$). The SI measure for luminance is called luminous sterance (L_V).

Luminous sterance (L_V). A measure of photometric brightness. It is the flux per unit area emitting from a light source. The SI unit for luminous sterance is candelas per meter squared (cd/m^2). Luminous sterance was previously called luminance.

Lambert (L). A unit of luminous sterance equal to $1/\pi$ candela per centimeter squared ($L = 1\pi\ cd/cm^2$). Also, a foot-lambert is equal to $1/\pi$ candela per foot squared ($L = 1/\pi\ cd/ft^2$). Also a meter-lambert is equal to a $1/\pi$ candela per meter squared ($L = 1/\pi\ cd/m^2$).

Foot-lambert (fL). A unit of luminous sterance equal to $1/\pi$ candela per foot squared ($fL = 1/\pi\ cd/ft^2$).

Luminous exitance (M_V). The measure of flux exiting a finite surface. The SI unit for luminous exitance is lumens per square meter (lm/m^2).

Illuminance (E). The measure of flux per unit area incident to a surface. The unit of illuminance is lumens per square meter ($1m/m^2$).

Luminous incidence (E_V). The measure of flux per unit area incident to a surface. The unit of luminous incidence is lux (lx).

Lux (lx). The unit of luminous incidence in lumens per square meter (lm/m^2).

Preferred SI Terms in Photometric Measurement

Luminous energy (Q_V)
Luminous flux (ϕ_V), in lumens (lm)
Luminous incidence (E_V), in lux (lx)
Luminous exitance (M_V), in lumens per square meter (lm/m^2)
Luminous intensity (I_V), in candelas (cd)
Luminous sterance (L_V), in candelas per square meter (cd/m^2)

The Inverse Square Law

This law of illumination states that the illumination of a surface due to a source point of light is proportional to the intensity of the source and inversely proportional to the square of the distance from the source to the surface (see Figure 10–5). For example, the following table shows how much and how brightly a 1-lumen source will light an area from different distances.

Distance from light source (ft)	Illumination (ft)	Area (ft^2)
1	1	1
3	1/9	9
5	1/25	25

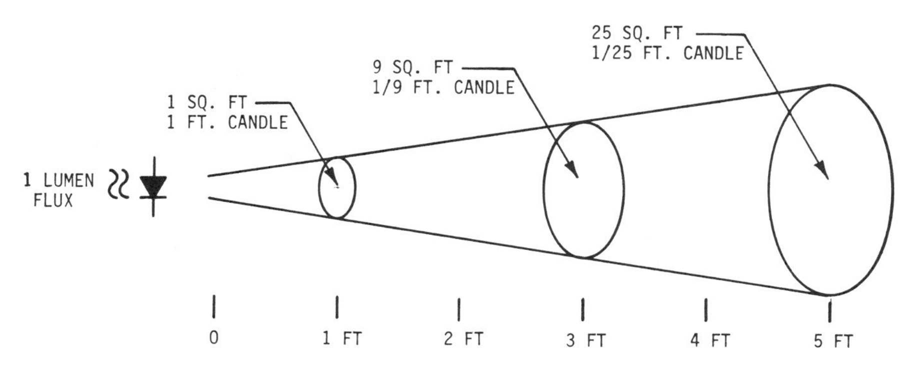

Figure 10–5 Inverse square law (*Mike Anaya*).

The Cosine Law of Illumination

The cosine law of illumination states that the flux radiated from an extended source in a given direction varies with the projected area of the emitter or the receiver in a plane perpendicular to the direction of the flux. In the illustration, an angle is made between a normal line (perpendicular to the surface) and the incident flux lines. The angle is used to calculate the effective area presented to the incident flux. The effective area is area $\times$ cosine $\angle\theta$. The illumination in lumens per square foot is equal to the area $\times$ cosine $\angle\theta \times$ incident flux (see Figure 10–6).

Terms Relating to Radiometry

The following terms are peculiar to measurement in all parts of the optical spectrum. The reader will note that descriptive words in radiometry such as prefixes will use the word radiant.

Watt (W). The basic unit of radiant flux. The watt is a measurement of power (energy) and relates to the photometric unit lumen (energy).

The watt in relation to the solid angle. You may recall that sphere contains 4π steradians. You may also recall that a solid angle is used as a parameter in the measurement of luminous intensity. Luminous intensity is measured in lumens per steradian (lm/sr). The parallel radiant intensity is measured in watts per steradian (W/sr).

Radiant energy (Q_e). Radiant energy is the generic term for radiant flux.

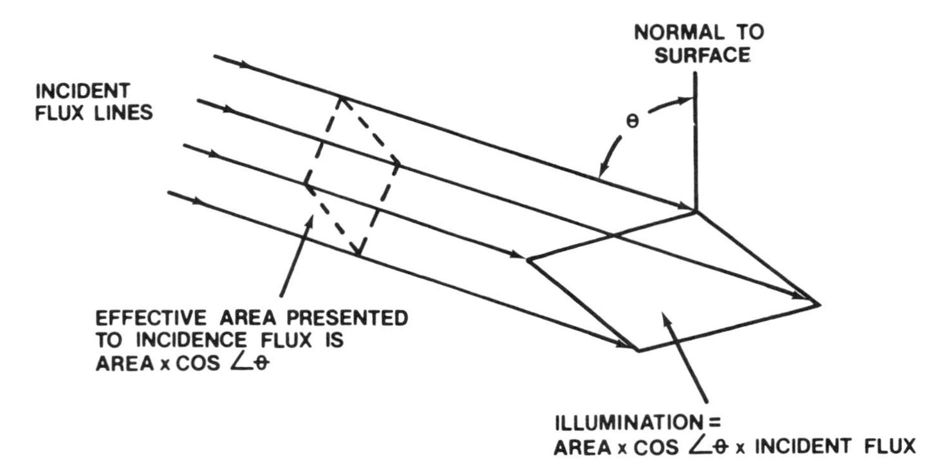

Figure 10–6 Cosine law of illumination (*Mike Anaya*).

Radiant flux (Φ_e). Radiant flux is the radiometric measure for radiant energy. The SI unit of radiant flux is the watt (W). Radiant flux is also called radiant power.

Radiant intensity (I_e). The measure of radiant intensity per unit solid angle from a point source. The unit of radiant intensity is watts per steradian (W/sr).

Radiance (N). Flux per unit solid angle per area of emitting surface. The unit of radiance is watts per steradian per meter squared ($W/sr/m^2$).

Radiant sterance (L_e). Flux per unit solid angle per area of emitting surface. The unit of radiant sterance is watts per steradian per square meter ($W/sr/m^2$). Radiant sterance was previously called radiance.

Emittance (W). Flux per unit area exiting from a finite source. The unit of emittance is watts per square meter (W/m^2).

Radiant existance (M_e). The measurement of flux per unit area exiting from a finite source. The unit of radiant exitance is watts per square meter (W/m^2). Radiant exitance was previously called emittance.

Irradiance (H). Flux per unit area incident to a surface. The unit of irradiance is watts per steradian per square meter ($W/sr/m^2$).

Radiant incidence (E_e). The measurement of flux per unit incident to a surface. The unit of radiant incidence is watts per square meter (W/m^2). Radiant incidence was previously called irradiance.

Preferred SI Terms in Radiometric Measurement

Radiant energy (Q_e)
Radiant flux (Φ_e), in watts (W)
Radiant incidence (E_e), in watts per square meter (W/m^2)
Radiant exitance (M_e), in watts per square meter (W/m^2)
Radiant intensity (I_e), in watts per steradian (W/sr)
Radiant sterance (L_e), in watts per steradian per meter squared ($W/sr/m^2$)

PHOTOMETER/RADIOMETER

A photometer/radiometer is a device that measures the intensity and sterance of a light beam. Any of the basic photometer/radiometer units consists of a light detector, an amplifier, a function selector, and either an analog or a digital readout/display. Light from some source, such as an LED or a laser is exposed to the photosensitive surface of a light detector. The light detector in modern units is usually a PIN silicon diode. The purpose of the PIN diode is to convert the light energy to electrical current. The current is proportional to the intensity of the light beam and varies with its wavelength. This wavelength must be carefully considered when recording optical-power measurements. The variance is named *spectral response.* The light energy in the form of electrical current is directed to an amplifier. The amplifier's gain is controlled by a range switch. A function selector selects the mode in which the meter will operate. The amplifier output is then directed into an ammeter meter movement or a digital display. The position of the meter movement or the displayed numerals is directly proportional to the intensity of the light at the power meter detector.

Light Suppression

When monitoring a light source, it is often necessary to utilize an external light attenuator to suppress background light. Background light, also called *ambient light,* may cause errors in power measurement especially with low power measurements. The ambient light may be handled in several ways. Before measuring a desired light source, a measurement of the ambient light is made, then the total source and ambient light is measured. Simple subtraction of ambient-light measurement from the total provides the source-light measurement. A second method is built into some power instruments. A calibration adjustment zeros the power meter, which in effect calibrates out the ambient light. A third method is to utilize an ambient light shade. The shade is placed over the detector and allows only the source light to reach the detector.

Light Attenuation

Just as ambient light affects the reading of an optical power meter at lower power, the opposite situation of high power is a problem. When measuring very high power, the optical range of a power meter may be exceeded without the operator being aware of it. In this event extreme damage to the detector and possibly the meter movement could result. To alleviate this problem an attenuator is placed in front of the detector in the path of the light beam under measurement. This limits the beam power reaching the detector. The attenuator is made of diffusing material and scatters the beam within its walls. Part of the beam is absorbed by the attenuator's surfaces. A small amount falls on the detector surface and is measured. The amount of light energy falling on the detector can be changed by adjusting the distance between the detector and the attenuator diffusing material. Shortening the distance provides less attenuation, lengthening the distance provides more attenuation. When external attenuators are used, the meter reading must be adjusted in relation to the attentuation factor.

Photometric Filtering

Optical power meters are often used to measure light energy that is similar to that seen by the human eye. To measure this light, the meter must respond to it the same way the human eye would. To accomplish this a photometric filter is placed in front of the light-source detector. This causes the meter to react the same as the eye would

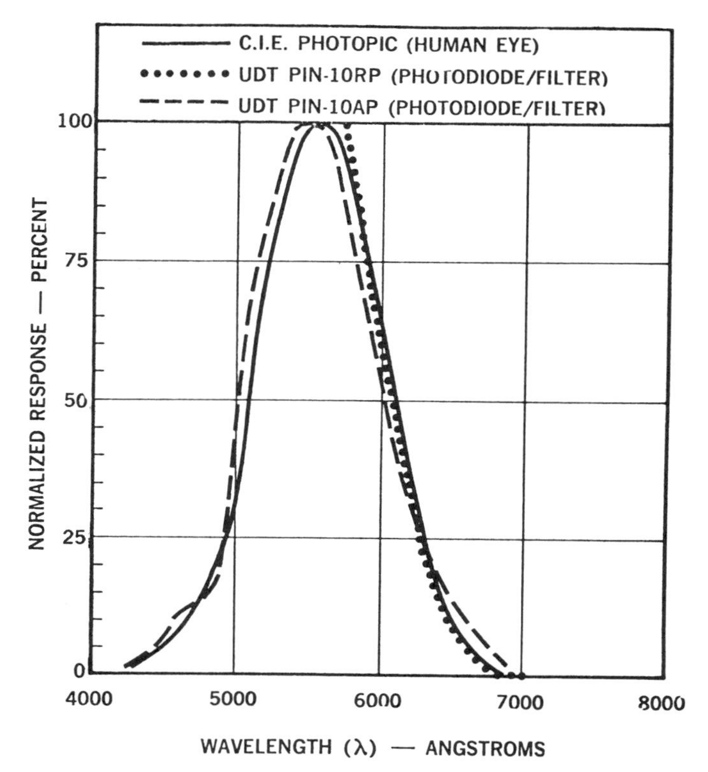

Figure 10–7 Photometric filter meter correction curve (*courtesy of United Detector Technology, Inc.*).

react to the light source. Figure 10–7 illustrates the CIE photopic curve in relation to photodiode filters. The photopic (human eye) curve is the solid line. Two PIN photodiode/filters are compared. These curves represent the wavelength response of the human eye and the filters. The reader will note that the difference is very slight. Most excellent power meters will have a selector switch that provides for photometric measurements.

Radiometric Filtering

Optical power meters are often used to measure light energy that is beyond the range of visible wavelengths (not photometric). To measure these optical wavelengths a radiometric filter is used (see Figure 10–8). The filter is placed over the detector and absorbs selected wavelengths of light while allowing other wavelengths to be seen by the detector with little or no absorption. When the filter is in place the meter responds to all wavelengths from approximately 400 to 1000 nanometers in the same level.

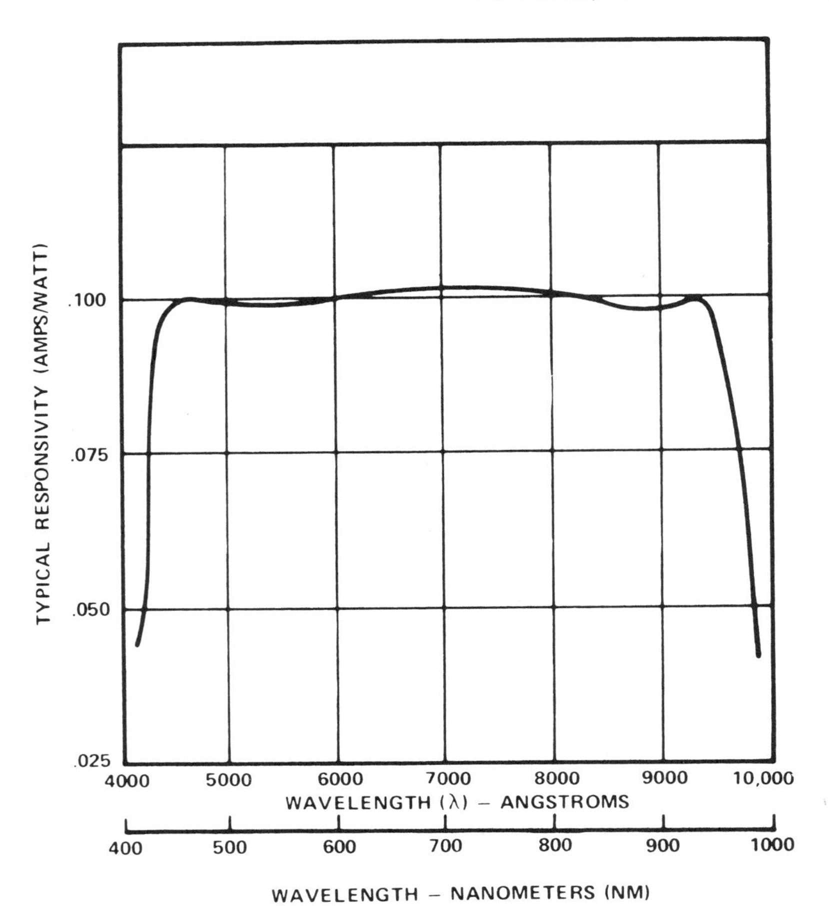

Figure 10–8 Radiometric filter meter correction curve (*courtesy of United Detector Technology, Inc.*).

Illuminance/Irradiance and Luminance/Radiance Measurements

Illuminance is the measurement of the luminous flux density (lumens/unit area) at a receiving surface as seen by a photopic sensor. The typical measurement unit in English-speaking countries is footcandle (lumen/square foot) and in most other countries is lux (lumen/square meter).

Irradiance is the radiant flux density (watts/unit area) as measured with a flat response sensor over given wavelengths. Illuminance/irradiance measurements are accomplished by setting the function selector at footcandles (illuminance) or microwatts (irradiance), screwing the appropriately marked filter (photometric or radiometric) onto the head assembly and directing the head at the desired light source.

Luminance (commonly termed *brightness*) is what the eye perceives, as specified by the C.I.E. curve. Luminance is luminous intensity of the source per unit source area projected in the direction of view (candelas per unit area). Radiance is the radiometric equivalent of luminance (watts/unit area of source/unit solid angle.)

Light Flash Measurements

Measurements can be made radiometrically in microjoules (microwatts-seconds) or photometrically in talbots (footcandle-seconds). Most photometers/radiometers are capable of measuring light flashes. Those that are capable accomplish this task in the following manner. Radiometric energy measurements are accomplished by attaching the radiometric filter to the sensor head, setting the function selector at microwatts and the range selector to the desired energy scale. Meter readings are then definable in microwatts-seconds. Photometric flash measurements are accomplished by attaching the photometric filter to the sensor head, setting the function selector at footcandle and the range selector on the desired energy scale.

Densitometry Measurement

Most photometers/radiometers can be used as a precision transmission densitometer for measuring the optical density of film, neutral density filters, colored-glass filters, and coated optics. Direct percent transmission readings can be obtained when the sample is inserted between the light source and detector head. Densitometric measurements can be made with a photopic detector response, spectrally flat radiometric response, or isolated wavelength response by the use of detachable filters.

Typical Portable Optical Power Meter

The UDT–40X optometer (see Figure 10–9) measures radiant power, radiant energy, illuminance, luminance, and integrated luminance over seven decades (logarithmically over six decades). A 1-cm^2 PIN silicon diode is used as the basic sensor. Two computer-designed, subtractive filters are provided for radiometric and photometric measurements. The radiometric filter provides a flat spectral response from 450 to 910 namometers. The photometric filter yields a spectral response comparable to the human eye. The relative reading function position "D" allows the user to set full scale at any value desired. The energy ranges integrate input light power from durations of 10 nanoseconds to several minutes. A manual reset button discharges the stored charge and resets the meter to zero. A logarithmic readout scale covers six decades of input light power without changing range.

Low-noise operational amplifiers are used to increase photodiode signal current to display levels. A taut band, mirrored scale, 4-inch meter movement displays the output. All the electronics are contained in a metallic shielded enclosure. The combination photodiode/operational amplifier is extremely stable and maintains its calibration independent of temperature and light-level extremes over many months. The instrument is supplied calibrated radiometrically and photometrically.

Figure 10–9 UDT–40X portable radiometer/photometer (*courtesy of United Detector Technology, Inc.*).

Typical Laboratory Photometer/Radiometer

The UDT–11A photometer/radiometer measures radiant power, radiant energy, luminance, illuminance, and spectral radiance (see Figure 10–10). This model is designed as a laboratory standard with accuracies of 1% throughout all ranges. The meter movement has an accuracy of 0.05%, ±1 digit-reading solution. Nine decades of light level can be measured from the lowest to the highest sensitivity. Construction of the model is of solid-state devices. The instrument can measure currents of 10^{-12} ampere. The range of the instrument is from 10^{-3} to 10^{3} microwatts. Its sensitivity is 10^{-11} watt, 10^{-5} footcandle, and 10^{-4} foot-lambert. The unit can be used as a standard photometer/radiometer or with different heads has variety of uses. Some of these are as a telephotometer, microphotometer, spectro-radiometer, and LED instrument.

In Figure 10–10, the UDT–11A is used along with a microphotometer. The microphotometer is basically a microscope with a 0.100-inch detector located as a target in the focal plane of the objective. The field of view is precisely defined by the detector area subtended at the objective. The objective resolves a spot size of 0.030 inch at a measuring distance of 1 inch. A computer-selected filter located in

Figure 10–10 UDT-11A photometer/radiometer (*courtesy of United Detector Technology, Inc.*).

the detector filter matches the C.I.E. (human eye) response to within ±2% of the total integrated area under the spectral curve. The microphotometer can be used as a microradiometer by simply removing the photometric filter.

LASER POWER METER

The purpose of a laser power meter is to measure the output power of a laser directly. It accomplishes this by coupling an extremely high-quality PIN diode with an integrating sphere. Figure 10–11 illustrates the integrating sphere and how it is set up to perform laser power measurements. In A of the figure a cross section of an integrating sphere is shown. The integrating sphere is accurately machined with a diffusely reflecting internal surface. An incoming light beam is dispersed through multiple diffuse reflections into a uniform intensity over the entire inner surface of the sphere. A partially transmitting diffuser, positioned on this surface and directly in front of an optical detector, accurately detects an exact percentage of the total input light. This energy level is then calibrated to accurately read the total input light beam.

In Figure 10–11, the test setup with the laser is shown. The integrating sphere is placed on a rack in line with the laser beam under test. An accurate power readout requires only that the entire laser beam being measured be directed into the input

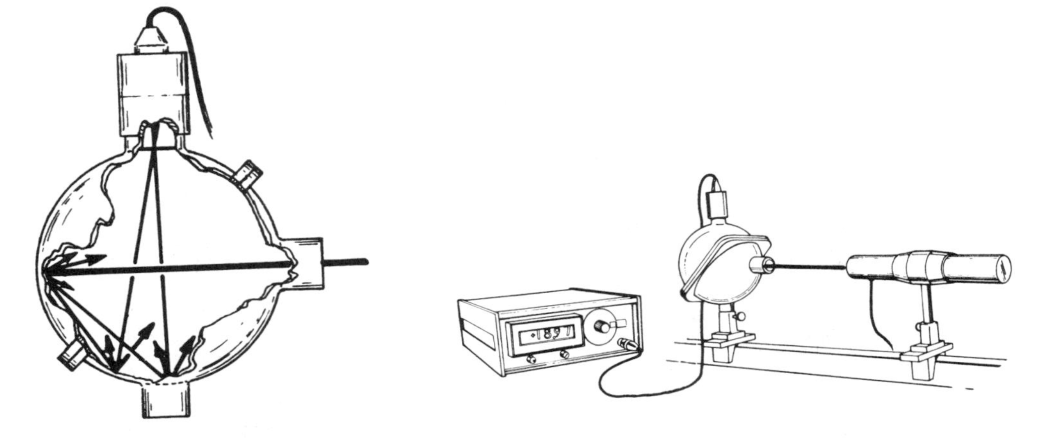

Figure 10–11 Integrating sphere and test setup for measuring laser power (*courtesy of United Detector Technology, Inc.*).

port. The integrating sphere does not have the interference problems encountered with the use of density filters. The detector is mounted on the top of the integrating sphere. Leads are connected from the detector to a detector-head input jack on the forward side of the power meter. The power meter indicates the laser power on a digital nixie display.

Measurement of laser power requires exacting safety precautions. These standards vary for individual wavelengths but for the visible region of 440 to 700 nanometers, class I lasers must not permit human access to radiation in excess of 0.4 × 10 watts, class II lasers must be below 1 × 10 watts, class III lasers below 0.5 watt, and class IV lasers are above 0.5 watt. In all cases various safety standards and precautions must be adhered to and displayed. An optional 7-millimeter-diameter aperture provides a direct power reading in accordance with proposed government input-aperture specifications.

Typical Laser Power Meter

A typical laser power meter is the United Detector Technology (UDT) Model 21 shown in Figure 10–12. The power meter is used along with the UDT2500 integrating sphere and the UDT223 detector filter assembly to measure laser power. The meter has a flat calibrated response over a wide range of laser wavelengths (450 to 900 nanometers). The meter can measure laser power over 12 full decades from 10^{-10} to 100 watts of input power. A digital readout requires no calibration curves. An offset adjustment provides a convenient means of meter zeroing and suppression of minor background radiation. Low-level power can be measured by removing the detector from the integrating sphere and placing it in line with the laser beam. Lower-level power is 10^{-10} watt to 1 milliwatt.

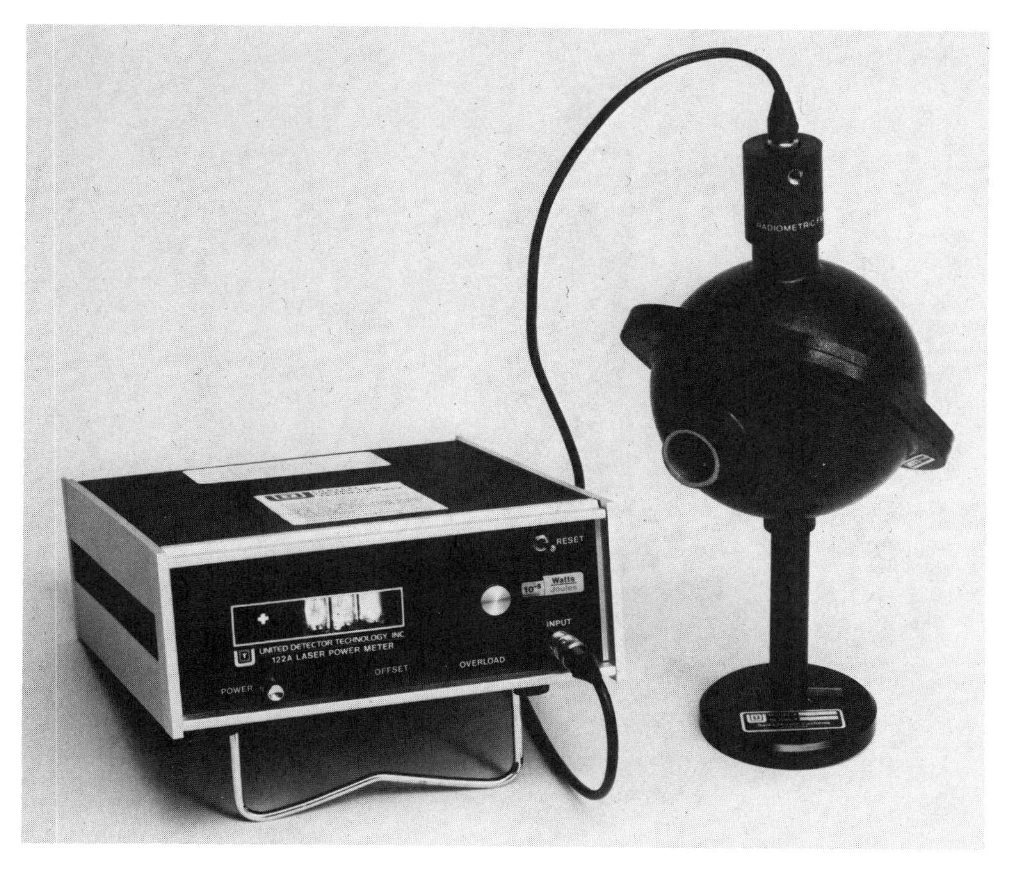

Figure 10–12 UDT model 21 laser power meter (*courtesy of United Detector Technology, Inc.*).

LED Display Measurement System

The United Detector Technology Model 111L is representative of an LED measurement system. It is pictured in Figure 10–13. The purpose of the LED display measurement system is to measure the total flux of green, yellow, or red LED displays in terms of luminous intensity. Single- or multiple-segment LEDs are mounted on a plate. The LEDs are energized by an external power supply. It is not necessary for the LED segments to have any exact alignment. The LED plate is mounted on a drawer that slides into the measurement module.

The measurement module consists of the LED enclosure and the detector head. The detector head is a black anodized optical cylinder containing detector, photometric filter, lens, and baffles. It is designed to position the detector at a fixed distance from the LED being measured. The sensor is a critically processed instrument-quality silicon detector with a circular active area of 1.0 cm^2. This high-performance detector is

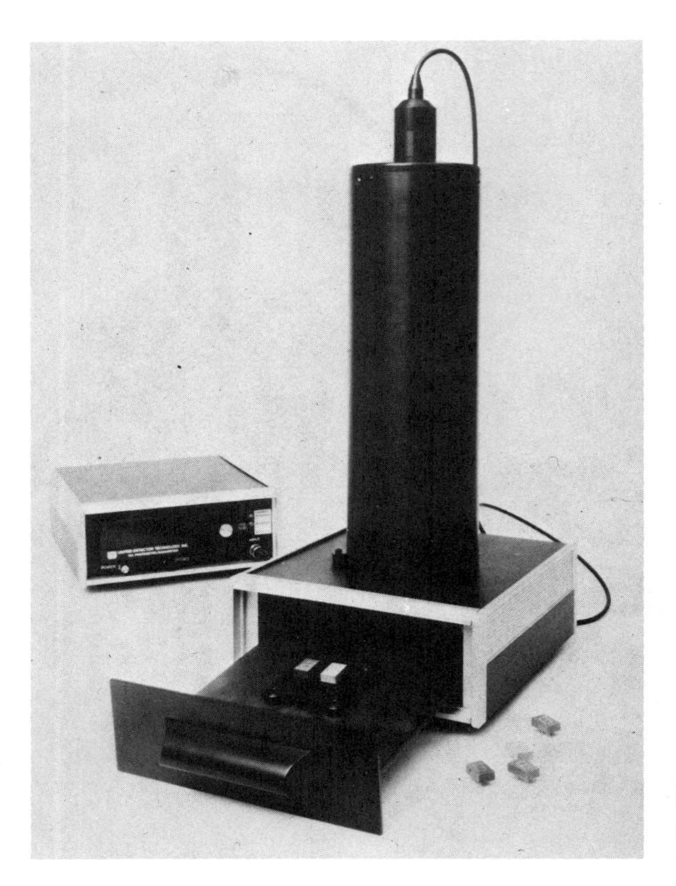

Figure 10–13 LED display measurement system (*courtesy of United Detector Technology, Inc.*).

specially selected for long-term stability, high responsitivity, low-noise uniformity, and linearity. Just below the detector is a red filter. The photometric filter is a computer-designed multilayered optical-quality glass filter. This filter corrects the detector's spectral response to that of the CIE curve and provides exceptional match for all visible LEDs. The lens defines the field of view of the system and focuses the energy emitted from the LED onto the detector. A series of baffles within the blackened cylinder eliminates the possibility of stray light causing erroneous measurements. The LED enclosure has a positioning module with provision for detachable mounting plates.

The control module contains the readout, electronics, power supply, and all necessary controls. A 3½-digit numerical display gives immediate direct readout in candelas. Direct-readout calibration channels are controlled by a four-position switch located on the rear panel. Each of the LED types, green (565 nanometers), yellow (585 nanometers), and red (665 nanometers), has its own dedicated calibration channel. A fourth channel is available for any other calibration required at the user's option. The particular channel in use is indicated on the control panel by a lighted "annunciator" of the same color as the LED being measured. Readings are displayed in absolute units and require no further reduction. A range switch provides system sensitivity over seven full-scale decades. A high-resolution full-scale offset control permits all stray light to be offset to zero. Outputs on the rear panel include analog and TTL compatible BCD.

PHOTOMETER/RADIOMETER ACCESSORY EQUIPMENT

Microscope accessory. The microscope accessory can measure brightness or radiance of small objects ranging from say 0.15 to 0.76 millimeter with a circular field of view.

Spot brightness accessory. The spot brightness accessory is a reflex viewing accessory used with an object lens to provide a variety of working distances and resulting image sizes.

Detector holder and aperture set. The detector holder and aperture set provides a convenient means of varying detector area without changing detector response.

Steradian shade. To fix the field of view to a specific geometry, the steradian shade should be used. The geometry is fixed at say 0.155 steradian. This permits the measurement of surface brightness (luminance or radiance) in terms of foot-lamberts, candelas per square meter, or microwatts per square centimeter steradian.

Light shade. The light shade is used in conjunction with detector/filter assemblies to limit the detector field of view in degrees thereby minimizing off-axis ambient light effects.

LED shade. The LED shade is a baffled tube that accurately defines the field of view, necessary when measuring LEDs.

Neutral density filters. The purpose of the neutral density filter is to extend the maximum power measurement capability of radiometers and photometers.

Laser beam spreader. The beam spreader extends laser power measurements capability.

Laser integrating sphere. The laser integrating sphere provides constant attenuation for high-power laser measurements.

LED integrating sphere. The LED integrating sphere facilitates laser power measurements. The sphere extends power measurements capability.

Laser attenuator. The laser attenuator provides constant attenuation for higher power laser measurements.

Photometric filter. When it is desired to measure light as the human eye would perceive it, the photometric filter should be used.

Cosine diffuser. The cosine diffuser is employed when making measurements of radiation at a surface.

Radiometric filter. The radiometric filter is used to provide a flat response when measuring radiation other than photometric.

Cosine receptor. The cosine receptor is employed to measure incident light over a wide angle.

Narrow bandpass filters. Optical glass interference filters with half-power bandpass at popular laser wavelengths.

Fiber optic probes. Probes used to make measurements in inaccessible areas. The probe is a flexible light guide constructed of glass-clad optic fibers sheathed in PCV with stainless steel end fittings.

Objective lens. Lens used to provide specific spot sizes.

Close-up objective. Provide a variable zoom at a specific degree viewing angle.

Lumilen. Lens used with a detector head for measurement of surface brightness.

Temperature filters. Filters used with a radiometer/photometer to determine color temperatures in Kelvin of incandescent sources.

SELF-CHECK QUESTIONS

Self-check questions allow students to evaluate how well they have learned what they are studying.

1. Define the scientific term *photometry*.
2. Define the scientific term *radiometry*.
3. What are the terms that deal with visible-light energy?
4. What are the terms that deal with radiant-light energy?
5. There are two types of light sources. Name these sources.
6. Define the term *solid angle*.

7. Name the five terms that relate to both photometry and radiometry.
8. Measurement in the visible part of the optical spectrum is the science of photometry. Provide a short definition of these terms as to their use in this science.
   (a) Candela
   (b) Lumen
   (c) Luminous energy
   (d) Luminous flux
   (e) Luminous intensity
   (f) Luminance
   (g) Luminous sterance
   (h) Lambert
   (i) Luminous exitance
   (j) Illuminance
   (k) Luminous incidence
   (l) Lux
9. State the inverse square law.
10. State the cosine law of illumination.
11. Measurement in the total optical spectrum is the science of radiometry. Provide a short definition of these terms as to their use in this science.
   (a) Watt
   (b) Radiant energy
   (c) Radiant flux
   (d) Radiant intensity
   (e) Radiance
   (f) Radiant sterance
   (g) Emittance
   (h) Radiant exitance
   (i) Irradiance
   (j) Radiant incidence
12. Define the device called a photometer/radiometer.
13. Why do engineers use a skill called light suppression?
14. Why do engineers require a light attenuator be placed in front of high-power light measurement apparatus?
15. What is meant by photometric filtering?
16. What is meant by radiometric filtering?
17. How are light flashes measured?
18. What are densitometers used for?
19. What is the purpose of a laser power meter?
20. Name several radiometric/photometric accessories.

Eleven

Safety

Safety education today has become an important phase of everyone's life. Under the 1970 federal Occupational Safety and Health Act (OSHA), the employer is required to furnish a place of employment free of known hazards likely to cause death or injury. The employer has the specific duty of complying with safety and health standards as set forth under the 1970 act. At the same time, employees also have the duty to comply with these standards. A full treatment of the subject of safety is far beyond the scope of this book, and there is ample justification for a full course in safety procedures, including first-aid treatment, for all electrical and electronic technicians. The intent of this chapter is to make the reader fully aware of the ever-present, invisible, and generally silent hazards in handling electrical apparatus and to point out some fairly common causes of electrical shocks and fires that can easily be overlooked.

Since the time when Benjamin Franklin flew his famous kite, it has become more and more apparent that electricity, even in its milder forms, is dangerous. We can therefore assume that good safety habits are mandatory for all who use, direct the use of, or come in contact with electricity. Electrical equipment is found in every place that the ordinary person may find himself or herself. For this reason it becomes the responsibility of each of us to be knowledgeable of electrical safety and to become our own and our brothers' keepers.

The most basic cause of electrical accidents, as other types of accidents, is carelessness, and the best prevention is common sense. However, knowing exactly what to do in an emergency is only achieved through formal education or experience. Unfortunately, the experience could be fatal, so it is more desirable to derive your knowledge through schooling. When an emergency happens it is often accompanied by panic that can cause the mind or muscles to become paralyzed. The antidote for this panic is to be safety conscious.

OBJECTIVES

After studying this chapter and completing the self-check questions at the end of the chapter, the reader will be able to:

1. Place safety responsibility on the right person(s).
2. Describe the phenomena of electric shock.
3. State the methods of rapid rescue.
4. Compare ionizing and nonionizing radiation.
5. Analyze the biological effects of radiation.
6. Determine the possible damage that radiation can cause on the eye.
7. State several classes of laser hazards.
8. Describe the problems of high-energy directed light.
9. Become familiar with eye and face protection from injurious light rays.

SAFETY RESPONSIBILITY

The supervisor is responsible for enforcing the rules of safety in the area under his or her direct supervision. Inspectors ensure that equipment is tested before it is released for use, and final testing should include safety precautions. Finally, the user of the equipment should be qualified in its operation so he or she knows when it is operating in the proper manner. No one who comes into contact with such equipment is exempt from some degree of responsibility. The average layperson may not believe they can get a severe shock from an electrical component disconnected from a power source, but even that could present a hidden danger.

ELECTRICAL SHOCK

There is a common belief that it takes a great volume of electricity to cause a fatal shock and that high voltage is the thing that provides the jolt to do the job. Imaginative stories about people being literally fried or jolted from their shoes are supplied freely by storytellers. Although there are true tales of this type, this is not the total picture. The bulk of electrical shocks come in small packages. Death from electrical shock happens most often from ordinary 60-hertz, 115-volt alternating-current house power. The effect, in general, is an instantaneous, violent type of paralysis. The human body contains a great deal of water and is normally somewhat acid and saline. For these reasons it is a fairly good conductor of electric current, which has no regard for human life or feelings.

Our brains are loaded with tiny nerve ends that provide a transmitting service to the muscles. Muscles, in turn, provide motion for the body functions such as the heart and lungs. Now, assume that an electrical impulse that was not called for told your heart to speed up or stop its pumping job. Or suppose that a similar impulse of electricity told the lungs to quit taking in air. These situations do occur and with a comparatively small amount of current flow. Depending on a multitude of complex variables, a small current such as 10 milliamperes can be very unpleasant. Currents no larger than 20 milliamperes can cause muscle tightening or freezing. Currents of 30 milliamperes can cause damage to brain tissue and blood vessels. And damage to brain and blood vessels can, of course, be fatal.

As you know, a decrease in resistance (according to Ohm's law) causes current to increase. Body resistance decreases when perspiring or otherwise wet. A great number of things can cause variation in resistance. The general health of the person is probably the most important variable. A body in good condition has a much better chance of recovering from electrical shock. The muscles, by being in better tone, can recover to normal from paralyzation.

The path of current flow varies the shock. For instance, if the current involves the brain or heart it is naturally more dangerous than another path. The length of exposure can also be a factor, as well as the size or surface area of the electrical contact. Large voltages can cause spastic action, but recovery can be rapid. Also, small currents can cause muscle paralysis. In the event of breathing paralysis, artificial respiration must take place as quickly as possible to prevent loss of body functions and damage to the brain because of lack of oxygen. A condition known as ventricular fibrilation (uncoordinated heart beats, both fast and irregular) can occur with high currents, say 50 to 100 milliamperes. This action will continue until something is done externally to restore regular heartbeat, that is defibrilation or CPR (heart massage).

Death can occur when currents reach 250 milliamperes. This does not have to occur, however, as rapid first aid can save a victim whose heart has actually stopped beating or whose lungs have stopped pumping momentarily. In many cases rapid action by knowledgeable persons can prevent body damage and save lives.

RAPID RESCUE TECHNIQUES

A special course in first aid will ensure the proper methods. A wrong method can be worse than no action at all. The first rule of thumb for an electrical shock victim is to remove the current path. This can be done by turning the power switch off if it is readily accessible. If not accessible, the person can be detached from the current source by using an insulator of some sort such as wood, rubber, cork, or plastic. Sometimes it is more suitable and sensible to remove the power source from the victim. Whichever is the case, isolation from the current source is by far the most important and first move that can be made.

Isolation procedures can cause problems. The rescuer may find himself or herself in the current path. Touching a person who is paralyzed to a current source provides the current source another path through the rescuer's body. Care must be taken to prevent this from happening. After isolation, artificial respiration and/or other first aid should be applied. In all cases speed is vital. The likelihood of an electrical shock being fatal increases the longer the victim remains without breathing. A person given artificial respiration in the first 3 minutes has a much better chance of survival than one who is given artificial respiration after 5 minutes.

SAFETY IN OPTOELECTRONICS

The biological effects of electromagnetic radiation to human beings fall into two categories. These are ionizing and nonionizing radiation. The first, ionizing radiation, is related to cosmic and x-ray wavelengths and to nuclear radiation (very small). The second, nonionizing radiation, is related to ultraviolet, visible, infrared, microwave, and radio wavelengths (very large). The larger a wavelength is, the smaller its frequency, and vice versa. To distinguish the exact position in the frequency spectrum where the two radiation categories (ionizing and nonionizing) fall is impossible. Each frequency and wavelength level has an overlap to neighboring areas. The dividing point between optical and microwave is usually established around 1 millimeter wavelength. The visible spectrum falls in the optical area in the range of 380 to 760 millimeters. The exact level is subject to disagreement and depends a great deal on whose book you are reading. In any event, ionizing radiation by scientific analysis is the minimum photon energy required to produce ionization in the atoms of biological tissue or around 103 nanometers (ultraviolet). The word *ionize* means to change atoms to ions. Nonionizing radiation is, of course, not involved with this process. The optical levels of the frequency spectrum are of the most significance in this study of optoelectronics. The writer does not wish to iterate that health hazards do not exist in other levels of radiation. If exposed to enough radiant energy of almost any type the damage to the human body can be bothersome, painful, critical, and perhaps final.

Since we are mainly concerned with the optical spectrum in this book, we shall limit the radiation safety discussion to nonionizing radiation.

BIOLOGICAL EFFECTS OF RADIATION

Modern developments in optical technology include fiber optics, the laser, and peripheral light sources such as electronic flashlamps and arc lamps have caused safety engineers some concern. The Division of Industrial Safety and OSHA have changed their standards to include these high-energy light devices. Beside the damage possible to the eyes, many of the strong light sources are capable of producing hazardous levels of ultraviolet radiation. OSHA standards are now set to include radiation that produces injury to the skin and the eyes. Prior to 1960, public use of microwave radiation was extremely limited. Changes in high-power radar equipment, microwave communications, industrial heating, and home microwave cooking have exposed the public to radiation that they previously had not encountered. The expansion of the fiber optics industry and the increasing use of new laser devices since the beginning of the 1970s has brought forth empirical data that have caused changes in safety regulations.

Biological effects of nonionizing radiation are dependent on the spectral region of the radiation and the duration of the exposure to the radiation. Furthermore, the damage to the eyes and skin is dependent on whether there was just a single exposure (acute) or daily exposure (chronic) to the radiation. Any discussion on permissible radiation exposure should include these factors. Quantities are usually expressed in exposure doses and exposure dose rate. Physical factors also affect the radiation hazard. These are: the radiation absorbed in, reflected from, and transmitted through the tissue. A final physical factor that must be considered is the spectral sensitivity of the particular tissue exposed to the radiation.

The biological effects of optical radiation to the human body are thermal (heat) and nonthermal. The eye is probably the organ that is most sensitive to injury. The injury to the eye is dependent on the region of the eye that absorbs the radiant energy.

Radiation and the Eye

The eyes, as we have stated, are generally considered to be the organ of the body which is most susceptible to damage by radiation. The parts of the eye that can be affected by radiation are the cornea, lens, eye fluid, and retina (see Figure 11–1). Different light radiation affects the individual eye parts. The damage to any of the parts occurs when the light is absorbed by the parts. The damage that takes place is dependent on the ranges of the exposure levels and the time of exposure.

In the optical spectrum, wavelengths for ultraviolet radiated frequencies range from 0.005 to 0.3900 microns. Short wavelengths of ultraviolet radiation are absorbed by the cornea. Long wavelengths of ultraviolet radiation are generally refracted at the cornea and the lens then absorbed in the retina. Some long wavelengths of ultraviolet radiation are absorbed by the cornea and lens.

Visible wavelengths of radiated frequency range from 0.3900 to 0.7500 micron. Visible wavelengths are generally refracted by the cornea and absorbed by the retina. Infrared wavelengths of radiated frequencies range from 0.7500 to 4000 microns. Near-infrared wavelengths are absorbed primarily in the ocular medium and at the retina and for that matter all parts of the eye. The near-infrared frequencies are the most dangerous to the eye. Far-infrared frequencies are absorbed at the cornea.

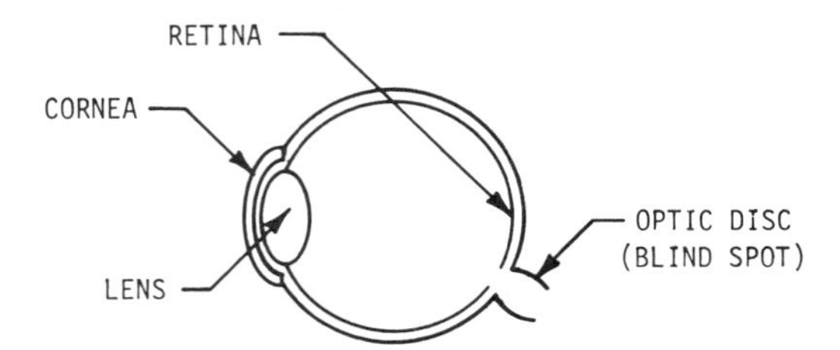

Figure 11–1 The eye and its parts (*Mike Anaya*).

ANSI Z126.1 Standards for the Safe Use of Lasers

The ANSI (American National Standards Institute) standard has divided lasers into classes which represent their possible hazards.

Class I lasers are at visible wavelengths. These are considered to be safe for continuous viewing at irradiance levels of 0.4 microwatt passing through the cornea.

Class II lasers are also at visible wavelengths. They are considered as those lasers that do not have enough power to cause retina damage during momentary viewing. Prolonged viewing will cause damage no matter what the irradiance levels are.

Class IIIA lasers will not generally cause damage to the eye under normal conditions, but will damage the eye when viewing the beam through optics. These are medium-power lasers in the range of 1.0 to 5.0 milliwatts.

Class IIIB are lasers which cannot cause hazardous diffuse reflection or fire. This medium-power class ranges from 5.0 to 500 milliwatts; they can cause direct eye damage.

Class IV lasers are a diffuse reflection hazard and a fire hazard. They are classified as high power and range from 500 milliwatts upward.

Lasers in general must be handled with extreme caution. General safety rules include prevention of direct sight at the beam, the avoidance of reflective objects such as mirrors, jewelery, and so on, and the access control to keep unqualified persons from harm's way. Lasers should only be used in well-defined areas. Appropriate signs must be posted in applicable places to alert people of the potential hazards. Whenever the laser is not in use, the laser room should not be accessible to persons who are not educated in laser technology. Accidents should be immediately reported. Even in minor eye accidents, an ophthalmologist should be consulted.

Problems of High-Energy Directed Light

In fiber optics and laser technology, the problems of radiation are compounded by extremes of density (high energy) and direction. Light beams in fiber optics are transmitted in glass fibers of 4 or 5 micrometers in diameter. Beam sizes of lasers are common in the 8- to 10-micrometer range. The hazards exist when an extremely strong beam of light is focused or is looked at directly. The beam, with its massive irradiant strength (often called power density) passes through the pupil and is directed in force upon the retina. Irreparable damage may be done in an extremely short period of time. If the beam is strong enough, damage can occur when the light beam is seen indirectly such as when bounced off or reflected from a smooth surface.

FEDERAL REGULATIONS

In Title 8, the Division of Industrial Safety provides general safety orders that are to be adhered to by industry.

Eye and Face Protection

Employees working in locations where eye hazards due to flying particles, hazardous substances, or injurious light rays are inherent in the work or environment, shall be safeguarded by means of face or eye protection. Suitable screens or shields isolating the hazardous exposure may be considered adequate safeguarding for nearby employees.

The employer must provide and the employee must use protection suitable for the exposure. Where eye protection is required and the employee requires vision correction, such eye protection must be provided as follows:

1. Safety spectacles with suitable corrected lenses, or
2. Safety goggles designed to fit over spectacles, or
3. Protective goggles with corrective lenses mounted behind the protective lenses.

The wearing of contact lenses is prohibited in working environments having harmful exposure to materials or light flashes, except when special precautionary procedures, which are medically approved, have been established for the protection of the exposed employee.

Design, construction, testing, and use of devices for eye and face protection shall be in accordance with American National Standard for Occupational and Education Eye and Face Protection, Z87.1–1968, except that integral lens and frame design will be allowed if the lens-frame combination provides unit strength as well as impact penetration, heat and flammability resistance, optical qualities, and eye zone coverage equal to or greater than is required by ANSI Z–87.1 1968.

Employees whose occupation or assignment requires exposure to laser beams shall be furnished suitable laser safety goggles which will protect for the specific wavelength of the laser and be of optical density (O.D.) adequate for the energy involved. The ANSI standard Z136.1 divides lasers into classes dependent on the radiation hazard.

SELF-CHECK QUESTIONS

Self-check questions allow students to evaluate how well they have learned what they are studying.

1. What is OSHA?
2. What is the most basic cause of electrical accidents?
3. Whose responsibility is safety?

4. Where do most electrical accidents occur?
5. When a person has an electrical shock, what are some of the things that happen to the body?
6. At what point does death occur because of an electrical shock?
7. What may an average person do to facilitate rapid first aid?
8. Compare the terms *ionizing radiation* and *nonionizing radiation.*
9. What are some biological effects of nonionizing radiation?
10. What parts of the eye are affected by radiation? What wavelengths are the most dangerous?
11. How are lasers classified as to their health hazards?
12. What can be done to protect the eyes in the workplace?

Glossary

Acceptance angle The critical angle, measured from the core centerline, above which light will not enter the fiber. It is equal to the half-angle of the acceptance cone.

Acceptance pattern The acceptance pattern of a fiber or fiber bundle is a curve of input radiation intensity plotted against the input (or launch) angle.

Active medium A collection of atoms excited to produce a population inversion.

Angstrom (Å) A wavelength value equal to 1/10,000 micron or 10^{-10} meter.

Antinodes The maximum amplitude point in standing waves.

Atomic lifetime The amount of time that an atom stays in an excited state before spontaneously falling to another, lower energy state.

Beam A slender shaft or stream of light or other radiation

Beamwidth (θ_o) The difference between the angles at which radiant intensity $I(\theta)$ is 50% (unless otherwise stated) of the peak value I_p.

Branch That portion of a cable or harness that breaks out from and forms an arm with the main cable or harness run.

Breakout The point where a branch meets and merges with the main cable or harness run or where it meets and merges with another branch.

Brewster window Glass windows placed at exact angles on the ends of a plasma tube.

Brightness Radiance. The amount of power radiated by a laser beam.

Buffer A protective material that covers and protects a fiber. The buffer has no optical function.

Bundle A number of fibers grouped together.

Bundle jacket The material that is the outer protective covering common to all internal cable elements.

Cable A jacketed bundle or jacketed fiber in a form that can be terminated.

Cable assembly A cable that is terminated and ready for installation.

Cable core The portion of a cable contained within a common covering.

Cable jacket The material that is the external protective covering common to all internal cable elements.

Cable or harness run That portion of a branched cable or harness where the cross-sectional area of the cable or harness is the largest.

Circularly polarized A wave is said to be circularly polarized when the electrical field traces out a circle in a plane perpendicular to the direction of wave propagation.

Cladding A sheathing or cover of a lower refractive index material intimately in contact with the core of a higher refractive index material. It serves to provide optical insulation and protection to the total reflection interface.

Cladding mode stripper A material applied to the fiber cladding that provides a means for allowing light energy being transmitted in the cladding to leave the cladding of the fiber.

Coherence Light waves which are in-step, in-phase, and monochromatic.

Continuous wave (CW) A laser whose output is continuous with no interruptions.

Core The high refractive index central material of an optical fiber through which light is propagated.

Corpuscle theory Theory of light in seventeenth century that suggested light consists of corpuscles that had an internal vibration of their own or in some way were controlled by waves or vibrations of the medium (such as air) through which they travel.

Coupler An optical device used to interconnect three or more optical conductors.

Coupling loss The total optical power loss within a junction, expressed in decibels, attributed to the termination of the optical conductor.

Critical angle The maximum angle at which light can be propagated within a fiber. Sine critical angle equals the ratio of the numerical aperture to the index of refraction of the fiber core.

Crosstalk Measurable leakage of optical energy from one optical conductor to another.

Crystalline Solid material that has a lattice crystal structure such as the ruby.

CVD fiber (chemical vapor deposition) A process by which a heated gas produces an oxide deposit to fabricate a glass fiber preform. The deposited glass becomes the core.

Decibel The standard unit for expressing transmission gain or loss and relative power levels. A decibel equals 10 times the log of the ratio of the power out to the power in [$dB = 10 \log_{10} (P_o/P_i)$].

Detector A device that converts optical energy to electrical energy, such as a PIN photodiode.

Diffraction Breaking up of light rays into bands when a ray is deflected by an opaque object or a slit.

Diffusion A scattering of light rays. A softening of light.

Directionality Not spread out in all directions. A narrow cone projected in one direction.

Dispersion The "spreadout" or "broadening" of a light pulse as it propagates through the optical conductor. Dispersion increases with length of conductor and is caused by the difference in ray path lengths within the fiber core.

Divergence Spreading out of a light beam.

Electroluminescence Electrical energy application to light-sensitive material causing emission of light (radiant) energy.

Emission Ejection of electrons from a surface by radiation.

Emission theory Theory of light developed about 500 B.C. by Greek thinkers. The theory suggested that all objects emit something which, when entering the eye, was caused by some part of the eye, therefore the brain could see.

End finish Surface condition at the optical conductor face.

Excitation mechanism Source of energy which excites the atoms in an active medium.

Exit angle The angle between the output radiation vector and the axis of the fiber or fiber bundle.

Extended source Light projecting from several points in space to illuminate a point from many directions.

Extinction A measure of how well a beam is polarized. It is the ratio of the power transmitted through an aligned polarizer and power transmitted through a polarizer rotated 90°.

Feedback Coupling of energy from output back into an active medium.

Fiber A single discrete optical transmission element usually comprised of a fiber core and a fiber cladding.

Fiber buffer The material that surrounds and is immediately adjacent to a fiber that provides mechanical isolation and protection. *Note:* Buffers are generally softer materials than jackets.

Fiber bundle A consolidated group of single fibers used to transmit a single optic signal.

Fiber cable A cable composed of a fiber bundle or single fiber, strength members, and a cable jacket.

Fiber cladding That part of a fiber surrounding the core of the fiber and having a lower refractive index than the core.

Fiber core That part of a fiber having a higher refractive index than the cladding that surrounds it.

Fiber jacket The material that is the outer protective covering applied over the buffered or unbuffered fiber.

Fiber optics A general term used to describe the function where electrical energy is converted to optical energy, then transmitted to another location through optical transmission fibers, and converted back to electrical energy. While the bulk of applications is in digital transmission data, fiber optics also can be applied to analog systems.

Fiber sheath A general term used variously to mean cladding, buffer, or jacket. Not to be used in military FO specifications.

Finite Measurable.

Footcandle Measure of intensity of illumination at all points of an illuminated surface 1 foot from a 1-candle source.

Frequency The number of times a wave occurs in a period of time.

Fresnel reflection losses The losses that are incurred at the optical conductor interface due to refractive-index differences.

Fringe A pattern of dark and light bands created by shining a monochromatic beam of light on a surface such as in thin films.

Fundamental mode A transverse electromagnetic mode (TEM) that oscillates in one mode only.

Gas loss A power loss, expressed in decibels, due to the deviation from optimum spacing between the ends of separable optical conductors.

Graded-index fiber A fiber whose index of refraction decreases with increasing radial distance from the center of the core.

Ground state The lowest energy state of an atom.

Harness A construction in which a number of multiple-fiber cables or jacketed bundles are placed together in an array that contains branches. A harness is usually installed within equipment or air-frame and mechanically secured to that equipment or airframe.

Harness assembly A harness that is terminated and ready for installation.

Hertz (Hz) Cycles per second of a radiated frequency.

Heterochromatic Of, having, or consisting of many colors.

Incoherent Light waves that are not instep, not in-phase, and nondirectional.

Index-matching materials Materials used in intimate contact between the ends of optical conductors to reduce coupling losses by reducing Fresnel loss.

Index of refraction The ratio of the speed of light in a vacuum to the speed of light in a material. When light strikes the surface of a transparent material, some light is reflected while some is refracted.

Infinite Lacking limits, extending beyond the bounds of measurement or comprehension.

Infrared Band of light wavelengths too long for response by a human eye.

Kilohertz One thousand cycles per second of a radiated frequency.

Lambda (λ) Greek letter depicting wavelength. Mathematically $\lambda = c/f$, where c is velocity in meters (300,000,000 meters per second) or (186,000 miles per second) and f equals frequency.

Lambertian A radiance distribution that is uniform in all directions of observation.

Laser Light amplification by stimulated emission of radiation.

Lateral loss A power loss, expressed in decibels, due to the deviation from optimum coaxial alignment of the ends of separable optical conductors.

Launch angle The angle between input radiation vector and the axis of the optical conductor.

Linearly polarized A wave is said to be linearly polarized when the electrical field oscillates along a line in a plane that is perpendicular to the direction of wave propagation.

Lumen Measure of luminous flux of one square foot of spherical surface one foot from a one candle source.

Maser Microwave amplification by stimulated emission of radiation.

Megahertz One million cycles per second of a radiated frequency.

Meridial rays The rays of light that propagate by passing through the axis of the fiber and travel in one plane.

Metastable state Energy states of atoms that have extremely long atomic lifetimes.

Micron (μ) A wavelength value equal to 10,000 angstroms or 10^{-6} meter.

Microwave Electromagnetic waves extending from 300 to 300,000 MHz.

Monochromatic Of, having, or consisting of one color.

Multiple-bundle cable assembly A multiple-bundle cable terminated and ready for installation.

Multiple-fiber cable assembly A multiple-fiber cable that is terminated and ready for installation.

Nodes The zero amplitude points in standing waves.

Numerical aperture (NA) The characteristic of an optical conductor in terms of its acceptance of impinging light. The sine of the acceptance cone half-angle equals the square root of the difference of the square of the index of refraction of the fiber core minus the square of the refractive index of the fiber cladding.

Numerical aperture, NA (10% intensity) The numerical aperture of a fiber or bundle is defined by: NA (10% intensity) = sin θ', where θ' is the angle where the measured intensity of radiation is 10% of the maximum measured intensity when plotting either the acceptance or radiation pattern.

Numerical aperture, NA (materials) The numerical aperture of a fiber or bundle is defined by: NA = $(n_1^2 - n_2^2)^{1/2}$ where n_1 and n_2 are the fiber core and cladding refractive indices, respectively, for step-index fibers. For graded-index fibers n_1 is the maximum index in the core and n_2 is the minimum index in the cladding.

Numerical aperture, NA (90% power) The numerical aperture of a fiber or bundle is defined by: NA (90% power) = sin θ, where θ is the angle between the axis of the output cone of light and the vector coincident with the surface of a cone that contains 90% of the total output radiation power, or where θ is the angle between the axis of the input cone of light and the vector coincident with the surface of a cone which contains 90% of the total input radiation power.

Opaque Material that blocks light rays.

Optical conductors Materials that offer a low optical attenuation to transmission of light energy.

Optical spectrum Wavelengths that operate in the light and sight spectrum.

Optics Branch of physics dealing with light and vision.

Optoelectronics Branch of electronics dealing with light.

Packing fraction (PF) The ratio of the active core area of a fiber bundle to the total area at its light-emitting or receiving end.

Panchromatic Sensitive to light of all colors.

Peak radiant intensity (I_p) The maximum value of radiant intensity, I.

Peak wavelength (λ_p) The wavelength at which the radiant intensity is a maximum. Unit: nanometers (nm).

Period The length of time designated to count the number of wave cycles. Usually the time is in seconds.

Phase Value of a wave in relation to another or to a specific time.

Photoconduction Change of resistance in a material when it is exposed to radiation.

Photoelectric Electrical effects caused by light or other radiation.

Photoelectrons Electrons that absorb light energy or photons.

Photoemission Light releasing electrons from a surface.

Photon Tiny pulses of light energy. An atom stimulated in an excited state emits photons of energy. A quantum of light energy.

Photovoltaic Generation of voltage with the use of dissimilar materials and light-sensitive material in response to radiation.

Plasma Ionized gas.

Plastic-clad silica fiber (PCS) A fiber composed of a silica-glass core with a transparent plastic cladding.

Point source Light projecting from one point in all directions.

Population inversion Movement of an atom from a ground energy state to an excited energy state.

Power density Irradiance.

Pulsed A laser whose output power has predetermined changes with time.

Quantum Unit of light energy dependent on wavelength.

Radiance The radiant flux per unit solid angle and per unit surface are normal to the direction considered. The surface may be that of a source, detector, or any other surface intersecting the flux.

Radiant energy Energy generated by the frequency of radiation of electrons. The energy of light.

Radiant intensity ($I = d\Phi/d\omega$) The radiant power per unit solid angle ω in the direction considered measured in watts per steradian (W/sr).

Radiant power (Φ) The time rate of flow of electromagnetic energy, measured in watts (W).

Radiant power ratio R is defined as the radiant power ratio Φ_2/Φ_1, where Φ_1, and Φ_2 are the measured power before and after specimen conditioning, respectively.

Radiation Process by which energy in the form of rays is sent through space from atoms and molecules as they undergo internal change.

Radiation pattern The radiation pattern is a curve of the output radiation intensity plotted against the output angle.

Ray A line of light extending between two points.

Ray angle The angle between a light ray and a reference line or plane.

Receiver optical An electro-optical module that converts an optical input signal to an electrical output signal.

Refraction The bending of a light ray when the ray moves through the intersection of two mediums such as air to water.

Repeater A device that converts a received optical signal to its electrical equivalent, reconstructs the source signal format, amplifies, and reconverts to an optical output signal.

Spatial coherence Phase correlation of two different points across a wave front at a specific moment in time.

Selenium Metal whose resistance is effected by light radiation.

Skew rays Rays of light that do not propagate through the axis of the fiber.

Source The source of radiant energy, such as a light-emitting diode (LED).

Spectral bandwidth (λ_{BW}) The difference between the wavelengths at which the radiant intensity I (λ) is 50% (unless otherwise stated) of the peak value I_p.

Spectrum Ranges of frequencies or wavelengths.

Splice Nonseparable junction joining optical conductor to optical conductor.

Spontaneous emission When an atom falls from an excited state to a ground state and in doing so emits a photon of energy, the photon is said to have been spontaneously emitted.

Sputtering Atoms jarred loose from a cathode by bombardment of ions in a plasma tube.

Standing waves A pattern created by waves reflecting back and forth within a laser cavity. These waves have the appearance of standing still if they are of the same frequency, amplitude, and moving in opposite directions.

Step-index fibers A fiber in which there is an abrupt change in refractive index between the core and cladding along a fiber diameter.

Stimulated absorption When an atom absorbs a photon and moves to a higher energy state.

Stimulated emission When an external source of power causes an atom to move from an excited state to a lower state of energy.

Tactile theory Theory of light accepted by Greek thinkers about 500 B.C. It asserted that eyes sent out invisible antennae and were thus able to feel or sense those things too distant to touch.

Temporal coherence Phase correlation of waves at a point in space at two instants of time.

Translucent Material that allows light to partially pass through.

Transmission mirror Mirror on the output of a laser that partially transmits the light beam and reflects the rest of the light back into the active medium.

Transmitter optical An electro-optical module that converts an electrical input signal to an optical output signal.

Transparent Material that clearly allows light to pass through.

Transverse wave A wave whose amplitude is perpendicular to its direction of propagation.

Ultraviolet Band of light wavelengths too short for response by a human eye.

Uniform lambertian A lambertian distribution that is uniform across a surface.

Unpolarized A wave is said to be unpolarized when the electrical field oscillates randomly in a plane perpendicular to the direction of propagation.

Velocity Rate of change of position in relation to time.

Visible light Electromagnetic wavelengths that can be seen by the human eye ranging from 380 to 770 nanometers.

Wave The motion of light as it travels through space.

Wave crests The part of a light wave that has maximum vibration.

Wavefront The forward side of a light wave.

Wavelength The amount of space occupied by the progression of an electromagnetic wave.

Wave trough The part of a light wave that has minimum vibration.

Bibliography

BOOKS

ALLEN, W. B., *Fiber Optics,* Plenum Press, London, 1979.

CLARRICOATS, P. J. B., Ph.D., *Optical Fiber Waveguides*, Peter Peregrinus Ltd, Instittition of Electrical Engineers, Stevenage, Herts, England, 1975.

GLOGE, DETLAF, ed., *Optical Fiber Technology*, The Institute of Electrical and Electronics Engineers, Inc., New York, 1976.

MIDWINTER, JOHN E., *Optical Fibers for Transmission*, John Wiley & Sons, Inc., New York, 1979.

PAMPHLETS

ANDERSON, JOHN D., "Fiber Optic Data Bus," Northrup Corp., Hawthorne, California.

BARK, P. R., and WEY, R. A., "Field Splicing of Optical Cables on Fiber Assessment with Optical Time Domain Reflectometry," Siecor Optical Cables, Inc., Horseheads, New York.

BARTON, DAVID M., "Driving High Level Loads with Iso-Lit Opto-Isolators," Litronix Inc., Cupertino, California.

BARTON, DAVID M., "More Speed from Iso-Lit Optical Isolators," Litronix Inc., Cupertino, California.

BIELOWSKI, W. B., "Low Loss Waveguide Communication Technology," Corning Glass Works, Corning, New York.

BISHOP, O. L., and SMITH, J. C., "Installation of a Fiber Optic System in an Electrical Power Station," ITT Telecommunications, Raleigh, North Carolina.

BLOOM, ARNOLD L., "C. W. Pumped Dye Lasers," Coherent, Inc., Palo Alto, California.

BORSUK, LES, "Fiber Optic Interconnections," ITT Cannon Electric, Santa Ana, California.

BORSUK, LES, "Introduction to Fiber Optics," ITT Cannon Electric, Santa Ana, California.

BOWEN, TERRY, "Low Loss Connectors for Single Optical Fibers," AMP, Inc., Harrisburg, Pennsylvania.

EBHARDT, C. A., "A T1 Rate Fiber Optic Transmission System for Electrical Power Station Communications," ITT Transmission Division, Raleigh, North Carolina.

EBHARDT, C. A., EDWARD, A. K., and MAGNER, R. G., "A T3 Rate, Repeatered Fiber Optic Transmission Systems," ITT Transmission Division, Raleigh, North Carolina.

EPPES, T. A., GOELL, J. E., and GALLENBERGER, R. J., "A Two Kilometer Optical Fiber Digital Transmission System for Field Use at 20 Mb/s," ITT Electro-Optical Products Division, Roanoke, Virginia.

FENTON, KENNETH, "Fiber Optic Connectors," ITT Cannon Electric, Santa Ana, California.

MARTINET, R. L., "A T3 Rate, Repeatered, Fiber Optic Transmission System," ITT Electro-Optical Products Division, Roanoke, Virginia.

MISKOVIC, EDWARD J., "Fiber Optic Data Bus," ITT Cannon Electric, Santa Ana, California.

NEMHAUSER, ROBERT I., ALEXANDER, GIL, and DUDA, RICHARD, "Radiometry and Photometry Once over Lightly," United Detector Technology, Inc., Santa Ana, California.

OLSZEWSKI, J. A., FOOT, G. H., and HUANG, Y. Y., "Development and Installation of an Optical-Fiber Cable for Communications," General Cable Company, Union, New Jersey.

SCHUMACHER, WILLIAM L., "Fiber Optic Connector Design to Eliminate Tolerance Effects," AMP, Inc., Cherry Hill, New Jersey.

SMITH, GEORGE, "Applications of Opto-Isolators," Litronix Inc., Cupertino, California.

SMITH, GEORGE, "LEDs and Photometry," Litronix Inc., Cupertino, California.

SMITH, GEORGE, "Multiplexing LED Displays," Litronix Inc., Cuperitino, California.

TAKAGISHI, DAVE, "Applying the DL-1416 Intelligent Display," Litronix Inc., Cupertino, California.

WEY, ROBERT A., and COLE, JAMES A., "Optical Fibers, Cables, and Links for Computer Systems Applications," Siecor Optical Cables, Inc., Horseheads, New York.

WHEELER, JOHN POST, "Design and Manufacture Considerations in Commercial Ion Lasers," Coherent Incorporated, Palo Alto, California.

Index

D

E

F

Q

R

S